D0982589

SYSTEMS APPROACH AND THE CITY

Systems Approach and the City

Edited by

MIHAJLO D. MESAROVIC
Professor of Systems Engineering
and Director of
Systems Research Center
Case Western Reserve University
Cleveland, Ohio

and

ARNOLD REISMAN
Professor of Operations Research
Case Western Reserve University
Cleveland, Ohio

1972

NORTH-HOLLAND PUBLISHING COMPANY – AMSTERDAM • LONDON
AMERICAN ELSEVIER PUBLISHING COMPANY, INC. – NEW YORK

Library of Congress Catalog Card Number 72-91320
North-Holland ISBN 0 7204 3069 0
American Elsevier ISBN 0 444 10410 0

Publishers:

NORTH-HOLLAND PUBLISHING COMPANY–AMSTERDAM
NORTH-HOLLAND PUBLISHING COMPANY, LTD.–LONDON

Sole Distributors for the U.S.A. and Canada:

AMERICAN ELSEVIER PUBLISHING COMPANY, INC.
52 VANDERBILT AVENUE
NEW YORK, N.Y. 10017

Printed in The Netherlands

PREFACE

This book represents the proceedings of the Fifth Systems Symposium held at Case Western Reserve University on November 9-11, 1970. The conference addressed itself to studies of the city using systems concepts and methodology. Urban problems were considered on all levels: the psychological effects of urban living on individuals, the socio-political problems in urban systems; the interaction of various services (health, transportation, etc.) and the role of the city as a sub-system in a region. The objective of the meeting was to provide a critical review of the state of the art, indicate current trends and potential for further development. The participants who came from throughout the United States represented the academic community as well as city administrations.

With the exception of one chapter (incidentally written by a political scientist) and minor sections of two other chapters the book presupposes no knowledge of mathematics nor of systems science. It is intended for students and professionals concerned with urban problems who are interested in what the systems approach can do in the urban setting and for systems scientists who are interested in the applications of their craft to urban problem solving.

TABLE OF CONTENTS

Chapter 7

AFTERTHOUGHTS ON FOUR URBAN SYSTEMS STUDIES

E. Cushen

Chapter 10

THE NORTHEAST CORRIDOR TRANSPORTATION PROJECT: FEDERAL FOLLY OR REGIONAL SALVATION

R. Nelson

Chapter 11

MODELS OF A TOTAL CRIMINAL JUSTICE SYSTEM

A. Blumstein and R. Larson

Chapter 12

SYSTEMS ANALYSIS OF CRIME CONTROL AND THE CRIMINAL JUSTICE SYSTEM

A. Blumstein

Chapter 13

URBAN GHETTO REVOLTS AND LOCAL CRIMINAL COURT SYSTEMS

I. Balbus

Chapter 15

SYSTEMS ANALYSIS AND SOCIAL WELFARE PLANNING - A CASE STUDY

A. L. Service, S. J. Mantel, Jr., A. Reisman

Chapter 16

DILEMMAS FOR SYSTEMS ANALYSIS OF URBAN PUBLIC PROGRAMS

S. Goldstone

Chapter 17

THE EXPERIENCE OF LIVING IN CITIES

S. Milgram

Chapter 18

HEALTH SYSTEMS - AN URBAN VIEW

C. Flagle

Chapter 19

SYSTEMS ANALYSIS OF URBAN AIR POLLUTION

F. F. Gorschboth

Chapter 20

POLLUTION AND ECOLOGY

Y. Y. Haimes

ACKNOWLEDGEMENTS

The editors are greatly indebted to the Thompson Ramo Wooldridge Foundation, for sponsoring the symposium and enabling the production of these Proceedings. The Symposium and the Proceedings could not, of course, have come about but for the efforts made by the contributors. Thanks are also due to Mrs. Mary Lou Cantini and to Miss Janet Snell who typed the manuscript; to Alex Forbes, who drafted the figures and to V. Sherlekar and Andrew Daughety who helped in the process of editing the final manuscript.

Chapter 1

INTRODUCTION *

M. D. Mesarovic

The objective of these remarks is to present a rationale for selecting the topic of our Symposium "Systems Approach and the City" and to discuss some of the viewpoints which led to the structuring of the conference as indicated by these proceedings. The subject matter of the various presentations is quite diverse, although they all deal with urban problems. What they have in common and how they are related will be emphasized.

First some comments on the systems approach. In the seventies, it can not be claimed that the systems approach is a novelty. One can be assured that there are perhaps more than a dozen different opinions as to what constitutes this approach. Qualifications are therefore in order. For the purpose of organizing this Symposium it is assumed that the systems approach is characterized by the following:

(1) When considering any particular problem one uses the broadest possible context. Whatever the system one focuses attention on, it is recognized that it represents but a subsystem of a still larger system. It is therefore, imperative to recognize that a study can not be limited to include the con-

*Opening remarks at the Symposium "Systems Approach and the City" held at Case Western Reserve University in November 1970.

cepts from a single traditional discipline, e.g., economics, sociology, architecture, the various branches of engineering, etc. On the contrary, there is a need for the blending of the tools from a large number of areas. This broad approach is quite apparent in the case studies to be found in several of the chapters of this proceedings.

(2) In describing the various factors of importance for the given urban problem, primary emphasis has been placed upon both the relationships between the factors as well as upon identifying the inner mechanisms which produce the observed behavior. The future, then, is assessed not on the basis of extrapolating past trends, but rather on the basis of understanding the effects determined by these inner mechanisms and how they affect the future. This is in essence where urban dynamics studies differ from the more traditional urban planning.

(3) Relationships between various factors, simply referred to as variables, are described in terms of decision-making and information processing concepts. It is necessary to consider carefully feedback affects, optimal response, information transmission, etc. The contrast of this method of description with the traditional ones will be particularly striking in several of the papers which consider the city as a cybernetic system. Of course, any of these variables have origin within some specific interpretation, but in the assessment of the interdependence, the variables are simply considered as a set of interacting objects (e.g., as a system). Such an approach has at least the following two advantages: (a) Methods of analysis developed in various specific fields can be used quite readily for the study of urban problems; (b) Since all subsystems are described within the same framework there is a better foundation for integrating the subsystems into a larger system - thus it encourages a broader view of the urban phenomena.

(4) Once the description of an urban phenomenon is posed in the system-theoretic framework the computer simulation and possibly quantitative methods of analysis can be used for the study of dynamics as well to investigate the logical consequences of the alternative courses of action - say, for policy evaluation. However, it should be emphasized that the usefullness of the system's description is not confined solely to the possibility of a quantitative analysis. The very description of a phenomenon in the systems-theoretic framework can provide important insight perhaps by the establishment of some important cause-effect relationships (even though these may be in but qualitative terms). A notable example of this is the recognition of the information-input overload as an important determinant of the psychology of urban living.

Let us turn to the city as an object of a study. An approach to classify various urban phenomena in a manner compatible with the systems viewpoint is to recognize four basic factors by which the urban phenomena can be described: (a) man, the individual, in the urban setting; (b) groups of people living or working together under urban conditions; (c) the natural environment, largely determined by the gergraphy of the area; and (d) the man-made or technological environment. The man-made environment itself can be classified into static, shelter type, and the dynamic, transportation services, etc. Such a classification is fundamental to the science of Ekistics of urban settlements as introduced by C. Doxiadis and presented in one of the chapters.

In conclusion it would be appropriate to leave you with one thought. Namely, that the city is a complex system capable of counter-intuitive responses which can be properly understood and controlled only if the interactions between the four basic urban factors as indicated previously are accounted for properly. In other words, the real question is not to better understand the individual subsystems (transportation, man, city government etc.) but to understand how they act together, how they are harmonized into this entity we call a city. At the danger of sounding a warning signal before we have embarked on the journey, I must express the opinion that the solution of many critical urban problems depends precisely upon our capability to recognize the complexity of the urban systems in an explicit manner so that the solutions are sought in reference to all relevant influences. A couple of examples here would be helpful.

In order to solve it's traffic congestion problem the city of London, England, has introduced a new traffic pattern considerably increasing the number of one-way streets and creating many new dead-end streets. In response to the request from the residents it also restricted the street parking to the residents only. The response to the change was very fast and not quite as expected. A working class neighborhood was transformed from a noisy, congested area with declining Victorian housing into a quiet urban retreat minutes away from the West End theatrical and shopping centers. First, the merchants complained about increased difficulty in deliveries. Latter, the residents showed concern because of the delays in abmulance, police and other services. But the greatest impact came from the action of developers and speculators. The rents went sky high, the upper middle class began to move in while the working class residents were forced to move further away which further affected trade and subsequently the entire pattern of living. The protests and complaints were quite loud and the City is considering what action to take. At any rate a simple action requiring minimal investment (i.e., just putting up the signs: "parking restricted to residents" and "one-way street") has greatly affected the social structure of the area. The action designed to help local residents was counter productive - it forced them away.

Another example of strong interactions between urban problems is discussed in the book by Edward Banfield "The Unheavenly City". He argues that the main problem of the urban ghettos is not racial but what he calls the low class mentality of ghetto residents. Living for today, with no expectations for the future, results in the behavior associated with ghetto residents. The problem is not racial, or even economic, but social and cultural and solutions ought to be sought along these lines. The problem can not be solved by money alone. The statement made by Norman E. Borlaug, who received the Nobel prize for enabling a dramatic increase in the rice production and in this way preventing the hunger and starvation of a large scale in certain parts of Asia, is of timely importance. "We have not solved the problem", he commented recently, "we have only delayed the world food crisis for another 30 years." The time has been gained but if the corrective actions in the population control area are not taken, the disaster can be even worse. These remarks are also quite appropriate for the urban scene. The short term solutions can only alleviate the problem temporarily, increasing its scale when the problem reoccurs. Long-range solutions have to be sought and the systems approach ought to make a major contribution to this process by providing a framework for interrelating different aspects and allowing for the investigation of the logical consequences of alternative policy actions. Systems engineers and systems analysts have claimed for years that their methods and approaches are applicable regardless of the specifics of the phenomena under consideration. The urban area might well provide an ultimate challenge and test for that claim.

Chapter 2

EKISTICS, THE SCIENCE OF HUMAN SETTLEMENTS

C. A. Doxiadis

We cannot acquire proper knowledge about our villages, towns, and cities unless we manage to see the whole range of the man-made systems within which we live, from the most primitive to the most developed ones -- that is, the whole range of human settlements. This is as necessary as an understanding of animals in general is to an understanding of mammals -- perhaps even more so. Our subject, the whole range of human settlements, is a very complex system of five elements -- nature, man, society, shells (that is, buildings), and networks. It is a system of natural, social, and man-made elements which can be seen in many ways -- economic, social, political, technological, and cultural. For this reason only the widest possible view can help us to understand it.

The author is an architect - planner and president of Doxiadis Associates, Athens, Greece, and Washington D. C.
This chapter is reproduced with permission from Science, Vol, 197, pp. 393-404.

There is a need for a science dealing with human settlements, because otherwise we cannot view these settlements in a reasonable way. Is such a science possible? The answer can be given in two ways. First, by observing that, in some periods of the past, people must have had such a science, which was probably written down only in ancient Greek times (in documents which have since been lost) and in Roman times (perhaps by the architect and engineer Vitruvius). Otherwise, how did people create cities that we still admire? Second, we are now convinced that man, in creating his settlements, obeys general principles and laws whose validity can be demonstrated. These principles and laws are actually an extension of man's biological characteristics, and in this respect we are dealing with a biology of larger systems.

It can be argued perhaps that we are dealing with a phenomenon with a ridiculously short life -- some tens of thousands of years, as compared with billions of years for the phenomena of microbiology and even longer periods for the phenomena of chemistry and physics. However, there is no way of proving that a certain period is too short, or long enough, for the development of principles and laws. In this case it is long enough to convince us of certain truths.

To achieve the needed knowledge and develop the science of human settlements we must move from an interdisciplinary to a condisciplinary science; making links between disciplines is not enough. If we have one subject we need one science, and this is what ekistics, the science of human settlements, has tried to achieve. Has it succeeded? The answer is that it is beginning to succeed, and that with every day that passes we learn more and more. How far have we come? How can we answer this question for any road we take if we know only the beginning and not the end?

In this article I try to demonstrate through a few examples the need for, and the existence of, a huge field of knowledge which man is trying to regain and develop in a systematic way. This field is a science, even if in our times it is usually considered a technology and an art, without the foundations of a science -- a mistake for which we pay very heavily. As I cannot present the whole case in a short article, I have selected a few points which can illustrate the validity of my statements made at the beginning of this article and the practical importance of this effort to achieve a science of human settlements.

2-1 THE PRINCIPLES

In shaping his settlements man has always acted in obedience to five principles. As far as I know this has always been true, and I myself have not found any cases which prove the opposite.

The first principle is maximization of man's potential contacts with the elements of nature (such as water and trees), with other people, and with the works of man (such as buildings and roads). This, after all, amounts to an operational definition of personal human freedom. It is in accordance with this principle that man abandoned the Garden of Eden and is today attempting to conquer the cosmos. It is because of this principle that man considers himself imprisoned, even if given the best type of environment, if he is surrounded by a wall without doors. In this, man differs from animals; we do not know of any species of animals that try to increase their potential contacts with the environment once they have reached the optimum number of contacts. Man alone always seeks to increase his contacts.

The second principle is minimization of the effort required for the achievement of man's actual and potential contacts. He always gives his structures the shape, or selects the route, that requires the minimum effort, no matter whether he is dealing with the floor of a room, which he tends to make horizontal, or with the creation of a highway.

The third principle is optimization of man's protective space, which means the selection of such a distance from other persons, animals, or objects that he can keep his contacts with them (first principle) without any kind of sensory or psychological discomfort. This has to be true at every moment and in every locality, whether it is temporary or permanent and whether man is alone or part of a group. This has been demonstrated very well, lately, for the single individual, by anthropologists such as E. T. Hall [Ref. 1] and psychiatrists such as Augustus F. Kinzel [Ref. 2], and by the clothes man designs for himself, and it may be explained not only as a psychological but also as a physiological problem if we think of the layers of air that surround us [Ref. 3] or the energy that we represent (Fig. 1). The walls of houses or fortification walls around cities are other expressions of this third principle.

The fourth principle is optimization of the quality of man's relationship with his environment, which consists of nature,

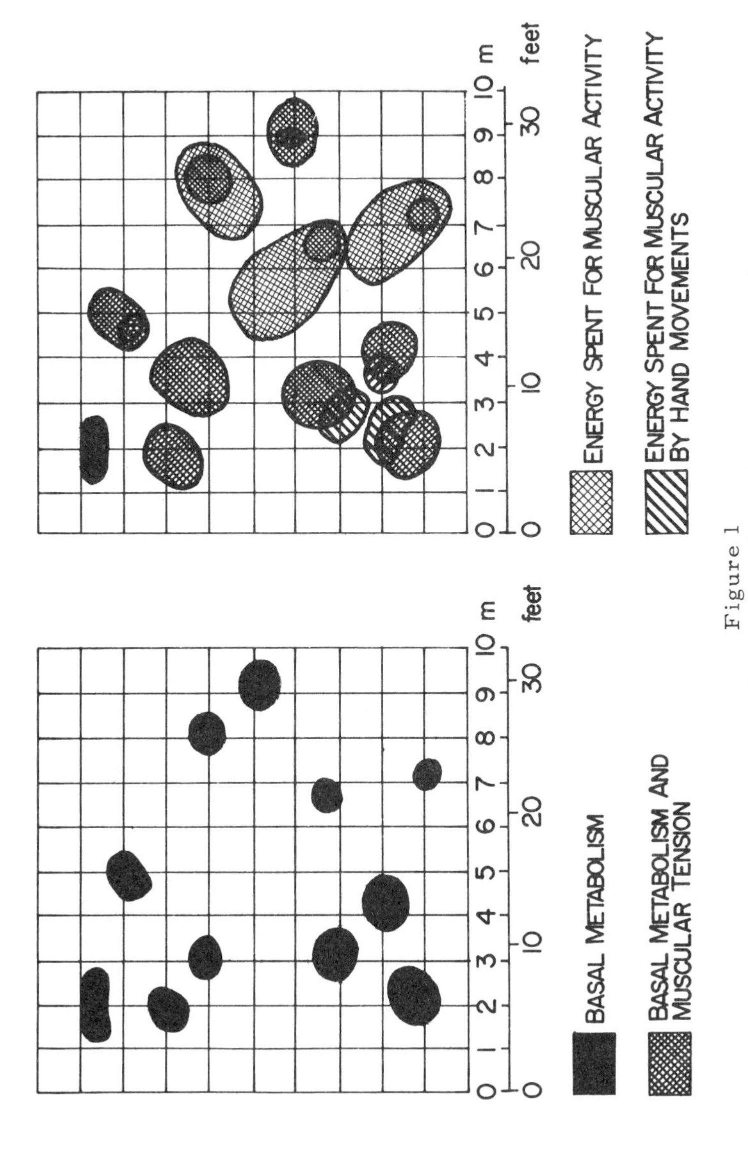

Figure 1

(Left) Static picture of a group of people as given in plans.

(Right) The real picture of the same group as given by energy measurements.

society, shells (buildings and houses of all sorts), and networks (ranging from roads to telecommunications) (Fig. 2). This is the principle that leads to order, physiological and esthetic, and that influences architecture and, in many respects, art.

Finally, and this is the fifth principle, man organizes his settlements in an attempt to achieve an optimum synthesis of the other four principles, and this optimization is dependent on time and space, on actual conditions, and on man's ability to create a synthesis. When he has achieved this by creating a system of floors, walls, roofs, doors, and windows which allows him to maximize his potential contacts (first principle) while minimizing the energy expended (second principle) and at the same time makes possible his separation from others (third principle) and the desirable relationship with his environment (fourth principle), we speak of "successful human settlements." What we mean is settlements that have achieved a balance between man and his man-made environment, by complying with all five principles.

2-2 THE EXTENT OF HUMAN SETTLEMENTS

Each one of us can understand that he is guided by the same five principles; but we are not aware of their great importance unless this is pointed out to us, and we make great mistakes in our theories about human settlements. This is because we live in a transitional era and become confused about our subject, even about the nature and extent of human settlements, confusing them with their physical structure ("the built-up area is the city") or their institutional frame ("the municipality is the city"). But human settlements have always been created by man's moving in space and defining the boundaries of his territorial interest and therefore of his settlements, for which he later created a physical and institutional structure.

When we view human settlements as systems of energy mobilized by man -- either as basal metabolic or as muscular or, recently, as commercial energy systems -- we get new insights. We see man spreading his energy thin in the nomadic phase of his history (Fig. 3), then concentrating in one area and using both energy and rational patterns when he organizes his village, where he spends more energy in the built-up part than in the fields (Fig. 4). Later we see him concentrating in the small city and using a wider built-up area, where he expends even more energy, and then, when more people are added, we

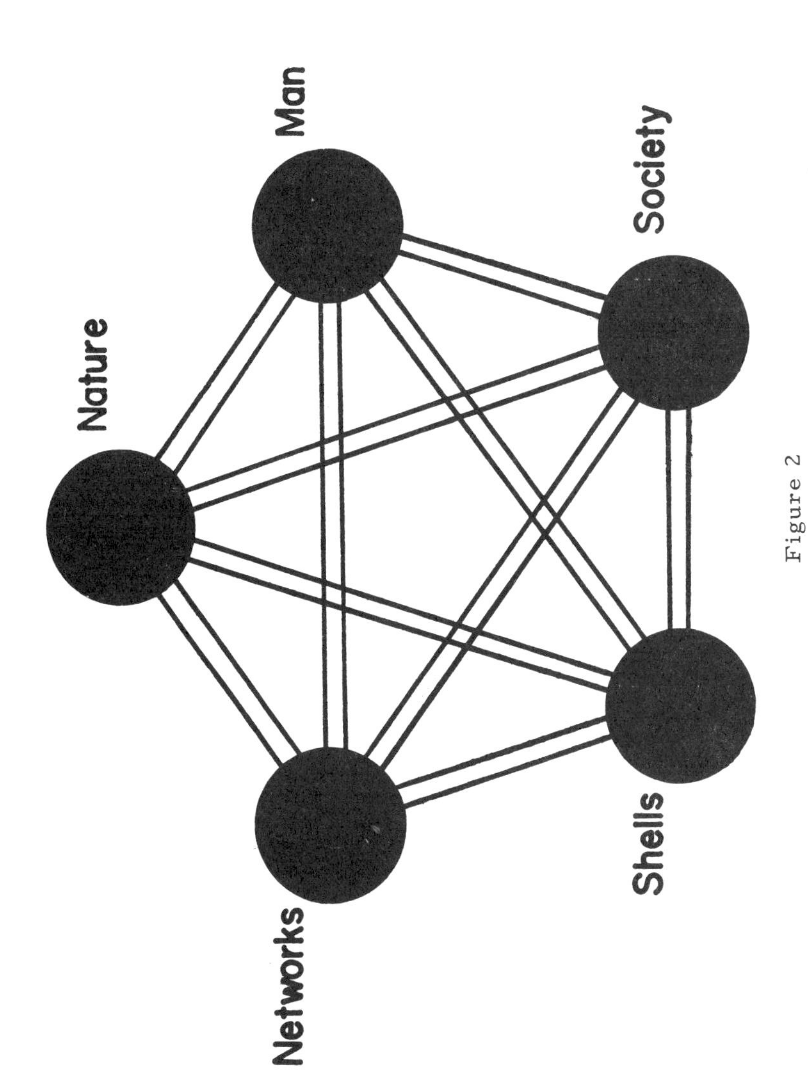

Figure 2
Fourth Principle: Optimization of the Quality of Man's Relationship with His Environment

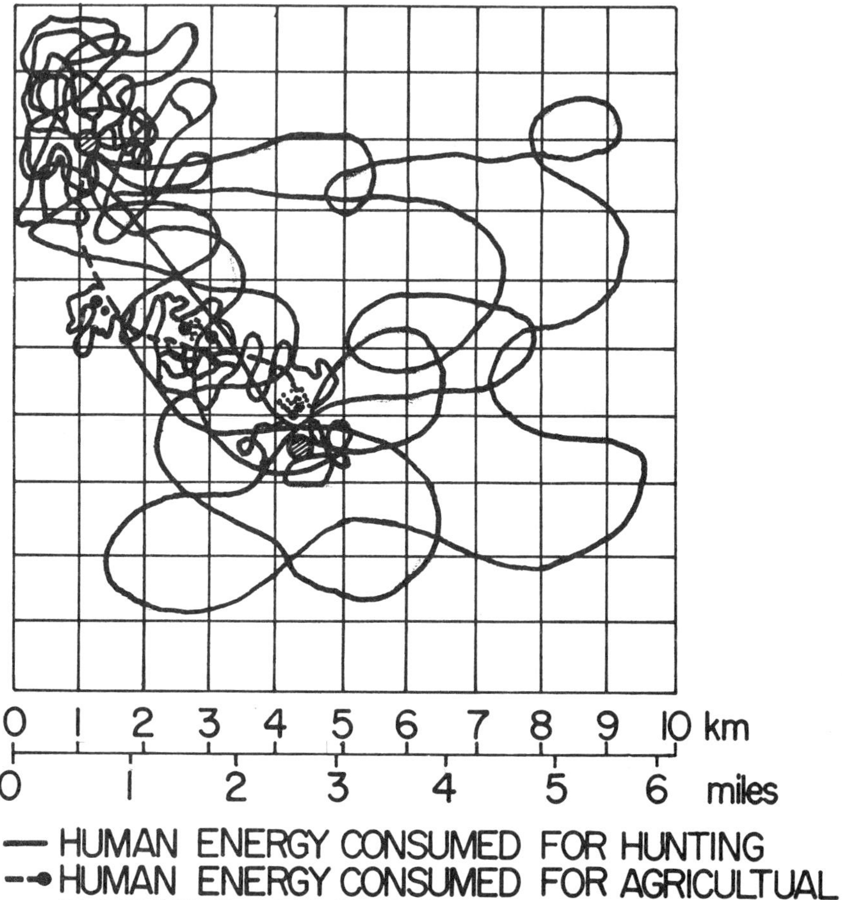

Figure 3
Energy Model for Hunters Who Begin to Cultivate the Land. Daily Per Capita Energy Consumption, 3000 Calories

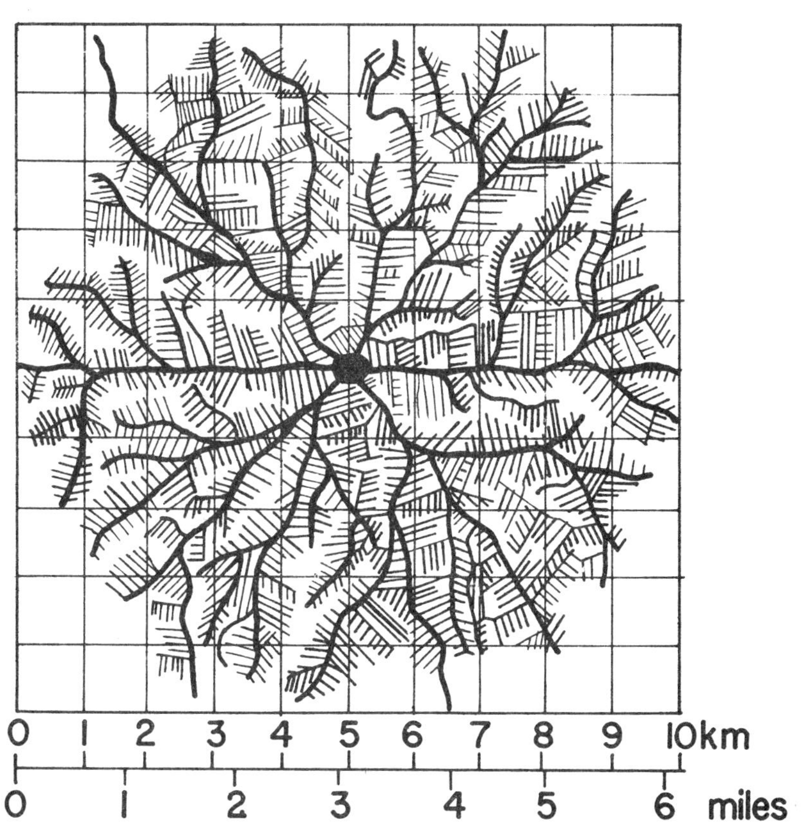

Figure 4
Energy model of a village. Daily per capita energy consumption, 8000 calories.

see him spreading beyond into the fields (Fig. 5). Finally, when he has commercial forms of energy available and can dispose much more energy without properly understanding its impact on his life and therefore without controlling its relationship to his settlement, man becomes completely confused by his desire for more energy. He suffers because, through ignorance, he inserts this additional energy into the system that he creates in a way that causes problems such as air and thermal pollution (Fig. 6).

Throughout this evolution there is only one factor which defines the extent of human settlements: the distance man wants to go or can go in the course of his daily life. The shortest of the two distances defines the extent of the real human settlement, through definition of a "daily urban system" (for a discussion of this process in urban settlements see "Man's movement and his city" [Ref. 4]).

In each specific case, the process starts with the circle whose radius is defined by man's willingness to walk daily up to a certain distance and to spend up to a certain period of time in doing so (the limit for the rural dweller is 1 hour, or 5 kilometers, for horizontal movement; the limit for the urban dweller is 10 minutes, or 1 kilometer). This leads to the conception of a circular city, and of a city growing in concentric circles (Fig. 7). When the machine -- for example, the motor vehicle -- enters the picture we are gradually led toward a two-speed system (Fig. 8), and then toward interconnected settlements (Fig. 9); then the road toward larger systems and the universal City of Ecumenopolis is inevitable [Ref. 5].

The idea that the small, romantic city of earlier times is appropriate to the era of contemporary man who developed science and technology is therefore a mistaken one. New, dynamic types of settlements interconnecting more and more smaller settlements are the types appropriate to this era. To stop this change from city (polis) to dynapolis [Ref. 6], we would have to reverse the road created by science and technology for man's movement in terrestrial space.

2-3 CLASSIFICATION BY SIZE

The changing dimensions of human settlements and the change in their character from static to dynamic, which gives them different aspects with every day that passes, makes the settlements

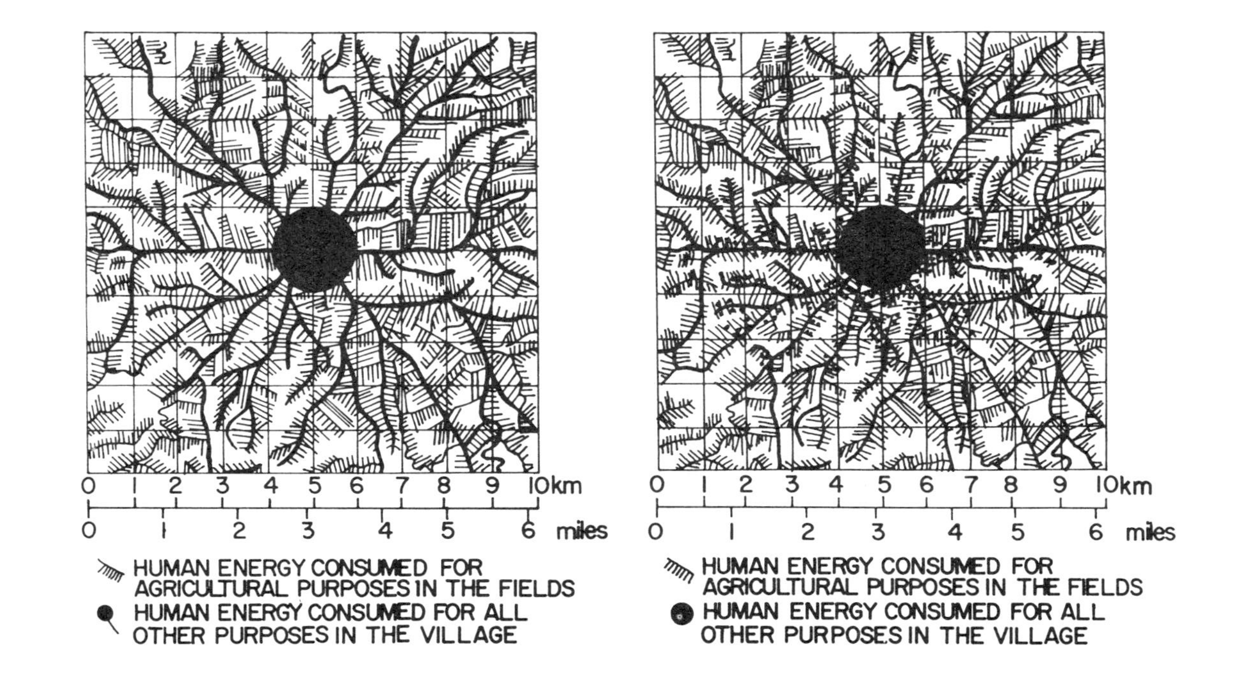

Figure 5

(A) Energy model of the central settlement of a system of villages. Daily per capita energy consumption, 12,000 calories. (Left)

(B) Energy model of the central settlement of a system of villages during the era of the automobile. Daily per capita energy consumption, 25,000 calories. (Right)

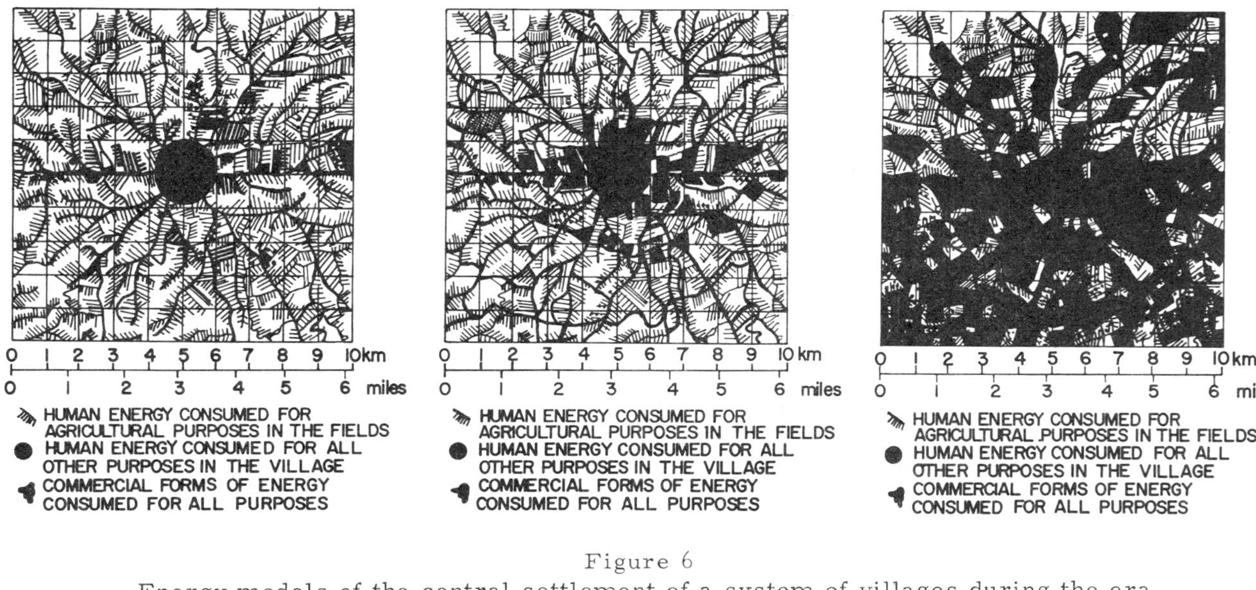

Figure 6

Energy models of the central settlement of a system of villages during the era of the automobile and of industry. Daily per capita energy consumption, (left) 33,000 calories; (middle) 45,000 calories; (right) 100,000 calories.

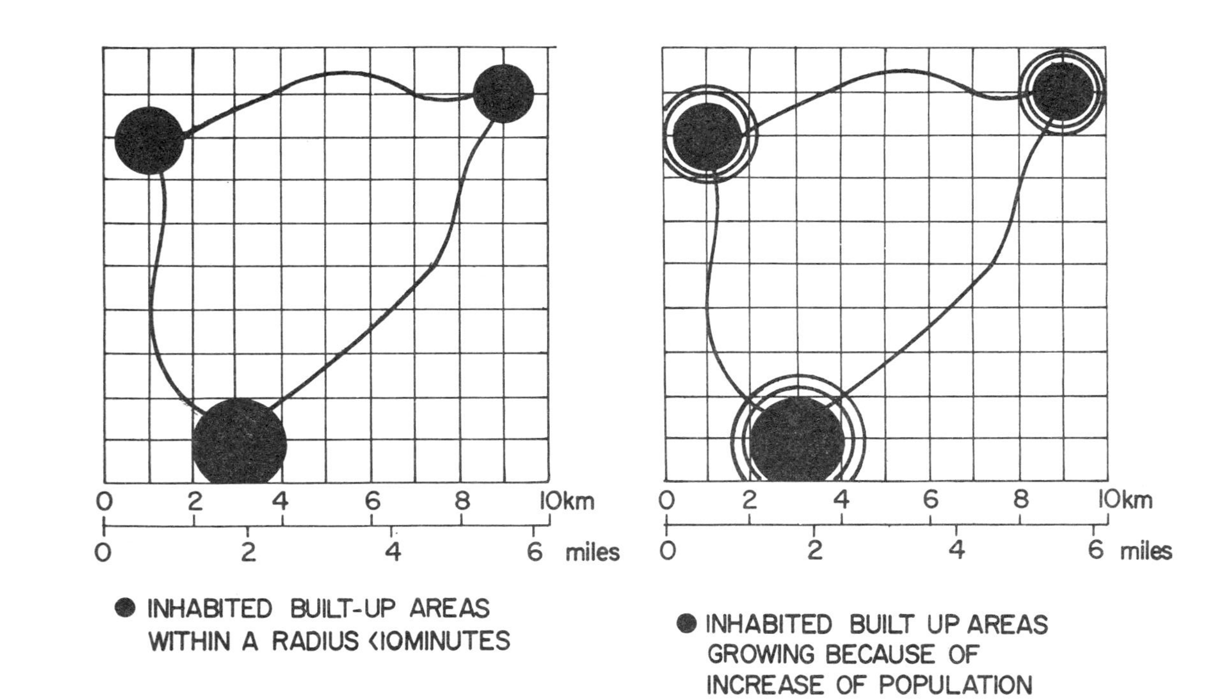

Figure 7
Growth of a system, pedestrian kinetic fields only.
(Left) Phase A; (right) phase B.

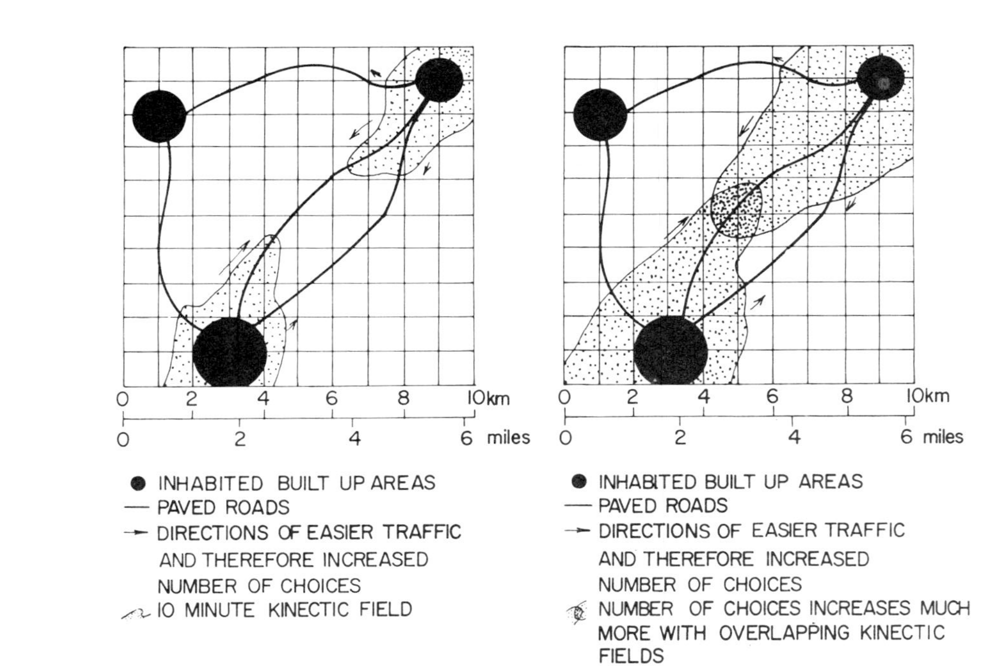

Figure 8
Growth of a system, pedestrian and mechanical kinetic fields.
(Left) Phase C; (Right) Phase D.

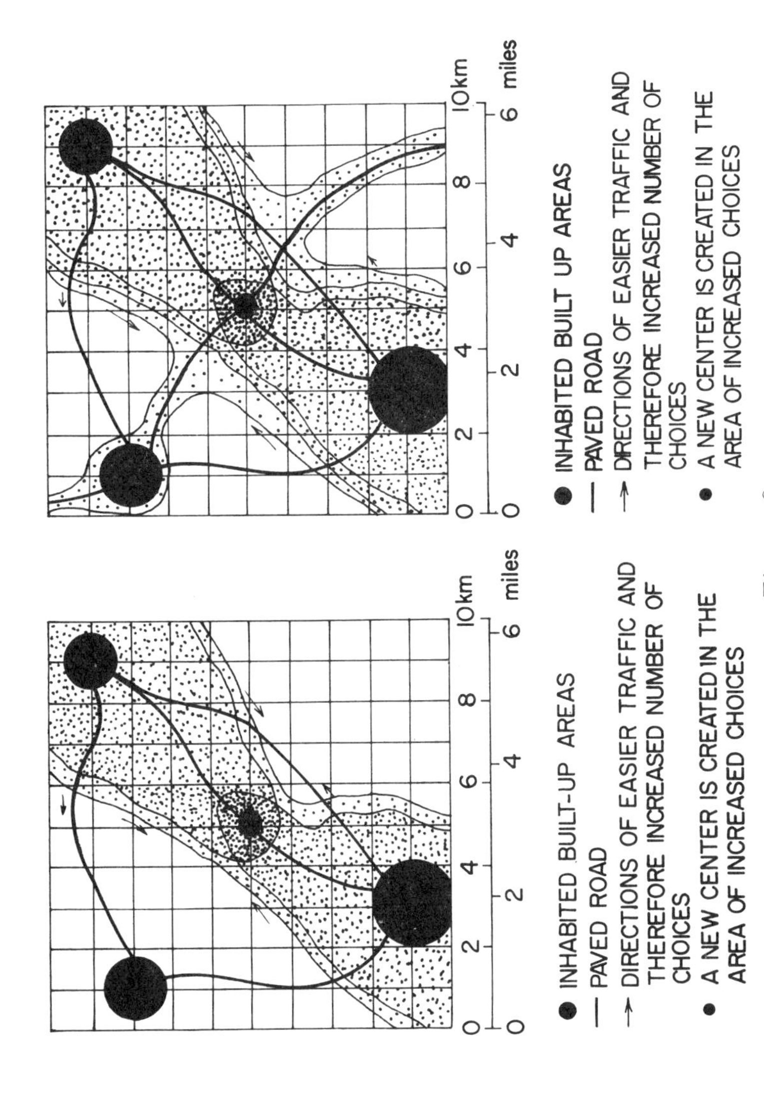

Figure 9

Growth of a system: pedestrian and mechanical kinetic fields. (Left) Phase E; (right) phase F.

confusing places in which to live, and people, instead of facing this new problem with realism, start trying to escape from the confusion. Some mistakenly support the utopian thought of returning to the system of the small city, but they do not define how this can be achieved without loss of some of the advantages that the great city has given us. Others, feeling that they cannot return to the small-city system, support the big-city concept but do not dare to face the big city's real structure; this is the attitude that leads to dystopia [Ref. 7] -- to the big city that lacks quality. But there is another road: to realize that the big city is an inevitable phenomenon, but that the quality of life within it is bad, and to try to improve the quality of that life. This is the only desirable and realistic road.

To discuss quality of life or any other important phenomenon in human settlements without referring to their size is impossible. The confusion caused by the use of terms such as small and big, town and metropolis, city and megalopolis is very great. If we want to avoid it, we must classify all human settlements by size in order to be able to understand them and assign them values. A small neighborhood with cars running through it loses its values, and a metropolis without means of very fast transportation cannot operate.

To achieve a proper classification, by sizes, of all human settlements, we should start with the smallest units. The smallest one is man himself as an individual. This spatial unit includes the individual, his clothing, and certain furniture, like his chair. The second unit is also very well defined; it is the space which belongs to him alone, or is shared under certain circumstances with a few others -- that is, his personal room. The third unit, the family home, is well defined also, as long as we have families. The fourth unit is a group of homes which corresponds to the patriarchal home of earlier days and probably to the unit of the extended family of our day; this is the unit that children need most, mothers need mainly because of the children, and fathers need, if perhaps not directly for themselves, because they are interested in the satisfaction and happiness of both mother and children. I have defined four units; of these the first three are very clearly defined, physically and socially, and the fourth can be conceived of as a social unit.

Beyond this point we do not have a clear-cut definition of any unit until we reach the largest one possible on this earth -- that is, the systems of human settlements of the whole planet. Thus we have five basic units, four at one extreme of our scale and one at the other. No other well-defined unit exists today, except for statistically defined units which are arbitrary, as may be seen from the differences in the official definitions from country to country. If we turn back in history we find, however, that, throughout the long evolution of human settlements, people in all parts of the world tended to build an urban settlement which reached an optimum size of 50,000 people and physical dimensions such that everyone was within a 10-minute distance from the center [Ref. 4]. There is no question that, for people who depend on walking as a means of locomotion, this unit is the optimum one from the point of view of movement and social interaction through direct contacts between people. Also, experience has shown that, for people who walk, it is a maximum one from the standpoint of esthetics; for example, creation of the Place de la Concorde in Paris cut from the total 3500-meter length of the Champs Elysees a length of 2100 meters, a distance from which one can reach, and enjoy, the Arc de Triomphe on foot. It is also perhaps an optimum one from the social point of view; for example, Pericles in ancient Athens could get a reasonable sample of public opinion by meeting 100 to 150 people while walking from his home to the Assembly.

Thus we now have four units at the beginning of the scale, one larger one somewhere beyond them, and one at the end -- a total of six. How can we complete the scale?

This can be achieved, for example, if we think of units of space measured by their surface and increase their size by multiplying them by 7. Such a coefficient is based on the theory, presented by Walter Christaller [Ref. 8], that we can divide space in a rational way by hexagons -- that one hexagon can become the center of seven equal ones. Similar conclusions can be reached if we think of organization of population, movement, transportation, and so on. Such considerations lead to the conclusion that all human settlements -- past, present, and future -- can be classified into 15 units [Ref. 6]. Thus the basic units are defined as units No. 1 (man), No. 2 (room), No. 3 (home), No. 4 (group of homes), No. 8 (traditional town), and No. 15 (Universal City), and a systematic subdivision defines the others. All these units can also be classified in terms of communities (from I to XII), of kinetic fields (for pedestrians, from a to g; for motor vehicles, from A to H; and so on).

2-4 THE QUALITY OF HUMAN SETTLEMENTS

We can now face the important question of quality in human settlements since we can refer to a specific unit by first defining its size. A small town, especially in older civilizations, can satisfy many of our esthetic needs for picturesque streets and squares, and this is why we like it. But most people want to visit it, not to become its permanent inhabitants, as they are guided by the first of the five principles discussed above and try to maximize their potential contacts in the big cities, in order to have more choices for a job, for education and health facilities, and for social contacts and entertainment.

In our era, which begins with London at the time it was approaching a population of 1 million, about two centuries ago, and in other areas later, we lost the ability to satisfy all five principles. Guided by principles 1 and 2 we reached the stage of the big city, but in these cities we do not satisfy the otherprinciples, especially principles 4 and 5, and we are not happy. We say that our settlements have no quality, and this is true in many respects, but we have to define what we mean. We need such a definition because we must remember that we now have much more water, of better quality, in our homes than man has had at any previous time, and we have much more energy available for conditioning our environment and for making contacts. A statement closer to the truth would be that our cities are better than the small cities of the past in many respects and worse in others.

Judgment about quality can be made in several ways in terms of the relation of every individual to his environment -- that is, his relation to nature, society, shells, and networks -- and the benefit that he gets from these contacts. We can measure his relations to air and to its quality; to water in his home, in the river or lake, and at sea (its quality and his access to it); and to land resources (their beauty and accessibility) and the recreational and functional facilities provided by them; and we can express judgments based on measurements of many physical and social aspects of the cities. Out of the great number of cases that I might cite I have selected three of the most complex ones.

We often talk about the greater contacts that the big city offers us, but we do not measure these contacts at every unit of the ekistic scale. If we do so we will discover that in units 2 and 3 (room and home) we have fewer person-to-person contacts than we had before, because of smaller families and new sources of information (radio and television); that in units 4, 5, and 6 (that is, in the dwelling group and neighborhoods) we have far fewer contacts because of the multistory building and the intrusion of automobiles in the human locomotion scale [Ref. 9]; and that in the larger units we have increased contacts because of the news transmitted to us by telecommunications media, the press, and so on (Fig. 10). In this way we see that we increase our one-way and, by telephone, two-way potential contacts with people and objects far away from our living area and decrease potential contacts with those close by. Is this reasonable for any of us, and especially for the children who cannot cross the street? This is a problem of quality of life seen in human terms. The answer to this problem is, I think, a city designed for human development [Ref. 10].

As a second case I have selected one which refers, not to the relation of man to his environment, but to the relation between two persons as they are related to their environment. If we take the case of the Urban Detroit Area, which has been defined by a 5-year study [Ref. 11] and covers 37 counties (25 in Michigan, 9 in Ohio, and 3 in Ontario), and rate the value of all its parts, taking as an example the esthetic value of its natural landscapes, and measure the number of units of esthetic value associated with places a person can visit within 1 hour, we find that the person who owns a car has access to 582 units from the centre of the city and to 622 from the outskirts. However, a person without a car has access to only 27 units -- that is, less than 1/20 the number of units to which the other person has access, even if his income is half as great. If we now remember that, in the past, poor and rich had equal opportunities to visit places by walking, we will see that modern technology has increased the gap between people relative to the choices they have for making contacts in their settlements (Fig. 11). If the Urban Detroit Area grows in a way which takes people farther apart, and if the wealthier ones move outward at a speed of 1.8 meters (2 yards) a day (Fig. 12), we can understand how critical is the situation we have created through the use of modern technology without an understanding of the whole system of the city and how we serve it.

As a third case I have selected the problem of complexity, about which we talk a lot and do very little. The great size of the modern city is not what causes the bad quality of our environment. Corporations have increased in size even more without

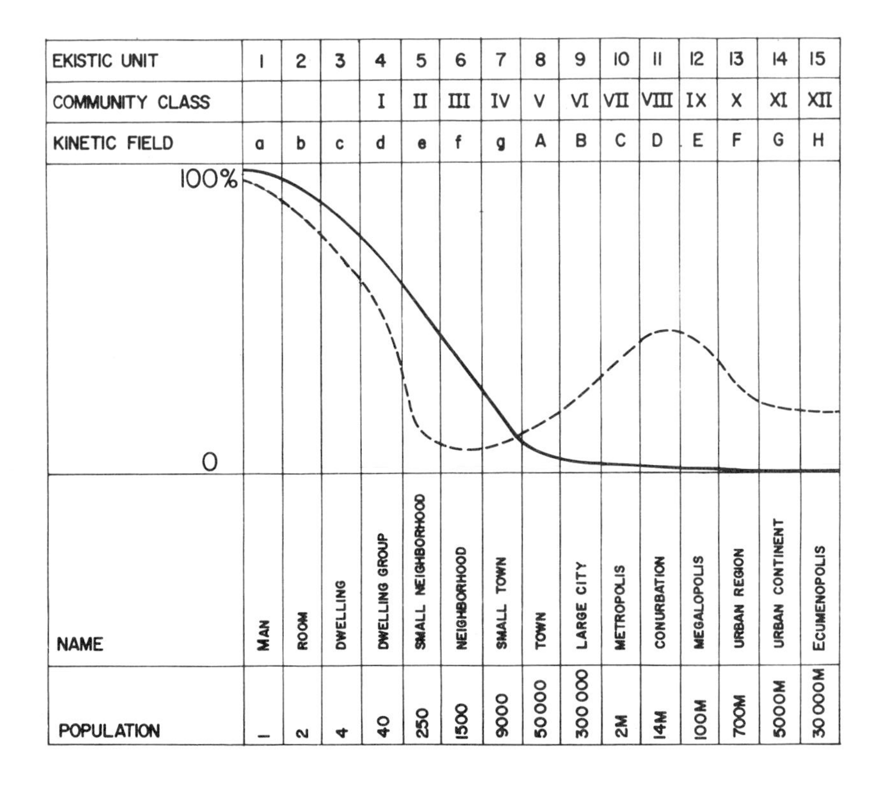

Figure 10

Contacts in the past and present in every ekistic unit. (Solid line) Past contacts, very much reduced beyond the unit of the town. (Dashed line) Present contacts. The greatest reduction is often in the small units.

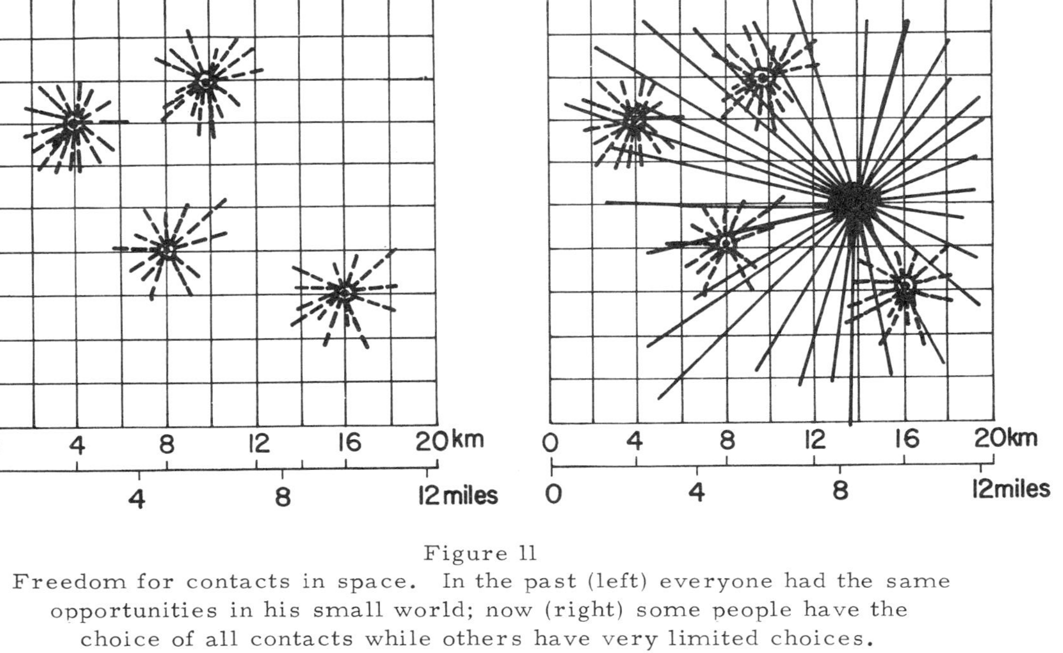

Figure 11
Freedom for contacts in space. In the past (left) everyone had the same opportunities in his small world; now (right) some people have the choice of all contacts while others have very limited choices.

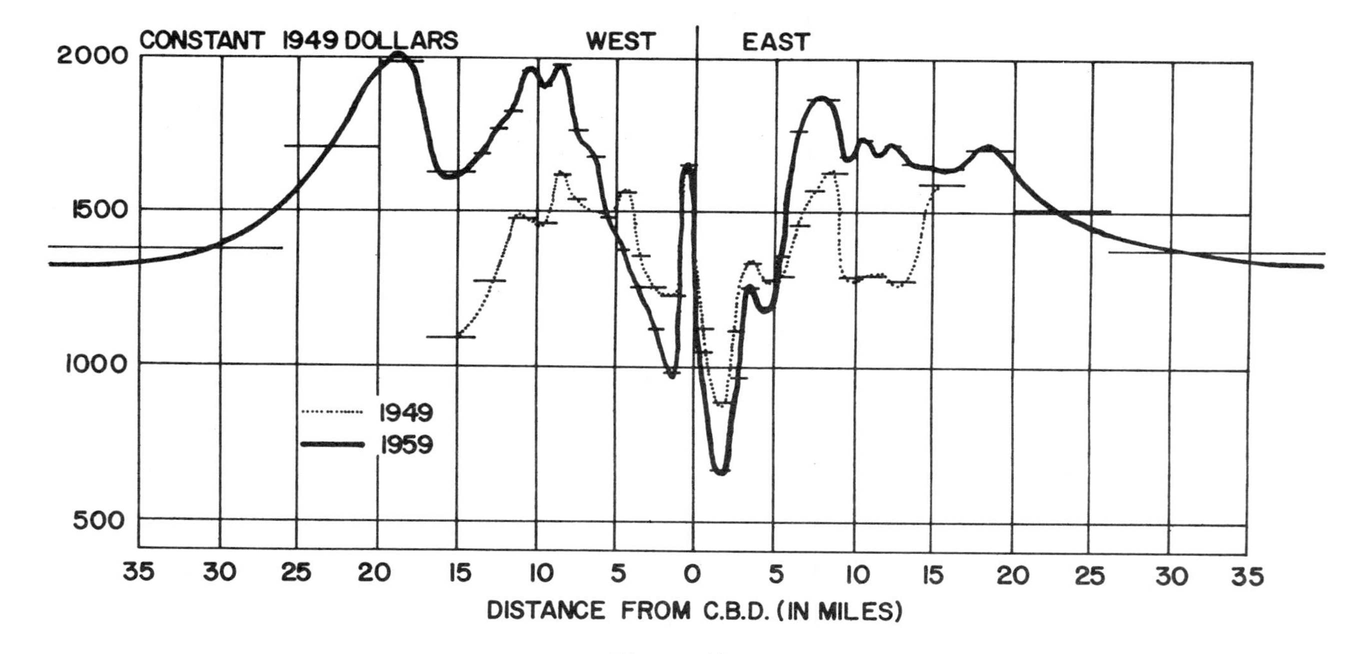

Figure 12

Outward movement of the higher-income groups in the Detroit Standard Metropolitan Statistical Area. The curves show the per capita income of people residing at several distances from the central business district (C. B. D.).

any loss in efficiency, and the armies of World War II were able to operate very efficiently despite their unprecedented size and rate of growth. The quality of our city, expressed, for example, in terms of a system of movement, is decreasing because we have not been able to reduce the increasing complexity by introducing a high degree of simplicity, as primitive man managed to do. The number of choices for primitive man in a space having no pattern of organization is the same as the number of persons in the space -- let us say 37. Since there is no structure in the system, the complexity equals the number of choices -- 37. When a structure -- social (family) or physical (wall of a compound) -- is built into the system (Fig. 13, top and bottom left), the number of choices remains 37 but the actual complexity is 15[6(compounds) + 9(Maximum number of persons within one compound)], and this means a coefficient of simplicity of 2.5. If this happens, then people learn to come together in large numbers and the same area may contain 75 people; that is, there are 75 choices (Fig. 13, bottom right) and a theoretical complexity of 75 but an actual complexity of 23 (9 + 14) or a coefficient of simplicity of 3.4.

In a similar way we find that the actual choices given an individual belonging to a group of 50,000 people, or living in a city of 50,000 population, theoretically number 50,000 (Fig. 14). These choices are reduced to 20,000 if 10,000 of the people live in the city and 40,000 live in the surrounding country (Fig. 15), and they are reduced to 5000 for a farmer living far out in the countryside (Fig. 16), as only a certain fraction of a man's time can be devoted to making contacts. What about the quality of contacts in the small village?

2-5 MORPHOGENESIS

The question now arises, if we know how to analyze and define quality, can we do anything to ameliorate conditions in cities whose quality is not high? The answer is that man has often faced many of these problems (not all) by giving his static settlements proper structure. By this I mean the settlements which were created up until the 17th century and which ranged in size from No. 2 units -- that is, from rooms which, once created, did not grow -- to No. 9 and 10 units -- large cities, very often surrounded by walls, that seldom grew. Peking is probably the only No. 10 settlement created before the 17th century. This is the structure which led to the shape and forms of the cities we admire today. It is time we tried to see how the changes came about; it is time we examined the morphogenesis of human settlements.

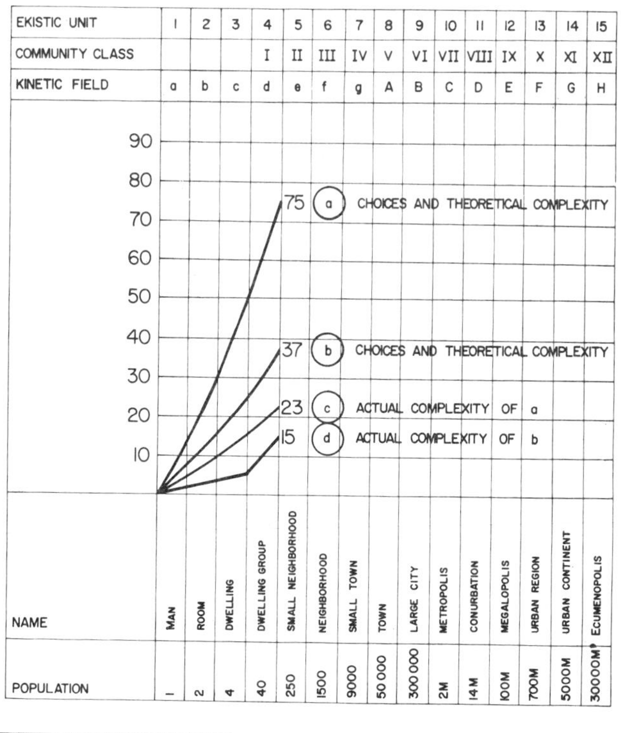

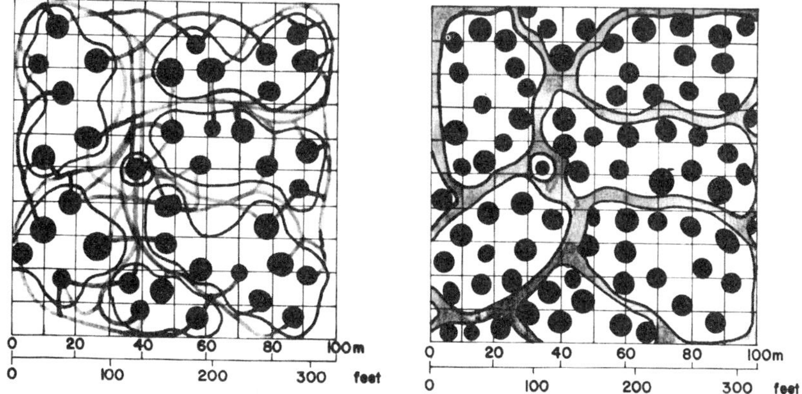

Figure 13: (Top) Complexity and Simplicity. (Bottom Left) Toward Organization of a dwelling group unit, showing first phase of organization: formation of dwelling groups, connections in certain areas, economy in use of space and time. (Bottom Right) Toward organization of a dwelling group unit, showing third phase at organization: order in function and structure, maximum economy in use of space and time,

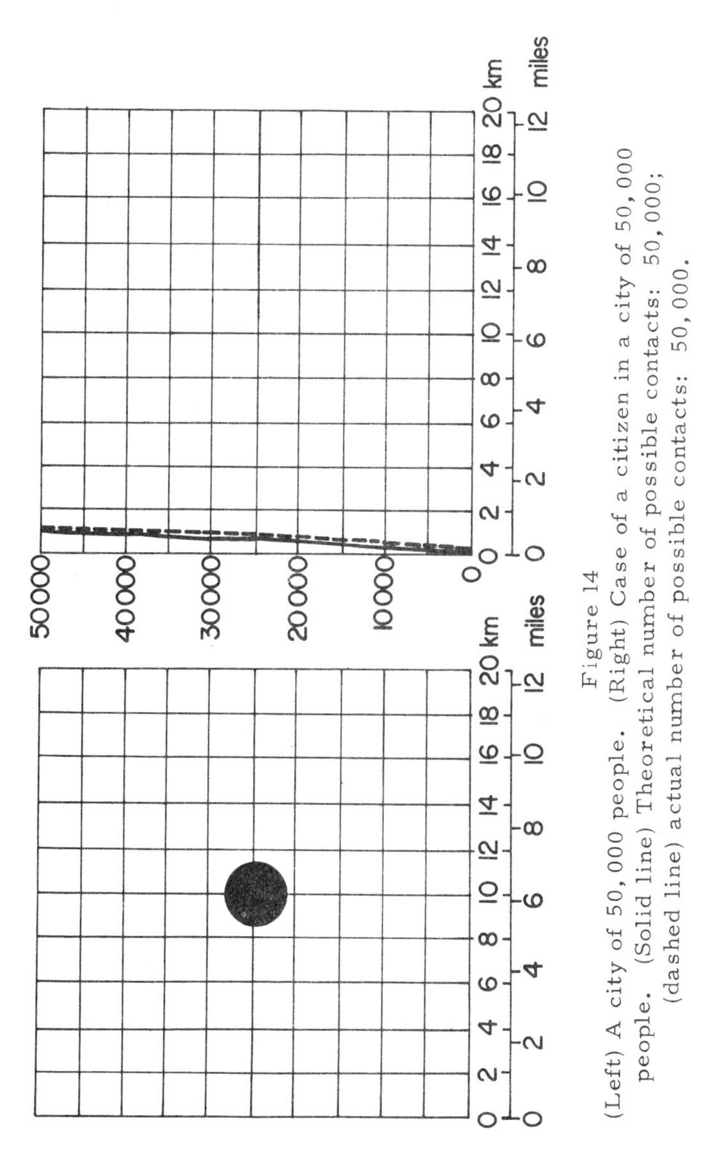

Figure 14

(Left) A city of 50,000 people. (Right) Case of a citizen in a city of 50,000 people. (Solid line) Theoretical number of possible contacts: 50,000; (dashed line) actual number of possible contacts: 50,000.

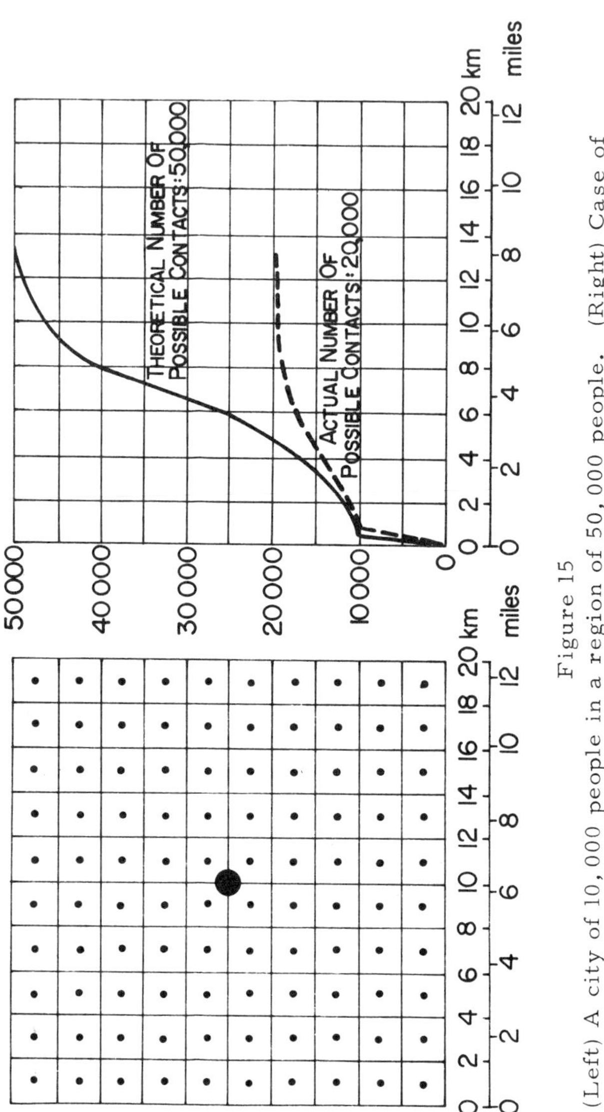

Figure 15

(Left) A city of 10,000 people in a region of 50,000 people. (Right) Case of a citizen in a city of 10,000 people in a region of 50,000 people.

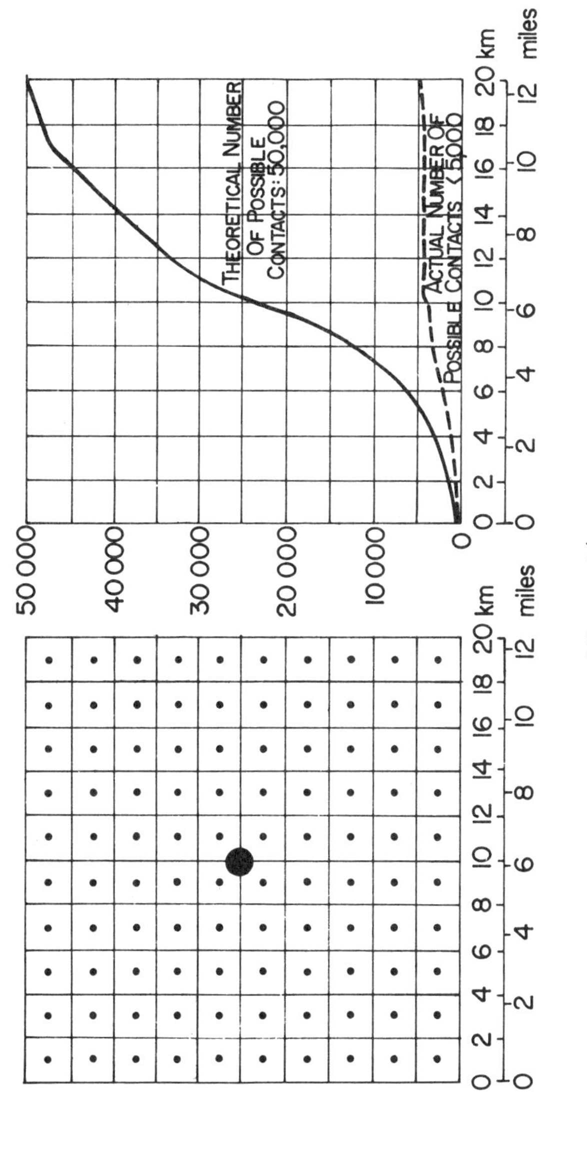

Figure 16

(Left) A city of 10,000 people in a region of 50,000 people. (Right) Case of a "peasant" in an outlying village of a region of 50,000 people.

Morphogenesis in human settlements varies with the type of unit we are dealing with. From the many types of units I will select the room, the No. 2 unit, and will follow its formation. We do not know how and when the formation of a room started. It probably started in many parts of the world, and probably the rooms had many forms and sizes. We have reason to believe that the first rooms were of moderate size (according to today's standards), but they may have been very small one-man, one-night huts similar, in a way, to those built and used by the apes [Ref. 12]. In any case the moment came when some primitive people had round huts and others had orthogonal ones, and when there were different types of roofs or, in some cases, no roofs at all. In at least one modern instance -- that of the Bushmen of the Kalahari Desert in southwest Africa -- there is no door to the hut; the Bushmen jump into it over the wall [Ref. 13].

Of great interest for us is the fact that, no matter how the first room started or how it was developed, the room always ends up, given enough time for the development of a composite settlement, with a flat floor, a flat roof, and vertical orthogonal walls. We can see the reasons for this. Man probably first builds the horizontal floor, so that he can lie down and rest, and walk without great effort or pain (the second principle). He then tends to build vertical orthogonal walls. The reasons for making the walls vertical and orthogonal are many: when he is in the room he feels at ease with, and likes to see, surfaces that are vertical relative to his line of sight (Fig. 17); he makes the walls vertical in conformity with the law of gravity; and by making them vertical and orthogonal he accommodates his furniture best (Fig. 18) and saves space when he builds two rooms side by side (Fig. 19). For similar reasons he needs a flat roof: a horizontal surface above his head makes him feel at ease when he is inside the room, and this construction enables him to use larger pieces of natural building materials and to fit one room on top of another without any waste of space, materials, and energy. In this way the form of the room is an extension of man in space (in terms of his physical dimensions and senses) and follows biological and structural laws.

Thinking in these terms, we reach the conclusion that the morphogenesis of the room is due to several forces derived either from man or directly from nature. When we move on to the house, the neighborhood, the city, and the metropolis we discover that several forces enter into the game, but their relationships change from case to case [Ref. 14]. The unit of the metropolis, for example, is too large to be influenced directly

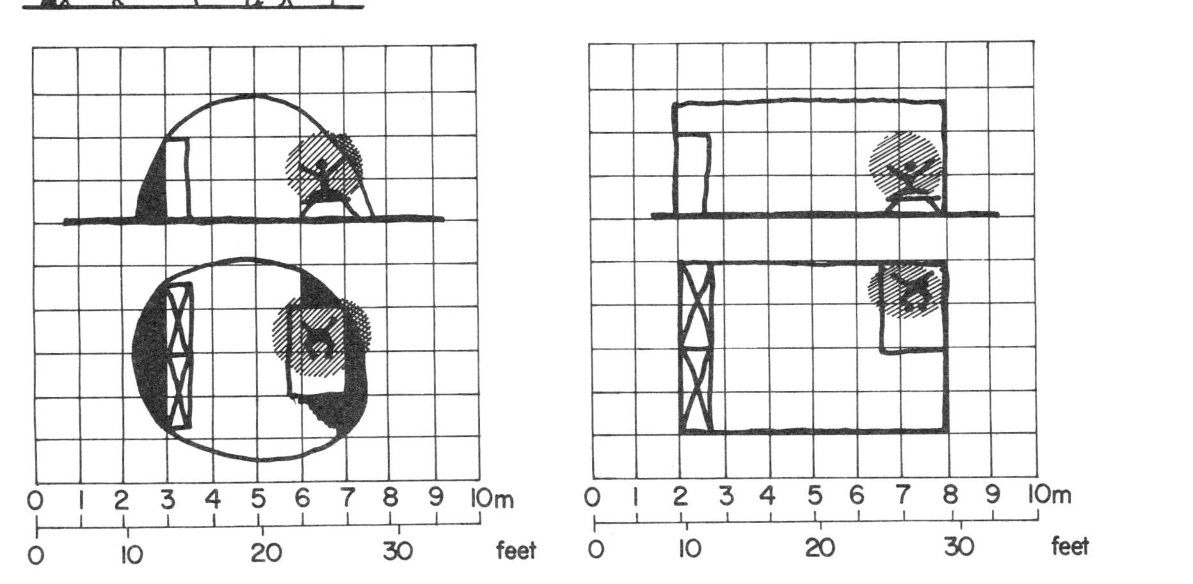

Figure 17
Formation of the walls. Walls have to fit the body and the senses of man.

Figure 18
Formation of the walls. Curved walls (left) lead to waste in the synthesis of furniture and room; straight walls (right) allow the most economic synthesis of furniture and room.

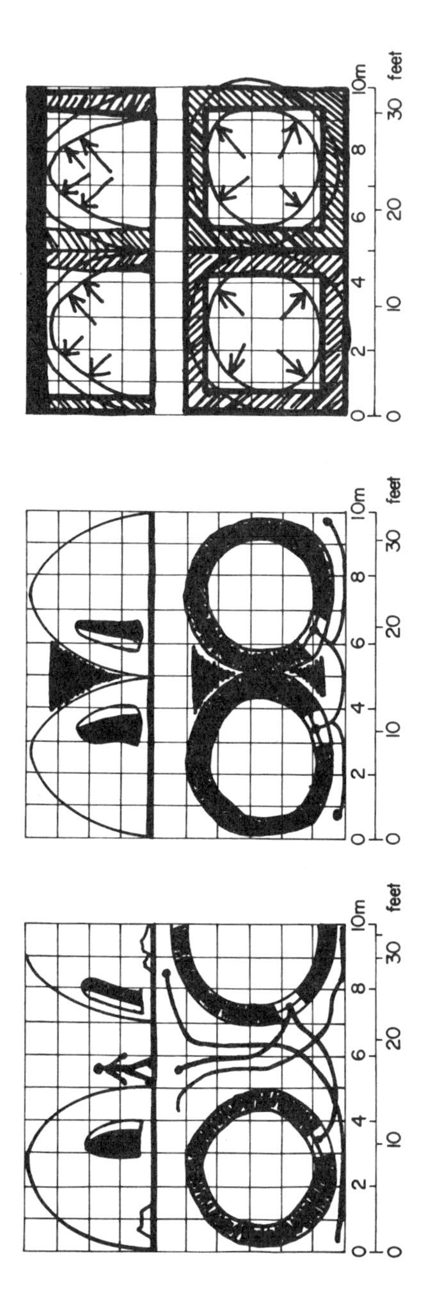

Figure 19

Formation of the walls. Two separate nonconnected rooms (left) can remain independent units, but people tend to bring them together. Two separate, connected rooms (middle) cannot remain independent units; they create many problem surfaces. Two connected rooms (right) tend to eliminate the problem surfaces; they tend to occupy a minimum total area.

by the unit man (again, in terms of his physical dimensions and senses) whereas it is influenced by the natural forces of gravity and geographic formation, by modes of transportation, and by organization and growth of the system.

Thinking in this way for all 15 ekistic units, we reach the following conclusion. The changing forces of synthesis which cause morphogenesis within every type of ekistic unit follow a certain pattern which, in terms of percentages, shows a decline of the forces derived from man's physical dimensions and personal energy and a growth of those derived directly from nature itself as a developing and operating system (Fig. 20).

Figure 20 can be understood, and will not be misinterpreted, if we keep in mind the following considerations.

First, it does not represent any specific case (a room in a desert house can be different from one in a mountain dwelling), but represents the average for all cases in each ekistic unit.

Second, the ratio between the different forces given for each ekistic unit in Fig. 20 is based only on personal experience which cannot be expressed by measurements at this stage. It is based on the assumption that all forces can be assigned equal importance. We have no way of proving that this is the case, but several trials prove simply that, by proceeding in this completely empirical way, we make the smallest number of mistakes. For this reason the shape of the surface representing the validity of each force (Fig. 20) can be considered to correspond to reality, while the ratio of one force to another is arbitrary.

What I can state here is that many years of experience as a builder of human settlements has proved for me the general validity of these diagrams in everyday practice for small-scale units and for several large-scale units, as shown in recent studies made in France [Ref. 15] and in the Urban Detroit Area study [Ref. 11, 16]. I can also say that the same diagram of synthesis is reasonably valid beyond the limits of the ekistic logarithmic scale, for units smaller or larger than the ekistic ones. Thus the ekistic logarithmic scale can be considered a basic tool for the study of synthesis in space, which is a basic characteristic of morphogenesis of human settlements. In nature, gravity, for example, plays an increasing role in larger units -- this is why large birds do little flying -- and a decreasing one in smaller units (Fig. 21). In this way we can understand the changing relationships between several types of forces which influence the formation of several types of organic and nonorganic systems in space.

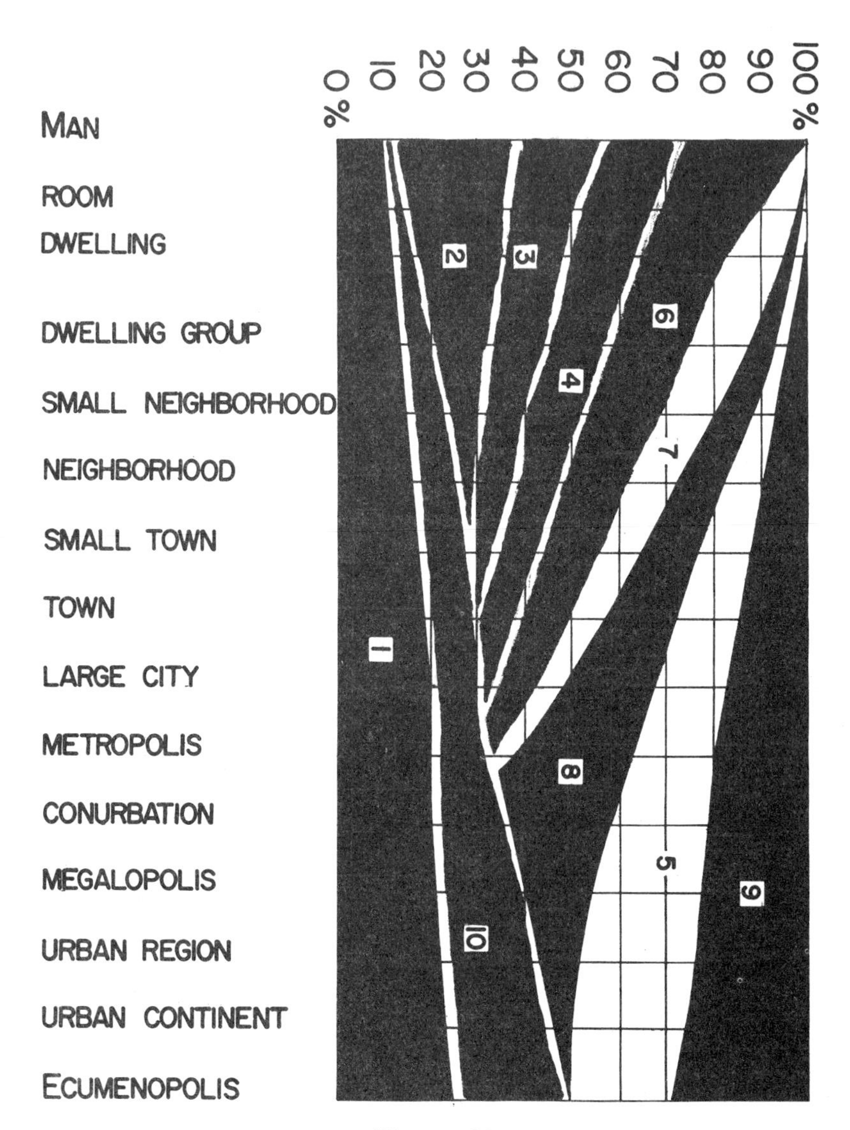

Figure 20
Probable validity of the forces of ekistic synthesis: 1, gravity; 2, biological; 3, physiological; 4, social; 5, movement; 6, inner structure; 7, external structure; 8, growth; 9, organization; 10, geographical.

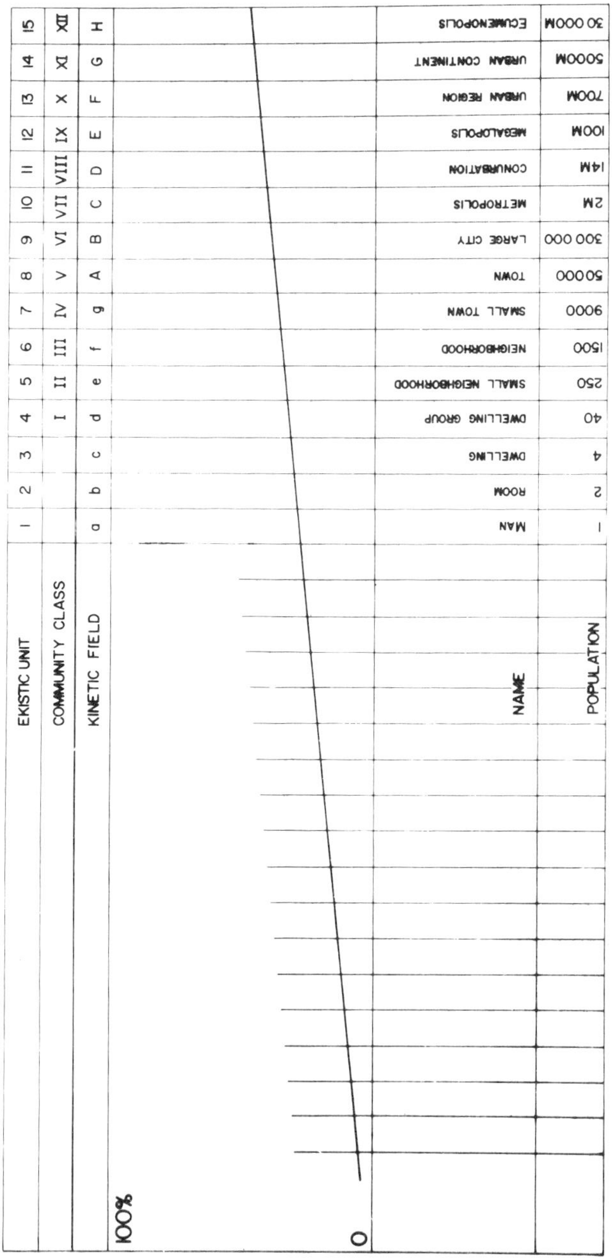

EKISTIC UNIT	1	2	3	4	5	6	7	8	9	10	11	12	13	14	15
COMMUNITY CLASS				I	II	III	IV	V	VI	VII	VIII	IX	X	XI	XII
KINETIC FIELD	a	b	c	d	e	f	g	A	B	C	D	E	F	G	H
NAME	MAN	ROOM	DWELLING	DWELLING GROUP	SMALL NEIGHBORHOOD	NEIGHBORHOOD	SMALL TOWN	TOWN	LARGE CITY	METROPOLIS	CONURBATION	MEGALOPOLIS	URBAN REGION	URBAN CONTINENT	ECUMENOPOLIS
POPULATION	1	2	4	40	250	1500	9000	50000	300 000	2M	14M	100M	700M	5000M	30 000M

100%

0

Figure 21

Influence of the Force of Gravity in Morphogenesis

2-6 TWO MYTHS

Another question now arises: If we can analyze the problem of quality and understand the morphogenetic process which should enable people to build properly and improve an undesirable situation, why are conditions so bad in our cities? The answer, apart from the fact that some problems are not related to the physical structure of the city, can run along the following lines.

1) Man, who understood the morphogenetic process for the small units, thought that the forces and laws valid for the small units were valid for the big ones that we build today, and this is not true.

2) New forces -- like motor vehicles -- have entered the game, and their impact on the city has not been understood.

3) Man did not seem able to learn about the new problems, and did not even seem interested in them, before the crisis came. He became confused, to the point of mistaking poverty for an urban problem, whereas it is simply a huge human problem which becomes more apparent in the urban areas because of the proximity of the rich, who have not been previously exposed to poverty, to the poor.

We can prove the foregoing three points in many ways, by considering some myths which still prevail in the minds of many people. I have selected two characteristic ones. I will start with the myth of the city of optimum size.

The city of optimum size. A long discussion is taking place throughout the world about the need to build new cities of optimum size, and proposals have been made by many experts and adopted in government policies, but no one can prove his case in a convincing way.

Some define optimum size as being related to the income of the people; but in the developing world, where the average per capita income increases by 2 percent a year (and by more in urban areas), what is the meaning of this optimum over a long period?

Others argue in terms of optimum numbers of people and of organizational and, more specifically, municipal efficiency, but they are not able to produce any convincing proof [Ref. 17]. Even if they could, comparisons of one city with another city have no meaning in a world where people no longer live in isolated cities but live in urban systems. But if I could prove that one city of

200,000 people had greater municipal efficiency than a city of 1 million, I must also prove that the people in the two cities were equally satisfied (otherwise what is the meaning of efficiency for them?) or that a system of five cities of 200,000 was an efficient as the city of 1 million, which is not the case.

Others base optimum size on organizational aspects such as one school or one hospital for so many children or people. But, in a world of changing ratios between age groups and of changing technical and managerial abilities, this line of thinking cannot lead anywhere. Such considerations are very useful for calculating needs which have to be satisfied in certain areas and periods but not for calculating the optimum size of the city. Technological calculations based on the means of transportation cannot be helpful either. Since speeds change continuously, how can we speak of an optimum distance? We can have an optimum distance expressed in terms of time, but this means a continually changing physical distance. Are we going to stop the development of technology?

In this changing world there is no optimum size for a city. The dynamic cities have no optimum size, but only an optimum speed of growth. And what this optimum speed of growth is, is a very complex question, the answer to which depends on many factors concerning the city itself and its relationship to the total space around it. For example, the answer for two dynamic cities, one 10 and one 30 kilometers from a metropolis, are completely different.

Is there no optimum size with which we can deal? The answer is that there is, because there is one relatively constant element, and this is man, insofar as his body and senses are concerned. I think that, for the foreseeable future, we can reckon with a man whose body and senses will not change. If this is so, we are led to the conclusion that there is a unit of space which will continue to serve his needs as it has done in the past; this unit is the circle that can be inscribed in a square 2 kilometers on a side [Ref. 4]. The importance of this unit is demonstrated by the growth of actual traditional cities and by the diagram of synthesis in space (Fig. 20), which shows that direct human forces do not go beyond the circumference of this circle. With traditional population, this unit contains 50,000 people.

The conclusion is that the optimum-size city is a myth. But any city can be divided into physical units of optimum size, and these may be used as a basis for planning that envisions an optimum number of people in a community. However, this latter goal is much more difficult to attain. I do not believe that we are ready for it, although we have the necessary arguments and data.

The static plan. Another myth which still prevails is that we can solve the problems of our cities through the conception, and official recognition, of a physical plan expressed by a two- or three-dimensional drawing. But our cities are growing organisms. They need a development policy leading to a development program which is expressed, in space, by physical development plans, but they also need economic, social, political, administrative, technological, and esthetic programs.

This does not mean that there are no areas where a physical plan can be final; if there were none, we would all be mentally ill. We need a room with constant dimensions, a home that gives us a feeling of permanency, a street and a square which do not change and which are esthetically satisfying. Such considerations lead to the question, to what extent can our environment be a constant one? The answer is that, if there is a unit of optimum size such as a room, a home, a community (up to the one of 1-kilometer radius), this can and should be constant. In this way we can face a world of changing dynamic cities by building them with constant physical units within which we can create quality -- units meant for a certain purpose and containing a certain desirable mixture of residences, cultural facilities, industry, and commerce. These would be designed on the basis of the long human experience which led to the natural growth of cities, such as Athens and Florence, or to the building of planned cities such as Miletus and parts of Paris, which we admire today.

We can design these small units if we understand the processes of synthesis and morphogenesis of the past and if we do not try to discover new patterns of life expressing nonexistent principles, just for the sake of changing the traditional ones. On the other hand, for the larger units and for the dynamically changing ones with which man has had no experience or a very bitter one, we must proceed in a different way. Not knowing what is going to be good or bad, we must use a completely different approach. We must build all possible alternatives and compare them in terms of the quality of life they offer their citizens. This approach is impossible in practice (we cannot play with the happiness and the incomes of millions) and would have been impossible in the laboratory even 20 years ago. But now we can build simulation models and compare them by means of computers.

To do this we have developed the IDEA method (the acronym stands for Isolation of Dimensions and Elimination of Alternatives). We first build all alternatives for the future of an urban system (this is possible if, through experience, we concentrate on the most important dimensions for every type of unit and every phase) and then eliminate the weakest ones. It is only in this way that we can avoid errors based on the mistaken belief that "I know," and can avoid the long period required for learning by trial and error, as primitive man learned.

This method certainly does not eliminate mistakes, but it reduces them to a minimum. Its application to the very difficult problem of the Urban Detroit Area [Ref. 11, 16] has demonstrated how useful it can be for large-scale areas for which there is no human experience at all.

Experience has convinced me that, if we can develop a science of human settlements and, through it, recognize the guiding principles, laws, and procedures of man's action regarding terrestrial space, we can build much better human settlements in the future. This will be, not through the repetition of past solutions, but through their synthesis within the new frame formed on the basis of the new forces that have entered the game. The physical features of future cities can be at least as impressive as those of the famous cities of history or of today. At the same time, the guiding principle of real freedom of choice for everyone, not for certain classes only, can be implemented for the benefit of every person, and thus man's cities of the future can be better and far more important for all their inhabitants than the famous cities of the past.

REFERENCES

1. E. T. Hall, The Silent Language (Doubleday, Garden City, N. Y., 1959); The Hidden Dimension (Doubleday, Garden City, N. Y., 1966).
2. Time 1969, 49 (6 June 1969).
3. Lancet 1969-I, 1273 (1969).
4. C. A. Doxiadis, <u>Science</u> 162, 326 (1968).
5. ______________, "The future of human settlements," speech delivered at the 14th Nobel Symposium, Stockholm, Sweden, September 1969 (Wiley, New York, in press).
6. ______________, <u>Ekistics; An Introduction to the Science of Human Settlements</u> (Oxford Univ. Press, London, 1968), pp. 27-31.
7. ______________, <u>Between Dystopia and Utopia</u> (Trinity College Press, Hartford, Conn., 1966).

8. W. Christaller, Die Zentralen Orte in Suddeutschland (Fischer, Jena, 1933).
9. C. A. Doxiadis, in Health of Mankind, G. Wolstenholme, Ed. (Churchill, London, 1967), pp. 178-193.
10. __________, Ekistics 1968, 374-394 (June 1968).
11. __________, Emergence and Growth of an Urban Region, vol. 2, Future Alternatives (Detroit Edison Company, Detroit, 1967).
12. G. Clarke and S. Piggott, Prehistoric Societies (Hutchinson, London, 1965), p. 75.
13. L. van der Post, The Lost World of the Kalahari (Penguin Books, Baltimore, Md., 1962), p. 25.
14. C.A. Doxiadis, Ekistics 1968, 395-415 (Oct. 1968).
15. Etude d'Organisation Urbaine Future de l'Espace Francais [published by Societe d'Etudes d'Urbanisme de Developement et d'Amenagement du Territoire (EURDA) under the direction of C.A. Doxiadis (French affiliate of Doxiadis Associates), as DOX-FRA-A5 for La Delegation a l'Amenagement du Territoire et a l'Action Regionale (Oct. 1969)].
16. C.A. Doxiadis, Emergence and Growth of an Urban Region; vol. 1, Analysis (1966); vol. 2, Future Alternatives (1967); vol. 3, A Concept for Future Development (1970) (Detroit Edison Co., Detroit).
17. W.A. Howard, Ekistics 1969, 312 (Nov. 1969).

Chapter 3

A SYSTEMS APPROACH TO URBAN REVIVAL

L. Alfeld & D. Meadows

3-1 INTRODUCTION

New programs touted as the answers to our many urban problems are announced in every morning newspaper. Every day somewhere in the U.S. a new urban program is initiated -- a program designed to eliminate poverty, to rehabilitate slums, to decrease traffic jams, to solve parking problems, or to alleviate some other aspect of our current urban "crisis". A local mayor promotes tax relief for new housing; a law enforcement officer seeks to hire hippies to fight against the spread of drugs in his community; a transportation department announces plans for a new mass transit system. To many Americans, the proliferation of new urban programs at the Federal, state and local levels in recent years must appear staggering.

In 1970 there were more than 600 Federally-funded programs, dispensed by 78 different agencies. 15% of our Federal budget went into these programs in 1970 and the dollar amount is increasing by 12% every year [Ref. 1, 2, 3] . These pro-

---

The authors are respectively, Director of Urban Dynamics Group and Assistant Professor of Management, Massachusetts Institute of Technology, Cambridge, Massachusetts. The research for this paper was supported by a grant from the Independence Foundation, Philadelphia, Pennsylvania.

grams include public assistance, highways, education, anti-poverty, food distribution, public health, urban development, public works, unemployment insurance, vocational rehabilitation, and business development. At state and local levels, a similar phenomenon exists.

Behind each program is someone who has identified a problem and initiated a program to solve it through direct action. These programs are based upon a common premise: by solving the city's problems individually, the city will become a better place in which to live and work. By any objective measure, this approach seems to have failed. Reports and statistics on urban crime, pollution, slums, traffic, taxes and welfare attest to this fact [Ref. 4, 5, 6, 7, 8, 9].

The quality of urban life in American cities has not been raised over the past decade. Most observers of the urban scene would agree that, relative to our expectations, the city is getting worse, not better. An ambitious employment program in Detroit, for instance, did not significantly reduce the rate of unemployment in Detroit [Ref. 10]. The almost endless supply of new housing in New York City has not alleviated that city's eternal housing crisis [Ref. 11].

If we can hope for no more than what vast expenditures of money and manpower have already accomplished, there may be little hope for our cities.

3-2 UNDERSTANDING URBAN SYSTEM BEHAVIOR

The symposium represented by these *Proceedings* has been organized to explore new approaches to urban problems, approaches which we all hope will not fall prey to the same deficiencies as those we so freely criticize. It is obvious that success will require new analytical approaches such as those to be discussed during this program. But an increase in the efficiency of present programs is not enough. If we were able, through the application of better tools to achieve a vast improvement in the operation of current urban programs, this still would not insure a better city nor the solution of current problems. Merely raising the operational efficiency of the city is not sufficient because the urban system does not seek to satisfy our goals; it operates to satisfy its own internal demands which are often quite apart from human aspirations.

It has become clear that the complex social and economic systems which are created from the interactions of individual efforts to achieve personal goals generally do not operate to further such goals. Large corporations tend to ignore individual contentment, government agencies often seem to possess a will of their own, cities do not reflect the values of their residents. Complex systems such as these are not structured to pursue objectives defined in personal terms.

This point is sufficiently important to warrant illustration. Many different individuals can take actions which appear to satisfy their own objectives; the aggregation of these efforts constitutes a complex social system. The goals which become implicit in that system will cause it to behave in some way very different from the expectations and the desires of its creators. The city is an excellent example of this perversity. Created so that the efficiencies of scale could provide better physical and economic well-being to its inhabitants, it has worked to trap many in poverty and to lower the physical and psychic health of those living in it.

Before we apply operations research tools to increase the efficiency of the city's individual components, we need a new theory which can effectively deal with urban behavior. That new theory must embrace a conceptual understanding of the city as a whole. It must facilitate analysis of the city's various activities as interrelated functions of a single complex system. Only when we are able to work within such a conceptual framework can we be certain that new urban programs will actually serve to solve urban problems.

Research over the past three years by the System Dynamics Group at M.I.T.[1] has been directed to the development and

---

1 System Dynamics is a theory of feedback system structure and a set of tools for representing complex systems and simulating their behavior. These tools were first developed at M.I.T. by Professor Jay W. Forrester in the context of industrial organizations. They were first presented in [Ref. 12] and are widely known by that name. Industrial Dynamics has become a misnomer, however, for in the past decade it has been applied to over a hundred different systems ranging from internal medicine to commodity price cycles and urban decay. Work reported here is an extension of research reported in [Ref. 13], Jay W. Forrester, Urban Dynamics, M.I.T. Press, Cambridge, Mass., 1969.

application of a theory to explain urban behavior. The primary key to understanding a city's total behavior is the concept of urban attractiveness. In this chapter we will define that concept, and describe an urban model based upon it. Through simulation analyses of that model, it has been possible to study the effects of alternative revival programs on the city as a whole. Analyses indicate that programs often have an effect exactly the opposite of that intended. Finally, we will introduce a new approach to the formulation of urban revival policies as a context for the development of more effective future urban programs.

3-3 THE THEORY OF URBAN ATTRACTIVENESS

The concept of urban attractiveness is familiar to everyone. We all make distinctions between urban areas based upon their appeal to us. The term "attractiveness" relates to our own preferences and biases in distinguishing among urban places. The sum total of all individual conceptions about the desirability of a given urban area constitutes the aggregate attractiveness of that area. It is this combination of influences which determines where we choose to live and work. On a larger scale, it simply defines the ability of a city to draw and to hold people.

Attractiveness is related to the movement of people to and from places. In our society people move between cities freely and without limit. Almost 5% of our population decides each year to move from one city to another; three times that many move from one community to another within a single metropolitan area [Ref. 14].

The decision to settle in a specific area may depend on many factors. It can involve job opportunities, the availability of housing, nearness to friends, climate, economic costs, quality of schools, racial attitudes and welfare benefits. Although attractiveness is a composite index and not a single objective measure, it has real meaning. People are quite consistent in their assessment of relative attractiveness as indicated in [Ref. 15].

A city's attractiveness is closely related to its population growth or decline. When an area is more attractive than its environment, its population will increase, as many people settle there in preference to other areas. There are, of course, time lags inherent in the perception of attractiveness. New job opportunities today will not mean new residents tomorrow. But over the long run, perhaps ten to twenty years, any increase in attractiveness relative to other areas will result in net migration into that city. That in-migration will continue so long as the area is relatively more attractive than the rest of the country. Only when this new attractiveness fades will net growth due to migration cease. Migration will, of course, continue, but new arrivals will be relatively balanced by those leaving the area.

What causes the attractiveness of a city to change? The answer to this question is the key to understanding a city as an urban system, the key to understanding why urban programs fail, and the key to restructuring a new approach to the solution of our urban problems.

3-4 GROWTH AND FEEDBACK

California is a very large area that has been a powerful magnet for migration for over a century. Each year many thousands more people move to California than move away from that state. In the last few years the population of California has grown steadily until today. 1 out of every 10 people in this country lives in California [Ref. 16]. It is obvious that such a disparity of population flow cannot continue forever; long before California is completely filled (or the rest of the nation completely emptied), people will stop moving there. Why? The answer is quite simple: most elements of an area's attractiveness are decreased by an influx of people. Schools and other city services become overloaded, new job opportunities eventually decrease, recreational facilities become crowded, travel becomes difficult, natural beauty is continually despoiled by an expanding population. All of these symptoms are becoming increasingly evident in California.

> So it is that in California one sees not only the consequence of unplanned, careless, or deliberately destructive past activity; one also gets the feeling that the worst is yet to come. There are times when the change without

apparent direction, and the growth without control, give the appearance of socially acceptable madness, of a human population irruption that may well end tragically both for the people and for the land [Ref. 17].

Increased attractiveness leads to population increases. And more people eventually decrease the attraction of any area! No city can escape from this law. Unsettled areas may become temporarily more attractive as initial population increases stimulate the creation of industry and urban amenities. Ultimately, however, attractiveness will be depressed by further population increases. Industrial expansion may enhance job opportunities in a city, but as these jobs are filled, the larger population drawn to the area will have succeeded in depressing other elements of the city's attractiveness. The opening of a new highway linking a city and a national seashore preserve will initially attract people who wish to take advantage of the new recreational opportunities. But migration stops when the highway between the city and the seashore is finally packed solid with cars.

This feedback phenomenon which limits attractiveness is not new; the same dynamics are at work in the lines which form before the tellers' windows at a bank, or in the lines of traffic in the lanes at a turnpike toll plaza. Each new arrival attempts to minimize his waiting time by joining the shortest lines. On the average, all lines tend to be of equal length and therefore, equal attractiveness. Each person attempts to maximize his own well-being and, in the process, creates a system whose internal dynamics keep all the people relatively equal.

If you accept this simple concept of attractiveness, one thing is clear: where migration is not restricted, it is impossible to raise the aggregate attractiveness of any one city relative to its environment for any significant period of time. This conclusion has profound implications for urban programs. <u>Any</u> program which successfully improves the city along some dimension will initiate a feedback response from the environment that, in turn, worsens other aspects of the city.

The failure of many current programs suggests that the urban system is too complex to analyze intuitively -- we cannot guess how the system will respond to the initiation of a new program. Individual parts of the system are sufficiently well known, however, to warrant construction of a formal model to examine urban behavior. The model then allows us to analyze system response to programs and reach conclusions about policy decisions. Alternative policies which change the component indices of attractiveness will affect migration patterns in different ways. We wish to look for policies which produce beneficial changes.

3-5 MODELING THE SYSTEM

A formal model which embodies the concept of attractiveness is necessarily complex, for the concept is a composite of many elements each influenced by feedback relationships. As an illustration of the feedback loops involved, Figure 1 indicates the relationship between housing and migration.

If new apartments are constructed in a particular city, occupancy levels fall tending to lower rental costs. As people are attracted by lower rents, the housing again fills up. The cost of housing rises until people stop flowing into the area. More housing may be built in response to higher rent levels until no more land is available for building.

Another typical relationship is illustrated in Figure 2, which shows the relationship between new industry construction and migration. Additional industry offers new jobs and thereby attracts people. People build housing. The availability of a larger labor force attracts yet more industry. This growth cycle also ends when the available land is completely filled, making it difficult to bring additional new industry into the area.

A third relationship is shown in Figure 3. If more housing is constructed, people are attracted to an area. A larger population places increased demands on urban services. Revenue requirements go up followed by tax increases. New housing construction is slowed because of higher taxes and fewer houses are built. The property base ceases to expand and taxes must again be raised.

There are time delays embodied in each of the above examples. The amount of time it takes for a response to be transmitted around any one of these loops is different in each case. New housing is generally sold during the year in which it is constructed. The time it takes to fill up the land within a city, however, takes a much longer time, perhaps a century or more, while the delay between the raising of taxes and the subsequent stagnation of the property base of any city varies greatly according to the depressent effect of new taxes and the demand for new construction in the face of such taxes.

It is relatively easy to describe and discuss each of these loops independently. Sufficient information is available in the literature and from the experience of those involved in urban systems to form an accurate picture of the system. Together, these relationships form a complex system, through which it is difficult, if not impossible, to intuitively trace the implications

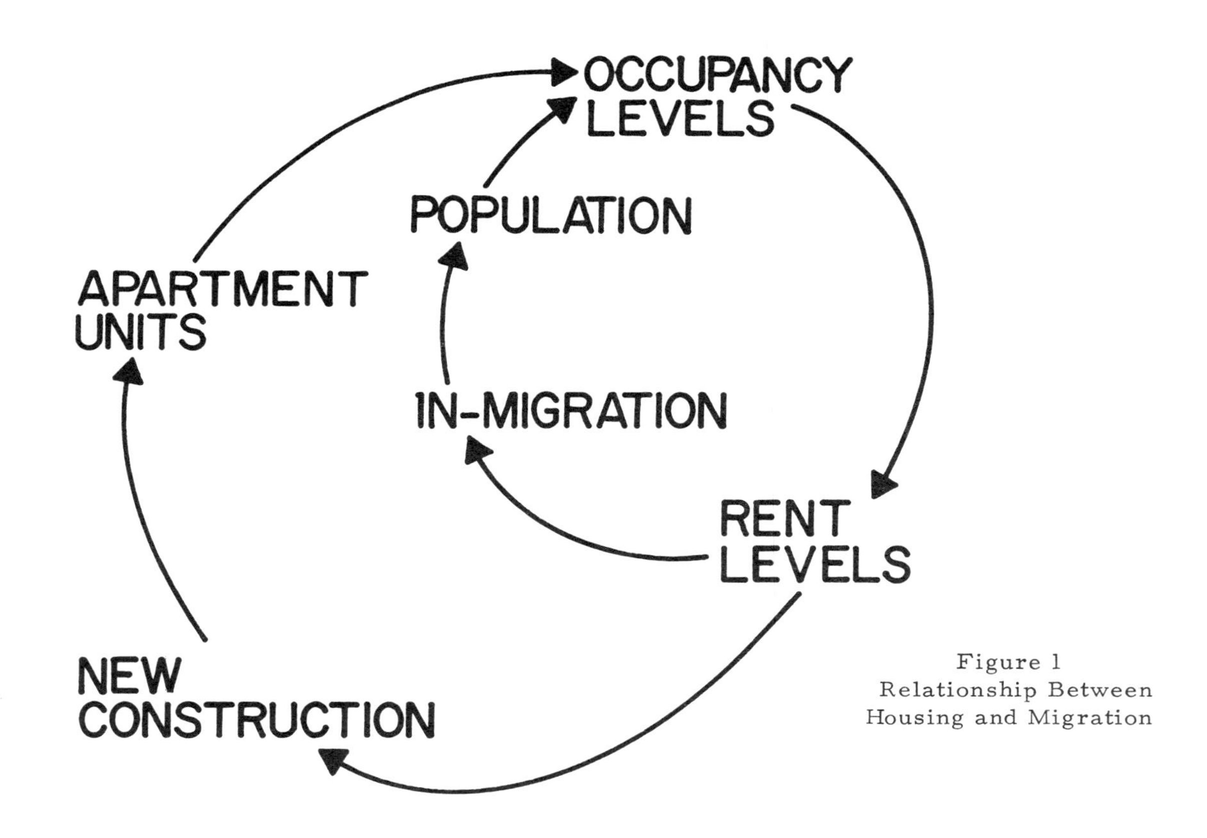

Figure 1
Relationship Between
Housing and Migration

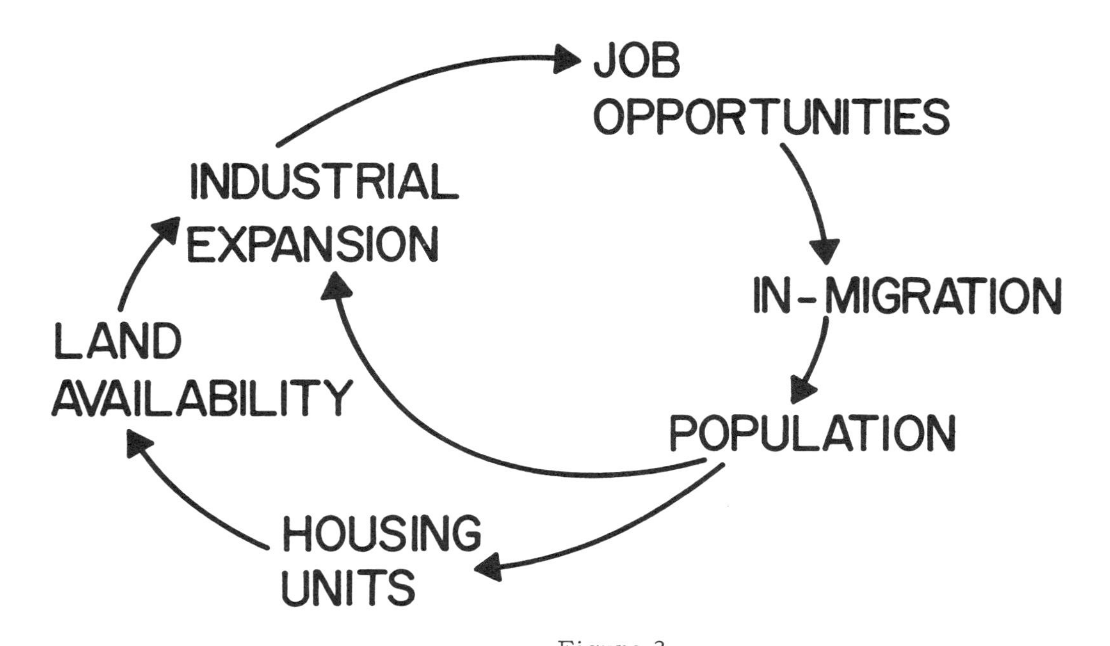

Figure 2
Relationship Between New Industry Construction and Migration

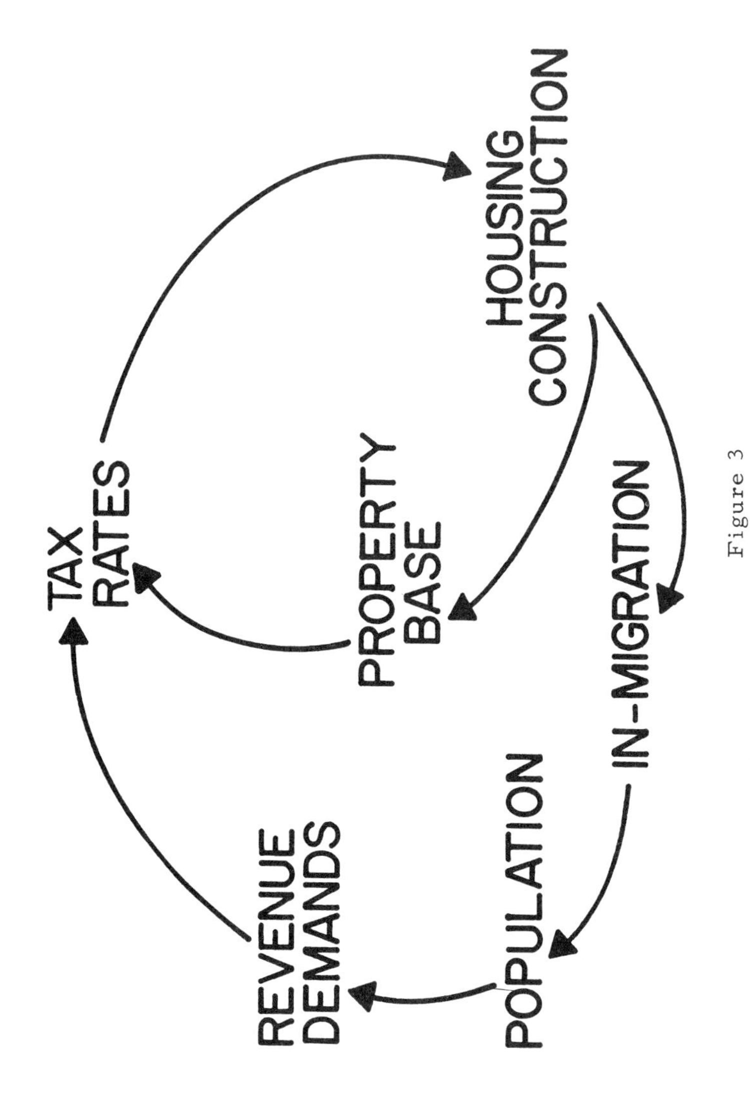

Figure 3
Relationship Between Construction Migration and Tax Rates

of a policy change. Building a model from many loops such as these, however, permits us to piece together the individual parts into a coherent whole. When the model is completed, it is then relatively easy to simulate the effects of any suggested change upon the behavior of the total urban system.

Without a comprehensive model, we are limited to tackling one problem at a time. Thus we will fail to understand the disturbing influences that any solution will propagate throughout the entire system. For example, one might reason that downtown traffic congestion can be alleviated by the construction of off-street parking facilities for commuter's automobiles. It may well be, however, that the increased availability of parking in the downtown area encourages more commuters to drive to work, thereby raising the total number of cars in the city. As parking garages reached capacity, traffic congestion would again begin to rise. The long run effect could be that the garages led to a change of commuter habits, causing them to rely more upon private automobiles and less upon mass transit. Downtown congestion problems would not have changed and the patronage of mass transit facilities would have been lessened. Although this example may be trivial, the point is valid. Long-term effects may be the opposite of short-term results. An ostensibly beneficial program may leave the city in a worse condition.

3-6 SYSTEM DYNAMICS AND URBAN DYNAMICS

System Dynamics is a method of systems analysis -- a philosophy of systems structure and a set of simulation tools for modeling a system's feedback loop structure and its behavior over time. It was created to represent the characteristics and implications of feedback loop structures such as those outlined above. These tools were first applied to industrial systems and the term "Industrial Dynamics" takes its name from a book by Jay W. Forrester which reported that work [Ref. 12]. The term has, however, become a misnomer. Over two hundred studies conducted since that first book have applied the methodology to problems ranging from internal medicine to commodity fluctuations and the management of research in the market economy.

One recent study resulted in the book titled Urban Dynamics [Ref. 13], which explored the complex feedback loop structure of the urban system.

Urban Dynamics is a book about the behavior of urban areas as they grow and eventually reach a point of equilibrium. It centers around the concept of an urban area as a system of interacting industries, housing and people. Normally, the internal dynamics of the sectors of a city cause it to develop, growing outward over the land area until the supply of vacant land is exhausted. Then the processes of aging often yield the symptoms of urban decay. As this process unfolds, the population characteristics and the types of economic activity within the city change. Unless some force for renewal is present, the filling of the land turns the city from growth to stagnation. Once the land is filled, new growth cannot occur as easily and, as the area ages without new houses and industries, it becomes even less attractive to new construction. This loss of attractiveness leads to further decline.

Urban Dynamics has been built around the concept of attractiveness. It includes those parameters which cause people to move to and from a city and those which cause the city to age over time. The behavior which we witness in American cities today is explained by the Urban Dynamics model. This model permits the study of programs which may be introduced into the city system to produce short-term changes in urban attractiveness.

The System Dynamics technique organizes the feedback processes of the system into a model and employs a digital computer to simulate the dynamic behavior of the system. The model's initial conditions can be set so that it grows over 250 years from a nearly empty land area to a large city. The model can also be initialized so that it starts with the equilibrium conditions that are reached at the end of the growth phase. The equilibrium model is used to explore the results of various policy changes upon the condition of the city over a 50-year time span. All such changes are represented in terms of the city's condition or attractiveness, relative to the condition of its surrounding environment. Although both the city and its environment change, only the relative differences in these changes are important to the model behavior.

The internal system of the model is composed of three primary subsystems representing business activity, housing and population. These are shown in Figure 4. Business activity is

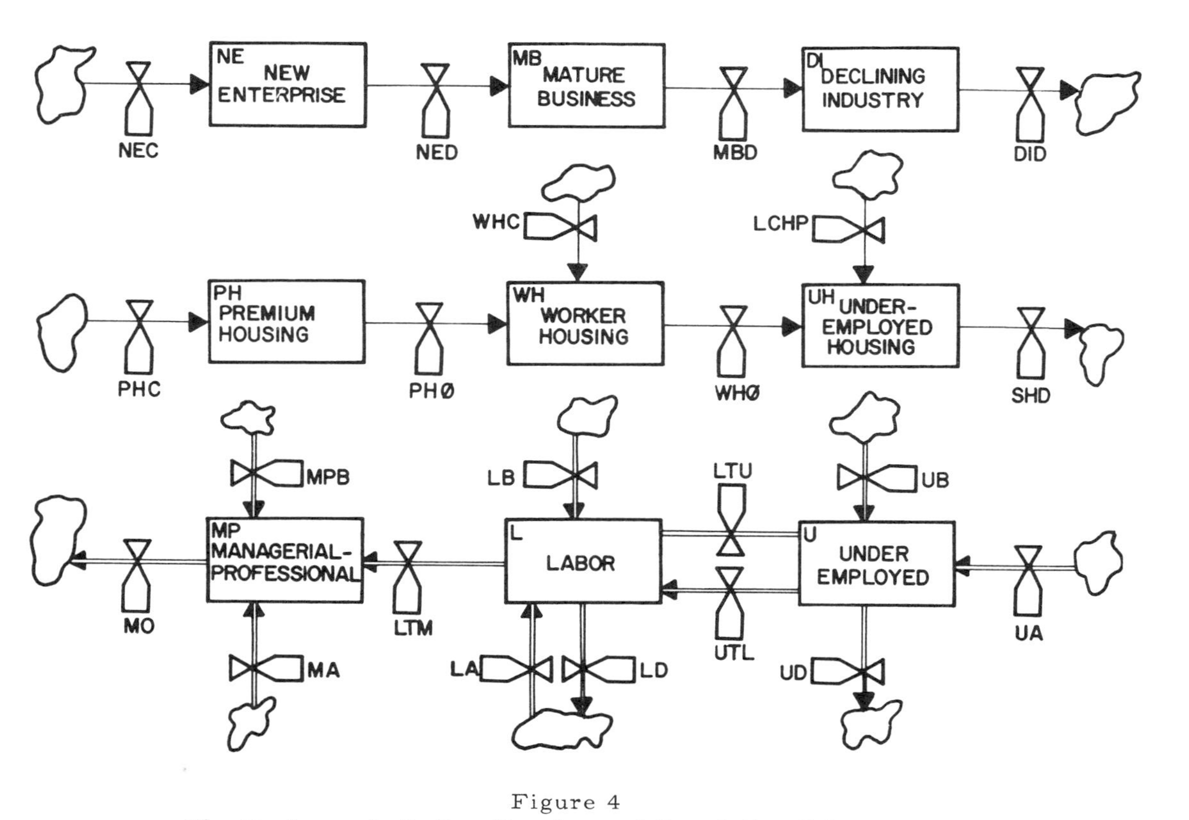

Figure 4
The Business Activity, Housing and Population Subsystems

divided into three categories according to the age of the buildings that house business activity.

The flow of business activity from one category to another depends upon both time and the condition of the entire urban system. Industry ages more quickly as taxes increase or as the labor supply drops. The middle subsystem contains three categories of housing which respond to the cycle of construction, aging and demolition. New housing enters the system on the left and eventually exits the system through demolition on the right.

The lower subsystem represents population and three categories of people are shown: manager-professional, labor and underemployed which correspond roughly to income distinctions in today's society. Note that in addition to flows of people between categories, people migrate directly between each category and the environment. As conditions in the cities change, each of these three categories of population will find the city either more or less attractive. They respond to this perception by entering or leaving the city. Thus, the composition of the population will change according to the conditions represented in the city.

Each business activity employs a certain ratio of managers, labor and underemployed. It is important to note that as businesses age, fewer jobs are available. Thus new enterprise is essential to the maintenance of high employment levels.

The categories which were described are called system levels. These principle levels are affected by rates of flow into and out of each of the levels. Each rate of flow is dependent at any moment upon the system levels. This network of levels and rates is connected by an information network. The model is too complex to show in its entirety, but Figure 5 shows how one flow rate within the system is affected by the levels of the entire system.

This represents the arrivals of underemployed to the city in response to their perceptions of changes in the aggregate attractiveness of the city.

With most of the important interactions between the various components of the urban system included in the Urban Dynamics model, the simulation runs are capable of explaining the growth and decay of cities, and the city's dynamic response to new programs.

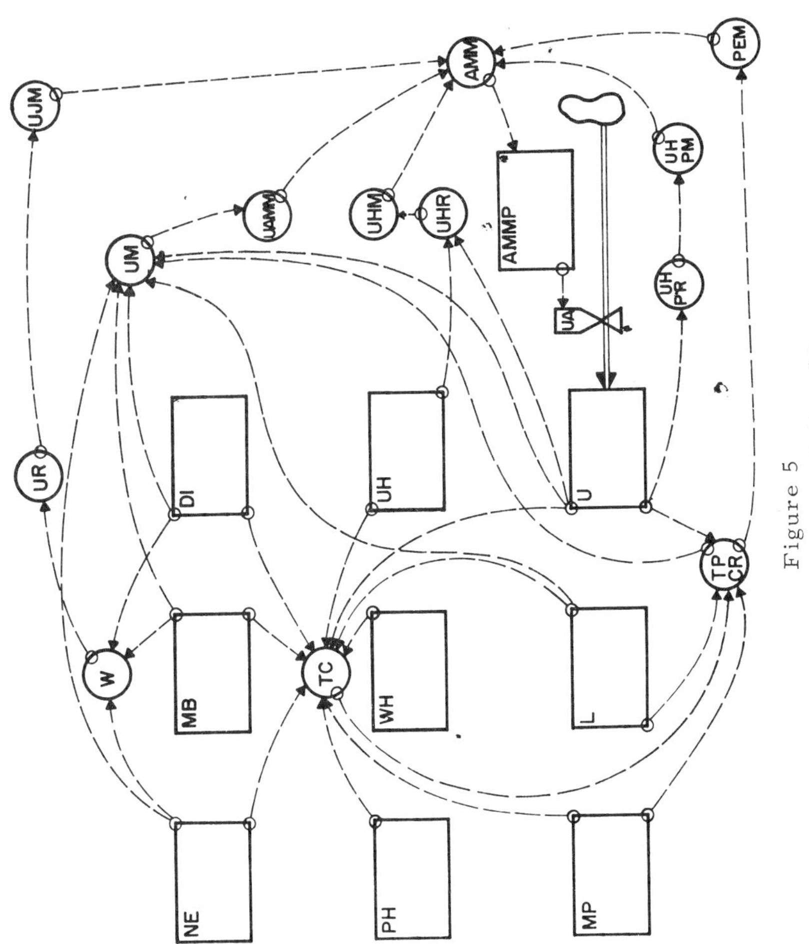

Figure 5
Subsystem Interrelationships

3-7 RAISING ATTRACTIVENESS

We are all aware that there is a growing pressure within our central cities for increased quantitites of low-income housing. Much of the housing stock in our central cities is rapidly deteriorating, while its cost is increasing [Ref. 18]. Residents of the central city, therefore, find themselves caught between an increase in price and a decrease in quality.

A common reaction to the problem is to construct low-cost housing. That program can be represented in the Urban Dynamics model. Consider, for example, a program of low-cost housing construction in the central city -- a program that attempts to add about 5% to the stock of low-income housing each year. This new housing should alleviate the city's housing problems; our brief description of the often counterintuitive behavior of the urban system, however, should make you wary of the ultimate results of such a program. We should expect that another problem within the system will be worsened. The problem we have chosen to attack -- that of low-cost housing -- may even suffer in the process. Both suspicions are correct.

Figures 6 and 6a give the results of the introduction of such a program on the city. The graphs show the outcome in rather vivid terms. Housing available to the underemployed begins to rise immediately as the program gets under way. Due to the increased availability of housing, more underemployed begin to be attracted to the city and the total underemployed population rises for the first 10 years. Filling the land available with low-cost housing lowers the attractiveness of the area for new enterprise, depressing the construction of new industry. The combination of increased numbers of underemployed and fewer new job opportunities acts to depress the overall attractiveness of the area, stopping additional in-migration of underemployed.

After 50 years, housing conditions have not materially improved. A scarcity of jobs has raised unemployment and lowered the upward economic mobility of the poor. A new equilibrium has been reached which is far worse than that which prevailed at the inauguration of the low-cost housing program.

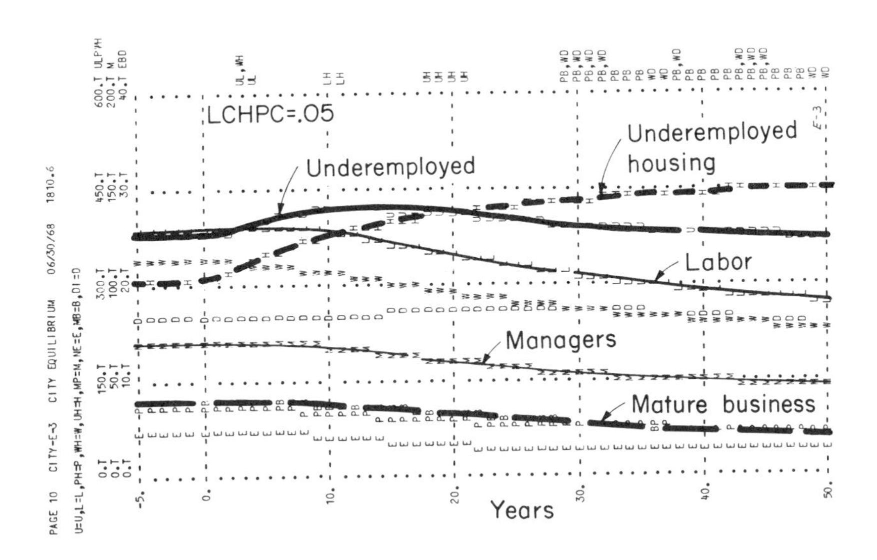

Figure 6

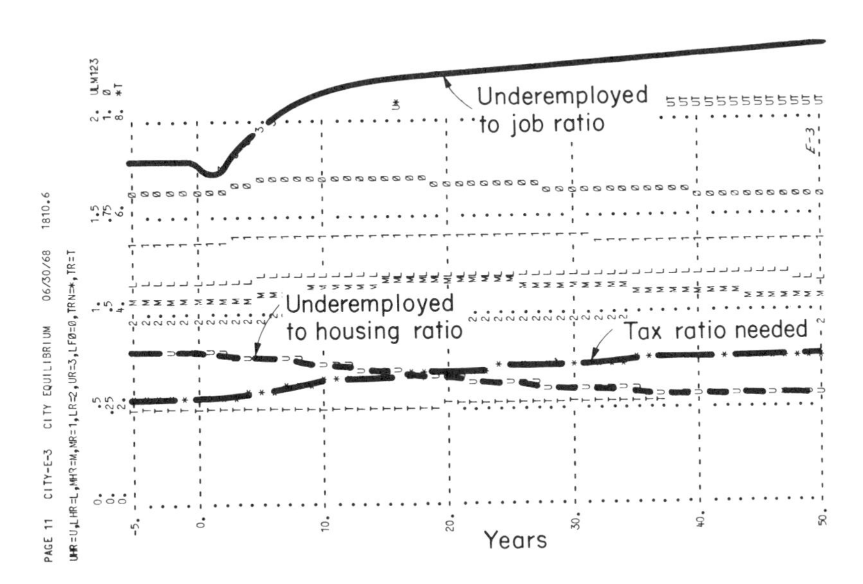

Figure 6a

Effects of Introducing a Lowcost Housing Program

American cities continue to form policy on the basis of short-term results. The vast majority of the programs which have been instituted to improve the city have had deleterious effects similar to those just illustrated. Is it any wonder, therefore, that urban problems are now reaching "crisis" proportions?

In the long run, most programs which succeed in raising the city's attractiveness ultimately succeed in increasing in-migration until the system is driven back down to its starting point. As a result, the attractiveness has not been changed, and many of the problems are often worse. This internal rebalancing of the system occurs automatically; as new pressures are introduced into the system they are distributed within the system. When one component of attractiveness is raised, another component often suffers. When additional housing is created, economic standards fall. It is impossible to increase the quantity and quality of everything for everyone. It is only through the explicit manipulation of the counterbalancing effects that we can gain a degree of control over the changing parameters of the urban system.

3-8 NEGATIVE COUNTERBALANCES

Efforts to raise attractiveness are defeated by negative counterbalances within the system. These are forces which act within the system to eventually stem the flow of in-migration. As the housing example showed earlier, the inevitable negative counterbalance may even show up as a decrease in that component of attractiveness which was first raised. A negative counterbalance can also appear elsewhere in the system to substitute for a natural attractive feature of an area. For example, in Los Angeles, the climate and the extensive beaches have not changed. However, the counterbalancing forces of smog, traffic, crowding and urban sprawl have seriously detracted from that area's natural amenities. This compensation within the urban system will always occur.

The principal lesson of Urban Dynamics is that such negative counterbalances need not be unexpected events chosen by the system. City administrations can decide which trade-offs they prefer, which amenities they wish to forego in gaining others. This decision will be politically difficult, but it cannot be avoided. If it is not made deliberately by citizens, it will be made implicitly by the system. When the system decides, it is rarely an outcome that anyone would consciously choose. We must understand that when we make a conscious choice to add housing to the city we are also making a long-term choice to increase unemployment, or traffic congestion, or crime, or raise the cost of living. Cities must develop comprehensive strategies for dealing with their problems, strategies which account for both the positive and the negative forces arising from any program.

Cities cannot close their doors to migration, though that would be a solution. The United States closed its doors to migration when it was threatened at the turn of this century with unlimited influx from countries around the world. Had the immigration laws not become more restrictive, the level of attractiveness in the United States could be no higher than any other country's. Its population would have grown sufficiently to exhaust its resources. Yosemite Park, one of the nation's most beautiful national parks was made more accessible to the public via a new road system. This ease of accessibility has attracted people to Yosemite. Today the crowds are so great that the park service is attempting to restrict entry to the park in order to save it [Ref. 19].

Some artificial barriers to migration between cities do exist. These are largely economic barriers. The barrier that exists between central cities and their suburbs, for instance, is an economic barrier. Residential zoning requiring large plots of land and large houses has restricted all but the well-to-do from entry into the suburban community. To a lesser extent, racial discrimination has also raised barriers to migration between city and suburb. I do not wish to argue the merits or fault of artificial restrictions to migration. It is enough to say that such economic barriers and discriminatory policies are not currently available to the central city. If the central city is to solve its urban problems, it must do so by explicit recognition of its inability to maintain an overall attractiveness substantially higher than that of the environment.

3-9 URBAN TRADE-OFFS AND ATTRACTIVENESS

In order to be effective, urban programs must involve trade-offs. Since we cannot hope to raise the attractiveness of the city we must instead concentrate upon shifts among the various components of urban attractiveness. We must identify the areas of city life which a city's residents deem most crucial. By de-emphasizing the less important aspects which its residents are willing to give up in return, the quality of urban life might be raised without raising the overall attractiveness of the city.

One of the basic urban problems is poverty. Efforts to raise the economic status of a city's residents are often given first priority. If programs to raise the economic well-being take the form of welfare payments, low-cost housing, increased job training, or special employment programs, they will attract others from outside the city. The question is then raised: what must the city give up if its relative attractiveness to outsiders is to remain unchanged? If new programs provide jobs for city residents, then the programs must take something away to keep newcomers from competing for those jobs. Unfortunately a city has few things that give it sufficient leverage to offset the attractiveness due to the availability of more jobs.

The list of possible negative counterbalances can be divided into two categories, economic and social. Economic counterbalances are those which make the city less attractive from an economic point of view. They increase the cost of living in the city. There are many devices for accomplishing this end; increased taxes, increased cost of living through higher prices, lower wage rates, and even automobile insurance laws can all tend to create economic counterbalances. Unfortunately, if the goal of a city administration is to raise the economic well-being of its residents, it can hardly rely upon economic counterbalances to discourage in-migration. Any gains that it might make through providing additional jobs would be eliminated by the counterbalances. Thus, the city's forced social counterbalances are those which decrease the quality of urban life. They can take many forms. A decrease in housing space, a poor transportation system, a low quality of urban services, a rising crime rate, fewer parks or higher pollution -- all are examples. Many of these social counterbalances are undesirable, infeasible or ineffective. Some of them have very little leverage in determining migration.

We must concentrate on those high leverage counterbalances which make up a significant portion of the composite attractiveness of a city, and which are politically practical in the sense that the residents of a city might be willing to trade them for increased economic affluence.

Housing, perhaps, has the greatest leverage; a lack of housing or a tight housing market will act as a deterrent to in-migration. Scandanavian cities, for example, are known for their continual housing shortage and lack of slums. A poor transportation system or a congested central city is another possibility, although perhaps less sensitive than housing. A third, and as yet unexplored area of counterbalance may be to put some self-imposed limits on the individual freedom of the residents of a city. These limits may take the form of curfews, repressive liquor laws, or some form of regimentation. Obviously, such a system could not be implemented without the full cooperation of the residents themselves. Such trade-offs, however, are difficult to discuss in the abstract. They must be tailored individually for each city. *The important point is that if the economic well-being of a city's residents is to be improved, such tradeoffs must be made.*

Urban Dynamics describes some of these trade-offs in analyzing the effects of various urban programs on the city. Model simulations show that programs for increasing housing, jobs training, financial aid, tax subsidies and increased employment all produce the type of urban failure illustrated earlier. They do not succeed in raising the quality of life in urban areas, because migration and the consequent negative counterbalances offset any gains they initially achieve.

Our research indicates, however, that some programs do succeed in making gains in raising the economic standards of the residents of the city. Programs which encourage the construction of new enterprises in the city to provide jobs for residents and at the same time discourage increased migration by limiting the amount of housing available in the city, tend to produce an increased upward mobility among the residents and to hold in-migration fairly constant. Part of this program is effected by demolishing slum housing and constructing new industrial parks in its place. Thus, the amount of housing is limited and, at the same time, the number of jobs is expanded.

Figures 7 and 7a indicate the major changes which result from this policy. Such a program involves a trade-off, of course: jobs for housing. It is a conscious choice, however, one that the residents of any city can make for themselves. The urban model permits us to analyze similar policies and programs which can have a beneficial effect upon the city. Combinations of programs can be structured which do not drive out the poor but rather force them up the economic ladder.

3-10 THE STRUCTURE OF URBAN PROGRAMS

Our staff at M.I.T. is now engaged in a continuing research effort to identify additional ways in which the city can make use of negative counterbalances to raise the well-being of its residents. Urban programs designed in this way will succeed in reaching their goals, because they are designed to operate within the system. They will take advantage of the system behavior to produce a new equilibrium point that is in harmony with the desires of its residents. Traditional urban programs have ignored the concept of equilibrium and negative counterbalances and, in attempting to raise the attractiveness of a city, have met with ultimate failure.

The adoption of policies such as are suggested by the Urban Dynamics analysis and the trade-offs which it requires will enable any city to become better off than it once was. Each of these cities, in a sense, can raise its attractiveness to its own residents and decrease its attractiveness to outside migrants by selectively focusing upon the internal trade-offs which its residents prefer. Urban pressures will not disappear for such pressures will always exist. What we have failed to realize before is that we can choose which set of pressures we are willing to tolerate in order to achieve the broader goals of society.

Through enlightened urban policies, it is possible for this country to move into an era of a new urban prosperity that would be quite different from the era of urban problems which we face today. To do so requires a concentration of will, both on the part of the city residents in adopting trade-off policies and negative counterbalances, and on the part of our national and state administrators and politicians in adopting policies which will further this process. At the center of such a process is the use of dynamic modeling techniques to evaluate new programs and

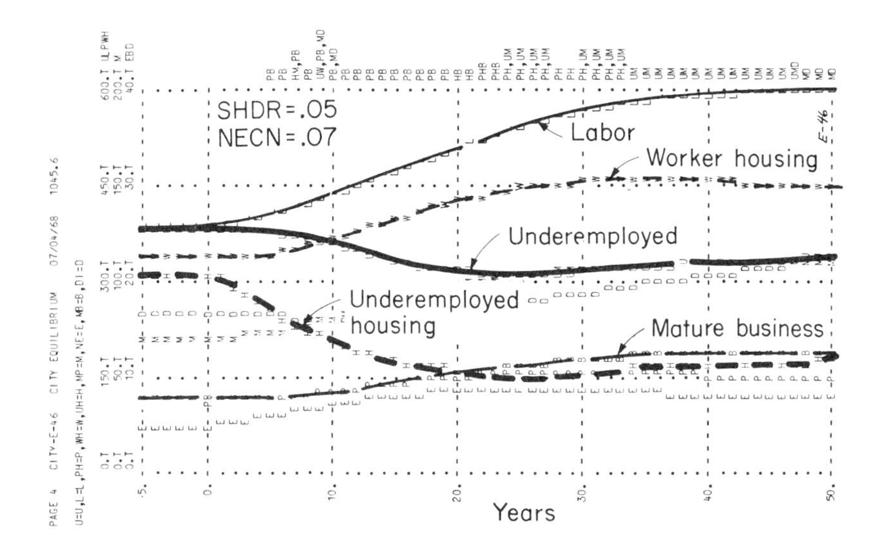

Figure 7

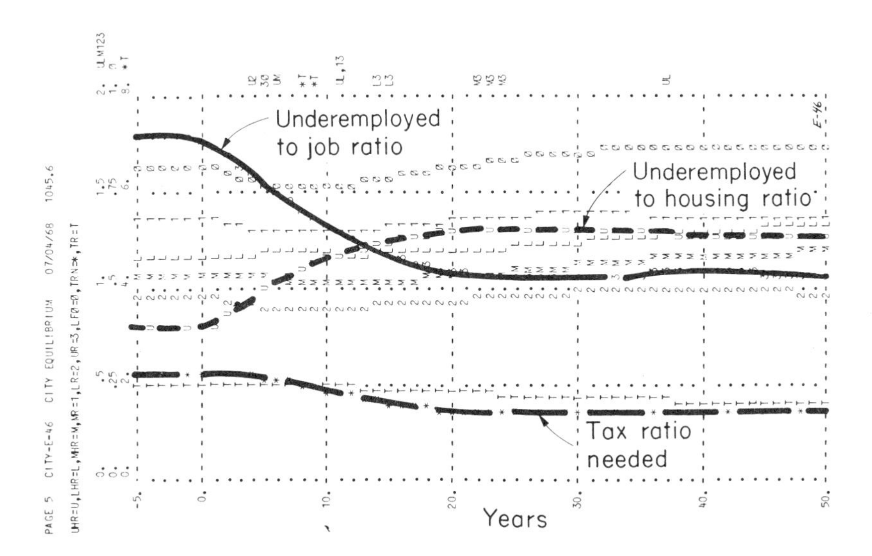

Figure 7a

Effect of Jobs for Housing Trade Offs

suggest ways for improving old ones. The application of the systems approach to the city promises tremendous benefits. It far offsets the price that we must pay to get them.

What we hope you will gather from this article is the understanding that efforts to raise the attractiveness of a city either through making it operate more efficiently or by the introduction of new programs, must be combined with specific programs of negative counterbalances if urban problems are actually to be solved. By working within the conceptual framework which were outlined, we should all now be able to produce solutions which have a far greater chance of success in addressing urban problems than those of the past. We can begin to look forward to the day when we can once again reclaim our cities as places in tune with our own aspirations and a true reflection of our own goals.

REFERENCES

1. Statistical Abstract of the U.S. 1969.
2. "Present Federal Programs", mimeo 150 pp., Laboratory for Environmental Studies, MIT, 15 November 1968.
3. Boston Globe, 18 October, 1970.
4. "Crime Expense", U.S. News and World Report, 26 October, 1970.
5. "Welfare Dilemma", Wall Street Journal, 17 September, 1970.
6. "Rapid Transit", New York Times, 1 June, 1970.
7. "Slums", Christian Science Monitor, 23 September, 1970.
8. "Local Tax Rates", Boston Globe, 6 September, 1970.
9. "Smog", Time, 10 August, 1970.
10. "Area Trends in Employment and Unemployment", U.S. Dept. of Labor, Manpower Administration, March, 1967-June, 1970.
11. "Decline of Bronx", New York Times, 13 October, 1970.
12. Forrester, Jay W., Industrial Dynamics, M.I.T. Press, 1962.
13. Forrester, Jay W., Urban Dynamics, M.I.T. Press, 1968.
14. U.S. Bureau of Census, Population Studies.
15. Lowery, Iras., Migration and Metropolitan Growth: Two Analytical Models, 1966.
16. U.S. Bureau of the Census; Population Changes, 1960-1970.
17. Dasmann, Raymond F., The Destruction of California, The MacMillan Co., New York, 1965, p. 21.

18. "Low-Cost Housing Shortage", Christian Science Monitor, 28 August, 1970.
19. "Siege and Conquest of a National Park: Admission of Automobiles", American West, Vol. 5, January, 1968, pp. 28-31.

Chapter 4

CHANGE AND EQUILIBRIUM IN THE URBAN SYSTEM

B. Harris

A systems view of urban problems should be much more than a catalogue of interactions or a platform from which to launch highly specific proposals for action or for research. Systems considerations as such are of little more than trivial interest if they do not provide major insights into problems through some general theoretic concepts. These theoretic concepts should have the property that they will sustain new deductive conclusions.

In this paper, I propose to discuss two or three concepts having to do with urban modeling, and more particularly, having to do with the relations between various urban policies and the goals of urban development in the context of system equilibrium. The principal topics which I will discuss have to do with problems of form, or morphology, problems of change, and problems of measurement.

The author is the chairman of Department of City and Regional Planning, Graduate School of Fine Arts, University of Pennsylvania, Philadelphia, Pennsylvania. This chapter is reproduced with permission from Highway Research Record Number 309.

Before taking up these topics, let me first comment briefly on my view of the city as a system. We know, naturally, that nearly every system is a subsystem or an element in some larger system, and frequently the degree of interaction with the external environment is so strong that the independent study of it is fruitless. Sometimes we have to distinguish different aspects of this central problem. For example, man is a self-contained biological system in many respects and can be studied as such. But it is almost useless to study a man as a social system, although he is a major element in any such system. Similarly, it may be argued that cities are far from independent of the national and the world economy, that their import-export relationships are powerful and even dominating and that, therefore, the economic life of the city is too open usefully to be considered as a system. I would agree that this is the case in respect of culture, technology, economic function, and national politics. On the other hand, as a labor market area, a pattern of settlement, a dense concentration of land development, and a site for daily social interaction, the metropolitan area functions as a coherent and identifiable system. From this point of view, the other considerations become a part of the long-term development which impinge upon, but are to a considerable extent independent of, the metropolis as a system. The balance of the paper indicates that I regard this second aspect of urban affairs as deserving systems study.

Let me now introduce a discussion of the form of metropolitan areas. By form, I refer to the patterned distribution in three-dimensional Euclidean space in a metropolitan area of artifacts, people, and their attributes. Form is not necessarily plainly visible since, for example, the relative distribution of occupational groups or religious groups might be a significant element of form but would not be immediately obvious to the observer. Form also includes in a sense flows and interactions, since these are attributes of both people who occupy an urban area and of their artifacts.

Urban form in its most general sense is an important object of policy manipulation, since it implicitly controls many of the aspects of the quality of life which people appreciate (positively or negatively). The cost, location, and quality of housing, the amounts of private and public open space, the length of the journey to work, the social environment, pollution or its absence, and public safety are all aspects of urban form which affect people's lives and which are more or less subject to public control. It is therefore important to know, amongst other things, how form is determined, so as to be able to improve the cost effectiveness of public policies. This needed knowledge implies some understanding of the urban system.

There are essentially two complementary ways of looking at the determination of urban form: static and dynamic. It is tempting and indeed useful to note the similarities amongst a wide variety of cities and to speculate that these similar forms represent the conclusion of an equilibrating process, an end state toward which, under present circumstances, many large urban conglomerations converge. If this were correct, the problem of emerging tendencies in urban form could be studied as a problem in general equilibrium, and this is a view to which I tend to subscribe.

This view is frequently counterposed to the picture of the metropolis as an evolving organism, in which the processes of change are more important than the states which exist at any particular time. In addition, it is suggested that the dynamic moving forces which motivate this process of change are so strong, so persistent, and themselves so changeable that the system never can achieve equilibrium. I agree that this view is true in its most literal sense, but I am inclined to believe that, in spite of external shocks and stimuli, any particular exemplar of the urban system is always tending toward equilibrium. With a proper definition of that equilibrium and with a proper understanding of lags, the equilibrating tendencies can be used to explain much of the dynamic picture.

These two different views of urban equilibrium play complementary roles in the evaluation of policy. In the theory of general equilibrium in economics, equilibrium is frequently identified with optimality, and it is probably very much worth our while to examine the extent to which this is true of spatial equilibrium. The opposite proposition also has considerable merit--that a study of the dynamic properties of systems operating over time will illuminate their anticipated behavior under a variety of policy assumptions. The static view neglects the path by which some desired equilibrium might be achieved, while the dynamic view tends to neglect ultimate objectives and focus principally on the immediate implications of policies. The dynamic view also turns out to be a very clumsy way of testing paths of arriving at desirable configurations.

A certain note of caution must be struck regarding the optimality implications of competitive spatial equilibria. The Hotelling problem regarding the location of two hot-dog vendors on a beach is the simplest possible example of a general and pervasive spatial problem. Competition leads the two vendors to locate side by side at the center of the beach, while social welfare would dictate their being located one quarter of the way from each end, where they would still equally divide the market. This example indicates that globally optimal solutions are not necessarily reached by "natural growth processes" as replicated in models, especially when there are indivisible units and spatial monopolies.

There are thus two very broad classes of urban models of spatial distribution. One class provides an equilibrium description of urban distributions in the static sense without arriving at this conclusion by way of an examination of the equilibrating process. Three examples of this type of model might be mentioned with varying degrees of explicitness in their definition of equilibrium. First, the famous Lowry Model of a Metropolis[1] defines an equilibrium distribution of population and service activities for Pittsburgh, based upon the transportation system, travel patterns, and the location of export or "basic" industry. The equilibrium implications of this model are hard to determine, but they reflect some stability in travel patterns. A second group of equilibrium models belongs to a class of gravity models used in the location of retail trade. These models if applied to a uniform distribution of purchasing power and a uniform class of commodities will, like central place theory, arrive at a distribution of equal-sized market areas. It can also be shown that this type of model tends, in a somewhat indirect way, to minimize total travel time to shop subject to certain constraints and to a stochastic distribution of shopping trip lengths. The equilibrium which exists is quite explicitly between spatially located supply and demand, and if an equilibrium were disturbed, it is implied that some centers would be more prosperous than others. Finally, we may mention the Herbert-Stevens model of residential location which is based on the Alonso theory of the land market and which uses linear programming to achieve a Pareto optimum which is also a behavioral, competitive, market-clearing solution and therefore a form of equilibrium.

At this point I will develop in brief the second case of static equilibrium above, since in one form at least it has a certain number of interesting properties. It connects ideas about the statistical behavior of users of a transportation system with ideas about the equilibrium of land use and location. At the same time, it may seem to produce an optimal situation from the point of view of the users. But it turns out that the equilibrium and the optimum are not exactly the same thing. This finding suggests

---

[1] See Reference at the back of the chapter.

that spatial behavioral models with equilibrium-seeking properties may not possess the same optimality properties that non-spatial economic models have.

The model in view here is one of the location of retail trade which has been extensively discussed in the literature, in an earlier form by Berry, Garrison, Huff, Carroll, and others. More recently somewhat similar ideas have been applied without any equilibrium properties in Lowry's Model of a Metropolis and the equilibrium model was developed simultaneously by Lakshmanan and Hansen and by me. Since the Lakshmanan and Hansen model is simpler and more directly related to the problem which I wish to discuss, I will use that rather than my own model. All of the models mentioned above produce results which are very similar to the results of central place theory: each marketplace is surrounded by an area of market dominance, and the areas of market dominance exhaust a plane. Unlike central place theory, however, these models, which are based on gravity models of trip interaction, admit of overlapping trade areas, and if the trade centers are of unequal size, the boundaries between their areas of dominance are neither straight nor equi-distant between the centers.

The model developed by Lakshmanan and Hansen assumes that we are dealing only with a uniform type of subregional shopping centers with floor areas in the vicinity of 500,000 square feet. Repeated applications of the model yield estimates of the number of trip-makers who will be attracted by each center as it competes with other centers. (Hypothetical center locations are an input to the model.) If the purchasing power which arrives at a center exceeds some predetermined average (say, $55 per square foot), then the center is expanded on the next iteration, and vice versa. Centers which become too small are dropped out and centers which become too large may be split in two if hypothetical sites are available. The outcome of this process is a form of equilibrium in which nearly all centers have an equal level of sales per square foot of floor area.

Lakshmanan and Hansen found, as a byproduct of their procedures, that the pattern of centers produced by this process also appeared to involve the minimum total miles of travel for the users of shopping centers. If this observation were absolutely correct, it would provide a useful consequence of the equilibrium aspects of the model. However, it may readily be seen that the equilibrium postulated in the model is primarily a producer's equilibrium. Sales at less than $55 per square foot are uneconomic and cause some firms to go out of business, while sales at over $55 per square foot on the average are excessively profitable and cause new firms to enter any particular center, thereby expanding its floor area and attractive power. There is a large element of consumers' or users' preferences involved in these equilibria--in that, owing to the convenience aspects of shopping as reflected in the gravity model of trip-making, it is impossible for all shopping to become concentrated in one center, and the distribution of centers becomes fairly even. This evenness produces the apparent optimality from the point of view of the user, but it must be stressed that there is no guarantee of such user optimality built into the model.

A simple way of viewing the paradoxical nature of this model may now be presented. Three assertions have been made. First, purchasers or consumers tend to behave as in a gravity model for any particular class of trips--say, food shopping. Second, producers achieve a spatial equilibrium by adjusting the size of their activity so that it can just precisely serve the level of activity which it will attract. Finally, it is hypothesized that this arrangement represents a minimum travel cost scheme from the point of view of the consumer. It is not difficult to show that only in very restricted circumstances can all three of these assumptions be true.

Consider the layout of market areas along a radial axis with a declining density gradient. It is apparent that if two centers are of equal size they will not have equal radii of service under the second assumption that the size of the center adjusts to the available market. If, on the other hand, they have equal radii of dominance, they will have unequal total markets and therefore be of unequal size. But we must note that, under the gravity model formulation of trip distribution, a market area boundary, defined as a line of equal probability, will be equidistant from two centers only if these centers are equal in size, since the interaction probabilities are generally proportional to the size of centers at equal distances. Finally, to a good approximation, it is evident that equal radii of market areas are necessary and sufficient for consumer travel times to be at a minimum. If, as a consequence of unequal sizes of centers, market boundaries are shifted towards one or another center, a substantial proportion of consumers will have imputed trips which are longer than those to the nearest center. This contradicts the hypothesis of consumer optimality.

Under this line of reasoning, we are usually constrained to give up at least one of our three original hypotheses. There is, however, one condition which seems to permit us in part to escape this tri-lemma. We can assume that centers are of equal size, but unequally spaced. At the same time, the lines of equal influence are perpendicular bisectors of lines joining the centers. Thus centers will have radii of influence which are shorter on the "up-hill" side of the density gradient, and vice versa, and will not be located in the center of their service areas. This seems to be the usual pattern of shopping center location, and Lakshmanan and Hansen seem to have been fortunate in their selection of potential sites, making it possible to arrive at a configuration which would approximately satisfy this set of conditions.

This solution, however, still contains a residual paradox. In the postulated configuration, the shopping centers are not necessarily at the centroids of their service areas. Within any one area, if a center could relocate and retain its customer allegiance, it could reduce total travel cost. But some customers would be disadvantaged in their choices of centers, and their consequent shift of allegiance would result in a change in center sizes. The equal size condition could no longer be maintained.

The model therefore permits all three assumptions to hold only on an isotropic, equal density configuration. It seems that this little example raises serious questions about the rationality and reality of the gravity model of trip-making, or of this family of retail trade models, or of the assumptions of optimality implicit in the equilibrium model. Such line of inquiry can thus be a powerful means for exploring certain aspects of models.

I should now like to turn our attention in more detail to a broad class of dynamic models which also have equilibrium and final state implications. This class of models is becoming very popular in metropolitan planning circles and goes under the general name of urban or regional growth models. The general form of such models is a system of differential or difference equations, not necessarily linear and sometimes quite large.

A recent publication, Urban Dynamics, by Jay W. Forrester, makes quite clear the structure of a system of simultaneous differential equations applied to urban phenomena. These systems of equations have the properties of embodying many feedback loops, of possibly providing contra-intuitive results (Intuition I), and or producing projections which for any particular phenomenon are not necessarily monotonically increasing or decreasing. All of these features have some considerable attraction as corresponding with our intuitive (Intuition II) views of the real world. Nevertheless, Forrester's presentation has a number of difficulties, most particularly in the nature of the assumptions regarding interregional change and the lack of detail regarding intra-urban distributions. Forrester also suggests that his ideas in their application to cities are altogether novel, although this is clearly not the case.

At least three major modeling efforts have been made in which an interacting set of models, used recursively in steps of five years or less, provide a much larger and richer mix of feedbacks than appear in the Forrester system. The argument is not essentially changed by the fact that these models are all based on difference equations rather than differential equations and that their results are cruder, but computationally more convenient. Models of this type include the EMPIRIC model for Boston (which also had a still-born companion in a differential equation formulation, POLIMETRIC), the Penn Jersey Transportation Study model package, and the Time-Oriented Metropolitan Model developed by Crecine for the Pittsburgh CRP and later further expanded. We might also include the Dyckman-Robinson model for the San Francisco CRP.

There is thus no shortage of relevant models of the dynamic type, but very little attention has been paid to their properties. I now propose very briefly to explore some of these by way of illustration of a number of points. First, I will look at the connection between equilibrium and dynamic models, and I will suggest that these ideas immediately provide another powerful means of examining both policy issues and the construction of the models themselves. Second, I will develop a particularly simple model of urban location patterns and examine the properties of its equilibrium solutions in slightly more detail. Finally, I will look at a group of statistical problems which arise in connection with these ideas and, indeed, in connection with a great deal of urban research.

The role of feedback and dynamic performance of systems in relation to homeostasis and equilibrium is complex, subtle, and not understood in sufficient details. For example, the models which I will discuss have linear feedback loops. If these loops were nonlinear or discontinuous, it is probable that in many cases the equilibrium tendencies of any particular system so described would depend on the initial state of the system as well as on its structural characteristics. Such dependency is common in biological systems and must exist in some social situations--indeed, quite commonly at least in any situation which has to do with matters of life and death. I will not, however, investigate any but linear and generally continuous systems.

Positive and negative feedback are distinctively different in their influence on dynamic systems. Positive feedback implies positive and self-reinforcing experiences, and consequently leads to growth and to extended exploitation of the environment. Negative feedback, on the other hand, leads to decline or to equilibrium-seeking. It is important to realize that positive feedback and exponential growth cannot continue to operate indefinitely. Systems possessing this characteristic ordinary encounter one of two modes of change which limit the growth. The ordinary or "liberal" solution results from a shift in relationships either internal to the organism or between the organism and the environment such that positive feedback is converted into negative feedback. Typically this happens when expansion is limited by the increasing cost of resources, or when the agglomeration economies begin to be offset by the diseconomies which result from congestion or pollution. The "radical" or less automatic solution arises when basic changes resulting from the growth of the system create conflicts of problems that necessitate new laws and new institutions. In the first case, the growing system reacts to changed circumstances. In the second case, either the system or the larger system in which it is embedded adapt by change or form. In society these changes of form are changes in institutions, laws, and social relations, and also in the response of technology.

As a consequence of this distinction, one of the first questions which we can logically and practically ask about models of urban systems is whether they generate any unlimited tendencies and whether in fact these tendencies correspond to those which can be observed in the real world. Unlimited growth, decline, concentration, or dispersion would in general seem to be contrary to our intuitive view of urban arrangements, but if they were realistic could in any case be expected to create various types of severe institutional stress. Systems which behave in this way have no equilibrium or homeostatic tendencies except when they have reached boundary conditions such as the concentration of national population in a single or very large city, ultra-high urban densities or uniform densities, or giant corporate monopolies. If an exploration of a model leads to the conclusion that it does not imply any normal equilibrium, it will be a matter of considerable delicacy to decide whether the abnormality lies in the construction of the model or in the true behavior of the system.

It seems to me much more likely that for well constructed models a set of equilibrium solutions will be available for most inputs of policies and environmental conditions. Such an equilibrium is, for example, displayed by the long-term solutions of the Forrester models of urban dynamics. More generally, various types of equilibria probably exist corresponding to no change (or a steady state turn-over of individuals, households, or firms), or to variously defined conditions of equiproportional growth. As I have suggested, in the case of linear models these equilibria are probably independent of starting conditions and rates of initial growth, but in the event that there are long lags such as may be identified with respect to the redevelopment and redeployment of the urban capital investment, the ultimate equilibrium might take a long while to achieve in any realistic growth situation.

When we explore the possible equilibrium positions of urban systems implied by dynamic models, we must take account of this and many other aspects of the relation between equilibrium and dynamic performance. Not only do actual physical investments tend to persist over long periods of time, but agglomeration economies, once established, may long outlast their original impetus. Thus, for example, urban financial centers are typically located near major cities' original port centers, even though these may no longer be in the central business district. Given this resistance to change, it is my view that cities probably tend toward their equilibrium position. This, however, may itself be constantly changing as a result of external impulses, and thus the homeostatic mechanism is aiming at a moving target. Tendencies which affect the rate of movement of the target are most particularly the rates of growth of metropolitan regions, the rates of change of economic function, the rates of increase or decrease of personal income, and the rates of change of technology--particularly in building, transportation, and communication.

It is attractive to consider that the manifest form in which cities are case is a joint product of two types of forces. It is manifestly a crystallization of the evolutionary tendencies embodying past history which we have just outlined, while at the same time the emerging basic functional form is determined current and anticipatory underlying equilibrium tendencies. Such a view might at the same time accommodate an explanation both of the convergent similarities of cities and of certain specific and evident differences. It is further attractive to compare this process, if it exists, with processes of biological morphogenesis and evolution. But these comparisons may be more dangerous than helpful, especially so long as our knowledge of both biological and metropolitan morphogenesis is so qualitative and so inadequately explored.

Aside from long-term speculations in the philosophy of science, the relationships which we have sketched between equilibrium and dynamics suggest that at any particular point in time the equilibrium which could be achieved for given environmental conditions, existing sunk capital, and policy determinations might tend to represent some sort of optimal arrangement. Actual anticipated development which takes into account short-run decisions which will result in capital investment and therefore foreclose some aspects of the long-term equilibrium would then be by definition less than optimal. In making use of this hypothesis we must constantly bear in mind the qualifications which were developed earlier regarding the possible mismatch between equilibrium and optimality. We must also recognize that avoiding currently attractive decisions which in the long run are less than optimal will usually impose costs either on government, on investors, or on users. Given all these qualifications, the equilibrium condition for dynamic systems may be extremely useful for the exploration of ideal future states and the policies which are related to their attainment.

In order to give this statement some realistic content, I should like to discuss briefly a modified version of the EMPIRIC model, originally developed by Donald Hill and his associates for application in the Boston region. This model, as I have already mentioned, is a multiple-equation, multiple-variable difference equation model which we shall consider in a modified form for simplicity of discussion. The dependent variable in the EMPIRIC model is a large set of area-specific and locator-specific rates of change--actually deviations from regional rates of change. The right-hand variable in these equations falls into four classes. First, the changes in all other locator quantities in a given area are assumed to affect the rates of change on the left. For example, if during a given period the volume of manufacturing in an area increases greatly, the rate of increase of residential location will be depressed. This is necessary for a difference equation formulation especially one with a time interval as long as ten years, but since it is not relevant to a differential equation model, we omit it from further discussion. The second class of variables defines the density of each locator variable in each area. In most cases it is anticipated that higher densities discourage additional location. In the original EMPIRIC model these densities appear in a concealed form in relation to zoning policy variables, but we will consider them explicitly. The third class of variables has to do with accessibility. Accessibility is a constructed variable which in this case is calculated as a sum, by weighting the locator volumes in all other areas by a declining function of time-distance from the area under consideration. While the distance functions are nonlinear, the weighting process is linear and the locators in various areas enter into the calculation in a linear way. Ordinarily, except possibly for conflicting land uses, the signs of the coefficients of accessibility are positive, thus differing from density. The fourth and final group of variables has to do with neighborhood qualities. These in principle may be both variables which are exogenous to the planning process, such as those having to do with slope elevation, micro-climate, and the like, and control variables such as water and sewer service and many other planned neighborhood characteristics.

Given this general description of the model, if we have N areas and M locators, we have MN equations for MN locators. Owing to the construction of the accessibility variables, all of the variables appear in all of the equations, or everything influences everything else, and there are MN feedback relationships involving all the variables. There are of course many less than $(MN)^2$ basic parameters in the model, owing to the manner in which the accessibilities are calculated. Here the network conditions (which are themselves policy variables) generate a large number of coefficients. In general, each equation will contain some positive and some negative coefficients so that for a properly selected vector of locator groups of length MN (with all elements positive), it may be possible to force all rates of change to zero. We shall look briefly at the conditions under which this might occur. There are in fact two different cases which may be of interest. If the left-hand sides of these MN equations are all set to zero, we have on the right-hand side a set of terms involving the locator groups and a set of terms involving neighborhood conditions. The dual problems are, then: first, given a certain set of neighborhood conditions, what would be the equilibrium distribution of locators; and second, given a desired distribution of locators, what would be the necessary configuration of neighborhood conditions? In both of these cases, certain mathematical difficulties arise which I will only sketch.

In the first case, there is almost certain to be a unique solution. Not only is the number of unknowns equal to the number of equations, but the combined neighborhood conditions provide a non-zero vector of constants. It seems unlikely on somewhat cursory examination that the (very large) matrix of coefficients applying to the locators would be singular. Such singularity could arise, however, in a case where the behavior of a locator is exactly similar to the behavior of any other locator (that is, where its coefficients are proportional to another locator), or indeed if any locator's coefficients can be defined as a linear combination of any other set of locators. From a certain point of view this might be taken as reason for reducing the number of locators to be considered. From a different point of view, however, it makes considerable sense to say, for example, that banking is indeed fifty per cent retail trade and fifty per cent business services and to analyze its behavior accordingly. This may be so even if it does not make very much sense to use the algebraically equivalent statement that business services is equivalent to banking doubled less retail trade. These problems though somewhat novel, are a perfectly legitimate field of inquiry in location models, and they suggest that other methods for dealing with this type of problem need to be explored. It is obvious, for instance, that for iterated solutions such as have actually been used for the EMPIRIC model and by Forrester, the singularity of a coefficient's matrix may not create the same type of difficulty.

Another problem which may arise in this first case is that certain constraints on the solution have not so far been built into our formulation of the problem. The first of these constraints is that none of the values of the locator groups shall be negative. The second is that the total volume of each locator must exactly match some predetermined or input value of population or business which has to be accommodated. The second problem can be converted into the first by eliminating M equations and M variables--for instance, replacing the Nth subarea variable for each locator by the predetermined total less the sum over all other areas. This calculated quantity itself cannot be negative. The existence of negative locator values in such a solution would indicate, in fact, that the system described has no equilibrium, since if the negative values are replaced by zeros, various rates of change will be non-zero.

The second class in this dual problem has to do with the circumstance when we have projected a possible pattern of equilibrium of locators and wish to know what public policies could bring this equilibrium about. We will not now discuss the subcase of the influence of transportation networks via accessibility on the equilibrium, since this leads into a problem which involves not only nonlinear functions, but also the combination of links into least-cost paths in a network. However, given a network configuration, it seems practical to ask what levels of other government services are necessary to ensure a certain pattern of development, short of direct controls. Where such controls are everywhere binding, the notion of equilibrium is no longer applicable. The mixed case where some controls are binding and others are not is most vexatious, not only as applied to solution methods, but also with respect to the observation of "natural" locational tendencies.

The first thing to be observed is that there are apt to be many more locator variables (and hence equations) than policy variables influencing location. This will always be the case if there are more types of locators than there are policy variables. Thus, the equation system for determining what policies are necessary may be over-determined. There are two general ways out of this dilemma. One of these is to make a least-squares fit of policies to the desired configuration. In this case, the RMS error could be interpreted as a measure of the lack of realism in the policy. The second means of dealing with the problem is to reduce the number of equations by combining locators. This could be done along the lines discussed in the preceding section or by any other reasonable procedure based on past locational behavior.

Whatever method is followed in solving this dual problem, the same difficulty regarding potential negative (or, more generally, unrealistic) policy values will probably be observed. In this case, the implicit advantages of an iterative scheme are not available, and it is necessary to conclude that the desired configuration cannot be achieved by way of influencing the behavior of the locators, but only by outright regulation. It seems likely that such regulation of locators implies some departure not only from equilibrium, but also from optimality. In other words, the imposition of a preconceived pattern of location may satisfy certain planner's goals, but does not necessarily best serve the interests of the locators.

I will now turn to my last major point in this discussion, which has to do with certain statistical implications of equilibrium models. In dealing with spatial location and perhaps even more with dynamic spatial processes, it often turns out that statistical problems are gravely complicated by aspects of multicollinearity. Statisticians sometimes argue that this problem should be avoided by reducing the number of variables, since it is "quite evident" that some of these variables must measure the same thing. This approach suggests the desirability of stepwise regression methods amongst others, but in my view is not entirely satisfactory. It seems to me that the preceding discussion leads directly to some conclusions which are at variance with this interpretation of multi-collinearity.

First, in order to clarify the situation somewhat, we must refer back to the earlier discussion of the uniqueness and linear independence of locational behavior. If, in fact, some locational behaviors can be represented as linear combinations of other locational behaviors, and if the system is approaching equilibrium, then the corresponding consequent locational patterns may be linear combinations of other locational patterns. Since the locational patterns enter uniformly into the variables of density and accessibility which make up many of the independent variables of this model, these variables will in turn be linearly dependent and the correlation problem is in principle not soluble. There may be some difficulty in identifying this case separately from the more difficult and more important case which follows. A simple way to deal with it, however, would be to component-analyze the locational patterns of all the locators over all areas. The locators themselves could then be replaced in the model by a set of component scores which would ordinarily be less than the number of locators. We have used this type of analysis to reduce the dimensionality of measures of accessibility, without sacrificing any of the information which is provided by taking a rich and detailed view of this set of variables.

The second and more serious difficulty arises out of equilibrium considerations. Even assuming that each locator entering into the model is truly independent, a correlation analysis might still break down. Consider the circumstances which arise where an urban system has either reached equilibrium or has approached it so that it may be said to be "tracking" equilibrium in a relatively uniform way. In this case, for each locator the rates of growth for areas, on the left side of the difference equations, are zero or uniform. In a regression analysis to determine the coefficients of these equations, the vector of correlations between the dependent and independent variables is zero. In this case, the equation for the coefficients yields only a trivial solution if the matrix of correlation coefficients for the independent variables is nonsingular. I think that we may justifiably generalize this situation slightly by saying that the closer an urban system approaches equilibrium, the more likely it is that an analysis of the rates of change will create a singular correlation matrix.

We may put the same problem in more intuitively attractive terms. The model which we have outlined depends on the interaction of factors which attract and repel the locators--for example, accessibility and density. For any particular size of city and location within it, there is some appropriate balance between these which means that their weighted algebraic sum is zero. (This discussion of course assumes a linear model.) If this condition is satisfied everywhere, the system is in equilibrium and has no impetus to change, yet this condition of a zero weighted sum of two or more vectors is precisely the condition for linear dependence in a set of variables. In practice in correlation analysis, we observe this phenomenon in the form of a very small determinant of the correlation matrix, followed by high and "unreliable" B values. Alternatively, if we correlate the dependent variable with component scores for the independent variables, we find that components with very small eigenvalues play a very large role in the analysis.

It is quite evident in this situation that throwing out variables is not the appropriate solution, although I hasten to add that the exact selection of appropriate methods is not altogether clear to me. However, it *is* clear that throwing out one of two highly correlated variables may be a disaster if in fact some phenomenon locational change is closely related, for example, to their difference. Since density and accessibility (as illustrative variables) obviously measure quite different phenomena, the statistician's original assumption of overlapping concepts and variables is no longer applicable. In other words, equilibrium provides an alternative explanation for collinearity.

From the preceding discussion it is quite evident that the stronger the equilibrating forces and the more responsive the system is to them, the less confidence statistical measures would give to the coefficients describing growth relationships. Let me put this point slightly differently. If I observed an SMSA in which accessibility and density were very highly correlated, I would take this as a confirming instance of my basic view of urban dynamics yet if I used these as variables in a correlation analysis of change in the same urban area, statistics would tell me that the influence of these variables is measured in a highly unreliable way. I must say that I cannot accede to this view, and I think that the problem of sorting things out is up to the statisticians. It is important because it is closely related to the predictive power of models.

There is a further extension of these ideas of statisticians with regard to the use of models of this type for projection. This is the idea that over time errors in the estimation of the model are likely to amplify. Thus, after a number of recursions of a difference equation model, it would be shown under certain statistical assumptions that the standard error of the projections is very large indeed. This approach assumes that each consecutive projection step is wholly dependent on the results of the preceding step. With a model which tends to equilibrium, this is not so, and errors tend to be self-correcting. The truly troublesome errors are those which arise from gros misspecification of the model and from unforeseen changes in tastes and technology.

In this brief case I have tried to develop in an illustrative way a cluster of ideas about how the relationships between dynamics, equilibrium, and optimality could be used to explore more fully our understanding of models and of urban phenomena. I think that the ideas here presented are perhaps somewhat naive and oversimplified, but I am confident that further exploration in greater depth would be more rewarding. I trust that these explorations will be widely undertaken.

REFERENCES

1. Alonso, William. _Location and Land Use--Toward a General Theory of Land Rent._ Cambridge: Harvard University Press, 1964.
2. Berry, Brian J.L. "The Retail Component of the Urban Model," _Journal of the American Institute of Planners_, Vol. 31, No. 2, May, 1965.
   Commercial Structure and Commercial Blight. Department of Geography Research Paper No. 85. Chicago: University of Chicago, 1963.
3. Carroll, J. Douglas Jr. "Spatial Interaction and the Urban-Metropolitan Description," _Papers and Proceedings of the Regional Science Association_, Vol. 1, 1955; also in _Traffic Quarterly_, Vol. 9, No. 2, April, 1955.
4. Fidler, Jere. "Commercial Activity Location Model," Publication TPOO-332-01, Subdivision of Transportation Planning and Programming, New York State Department of Public Works. Albany, January, 1967.
5. Harris, Britton. "Basic Assumptions for a Simulation of the Urban Residential Housing and Land Market," Institute for Environmental Studies, University of Pennsylvania, July, 1966. Mimeo.

"A Model of Locational Equilibrium for Retail Trade." Paper presented at a Seminar on Models of Land Use Development, Institute for Urban Studies, University of Pennsylvania, October, 1964. Mimeo.

6. Herbert, John, and Benjamin H. Stevens. "A Model for the Distribution of Residential Activities in Urban Areas," Journal of Regional Science, Vol. II, No. 2, 1960.
7. Hill, Donald M. "A Growth Allocation Model for the Boston Region," Journal of the American Institute of Planners, Vol. 31, No. 2, May, 1965.
8. Huff, David L. Determination of Intra-Urban Retail Trade Areas. Real Estate Research Program, Graduate School of Business Administration, Division of Research. Los Angeles: University of California, 1966.
9. Lowry, Ira S. A Model of Metropolis. Memorandum RM-4035-C. Santa Monica: The RAND Corporation, August, 1964.

   Seven Models of Urban Development: A Structural Comparison. P3673. Santa Monica: The RAND Corporation, September, 1967.
10. Lakshmanan, T.R., and Walter G. Hansen, "A Retail Market Potential Model, Journal of the American Institute of Planners, Vol. 31, No. 2, May, 1965.
11. Muth, Richard F. "The Spatial Structure of the Housing Market," Papers and Proceedings of the Regional Science Association, Vol. 7, 1961.
12. Reilly, W.J. The Law of Retail Gravitation. W.J. Reilly Co., 1931.
13. Robinson, Ira M., and Harry B. Wolfe and Robert L. Barringer. "A Simulation Model for Renewal Programming," Journal of the American Institute of Planners, Vol. 31, No. 2, May, 1965.
14. Seidman, David R. "An Operational Model of the Residential Land Market," paper presented at Seminar on Models of land Use Development, Institute for Environmental Studies, University of Pennsylvania, October, 1964. (Mimeo.)

    The Construction of an Urban Growth Model. Delaware Valley Regional Planning Commission Report #1, Technical Supplent, Volume A. Philadelphia.
15. Voorhees, Alan M. "A General Theory of Traffic Movement," The 1955 Past Presidents' Award Paper. New Haven: Institute of Traffic Engineers.

16. Wingo, Lowdon, Jr. Transportation and Urban Land. Washington: Resources for the Future, 1961.

Chapter 5

THE CITY AS A SYSTEM: A POLITICAL-ADMINISTRATIVE VIEW

E. S. Savas

The first paper portrayed one particular view of the city as a system. In this paper I would like to present--from a totally different vantage point--another view of the city as a system and discuss politically feasible policy that emerges from this examination.

Elsewhere [Ref. 1] I have described city government as a classic cybernetic system, with goal-seeking, disturbances, an action element, and information feedback. In other words, city government is a very large, complicated, sophisticated controller. In common with other controllers that have such attributes, this controller, too, suffers from component failures, gear slippages, time lags, loose connections, calibration errors, and nonlinear effects. In government these are, respectively, incompetents, poor coordination, bureaucratic delays, poor communications, absence of shared value judgements, and conventional responses to unconventional problems.

The author is the first Deputy City Administrator in the City of New York.

Alas, our urban systems exhibit some fundamental design problems. For example, over the last century we have constructed elaborate, time-consuming, costly bureaucratic systems of checks and balances designed to assure that municipal governments get fair value in their purchases and to protect against corruption in contracting for supplies and equipment. However, the consequence is a long delay in securing bids, ordering goods, and paying bills: Requests are prepared and submitted to bidders on an approved list. Sealed bids are received and opened ceremoniously, bids are awarded, purchase orders are prepared and issued, goods are received, several different agencies check to see that the right goods were delivered in good condition to the right place at the right time, payment is authorized after a proper invoice is received and cross-checked, and finally a check for payment is grudgingly issued by the city government.

The result of all this red tape is that many potential vendors refuse to do business with the city, while those who do charge higher prices in order to make up for the additional expense of dealing with it. Thus, a strategy intended to increase competition and reduce the cost of goods has the precisely opposite effect of reducing competition and increasing the cost of goods! Furthermore, the agency which originally required the supplies sees its program languish while it waits for them to arrive. No matter how well-intentioned the original design, it does not function well. And what is worse, there is no effective feedback to cause modification of the system.

In fact, we have too many open loops in our systems. Look at the case of a city which provides municipal refuse collection service. When that service is paid for indirectly, by real-estate taxes, there is no incentive to reduce the amount of refuse that is generated; whether one produces a lot of refuse or a little makes no difference, for it is removed free of charge. Thus, we encourage indiscriminate production of waste in our "effluent society" at the same time that we are running out of land for waste disposal. To repair this portion of our malfunctioning system, we ought to impose a disposal tax--collected at the manufacturing source--on all inedible products, with the tax proportional to the difficulty of disposal, or else charge the consumer directly, by the pound, for the waste he nonchalantly bequeaths to his municipality.

The present system provides no regulatory feedback link to reduce the demand for refuse service or for that matter, to increase the efficiency with which the service is provided.

Similarly, on a larger scale beyond the city in size, the consumer of a product does not pay for its full cost. Someone else bears the external cost of the pollution that is generated as a by-product in the manufacturing process.

Look also at the following bizarre sequence in our economy: we pollute our air when we burn fuel to generate electricity which is used to produce aluminum which is formed into cans which litter our landscape. The public suffers at both ends of this destructive chain of events.

Incidentally, geophysicists are excited because they've finally detected continental drift. This comes as no surprise to the public official, who has already observed that the island of Jamaica, a huge exporter of aluminum ore, is gradually drifting onto the United States and covering us, in the form of a thin layer of aluminum beer cans!

Let me digress for a moment to mention an interesting discussion that I had with the cab driver who brought me in this morning from the airport. It was about homeostatic systems (!) He complained about the rising burglary rate in the suburbs, and I remembered reading that it was increasing more rapidly than in the center cities. Then it occurred to me that what we were witnessing was a fundamental homeostasis: during the day, suburbanites come into the city, work, and bring money back to the suburbs. At night, burglars--presumably from the city--go into the suburbs and bring some of it back. This basic balancing of the regional economy derives from simple cost-effectiveness considerations: even without a degree in the subject, a burglar readily calculates that he is better off working the suburbs; the number of policemen per unit area is less and his dishonest reward is likely to be greater than in the city.

Another major design defect in our urban system can be found in the civil service subsystem (which has features in common with the tenure system in universities). It acts as a fly wheel providing enormous inertial stability for the system. (At least that's how a mechanical engineer would describe it. A chemical engineer would say that the system is like a large settling tank with a long residence time so that a change in the input composition doesn't show up in the output for a long time.) Now this subsystem was designed to promote quality in the public service by providing security for the individual and freedom from external influences. Unfortunately, this has come to mean freedom to be unresponsive to the changing needs of society. (And we all know what happens to an organism which fails to adapt to a changing environment. For examples, see any big-city school system or the dinosaur exhibit at a local museum.) The problem shows up all over the country in the form of uncivil servants going through preprogrammed motions while awaiting their pensions. Too often the result is mindless bureaucracies which appear to function solely for the convenience of their staffs rather than for the public whom they ostensibly serve. When Servan-Schreiber alerted Europe to the challenge posed by America's managerial prowess, [Ref. 2] he wasn't thinking of our civic ineptness.

The systems that have emerged haphazardly in our cities are very vulnerable. Just remember that one of the most important functions of government is to provide (or regulate) those services which "by their very nature" are monopolies, and so one sees governments providing police, fire, sanitation, and ambulance services, for example, while regulating services such as power and communications. This means that many agencies in city government are monopolies, and their staffs exercise monopoly power. When a department is a malfunctioning monopoly, which no longer serves the public interest but its own, the public is left angry and helpless before it. We have unwittingly created a system in which our citizens are at the mercy of our public employees.

In a complex society, any one of a thousand organized groups can shut down the system. Government, with its monopoly power over vital services, is uniquely vulnerable, for any group within the government can utilize that monopoly power to its parochial advantage, and the public be damned. Strikes by public employees in big cities around the world have demonstrated this. Reliability theory tells us that in this sort of situation one should try to build redundance into the system. Redundancy is what an engineer would call it. An economist would use the term competition to describe the very same concept and a control engineer would call for more minor loops. Redundancy, competition, or minor-loop control will increase the reliability of the system and decrease the vulnerability of our urban governments to the abuse of their inherent monopoly power.

Increased reliability can be achieved by reducing the scope of some of the governmental monopolies: for example, contracting with private agencies for certain protective services and permitting private firms to compete for the opportunity to collect refuse, clean streets, and repair roads.

Another, complementary, approach to increased reliability is to reduce the scope of the municipal monopolies. This can be achieved by moving toward neighborhood government, i.e., putting decision-making at a level closer to the individual community and thereby providing more rapid response and more effective performance by city government at the local level.

The systems approach to city problems is usually applied to functional subsystems such as health, education, criminal justice, environmental protection, etc. I think that the systems approach can be applied even more profitably to the geographic subsystems of the city, the neighborhoods, and also to the geographic total systems, the metropolitan regions. In other words, the challenge is to construct minor-loop control systems at the neighborhood level to make our cities function better.

Here is one way to view the urban problem: The urban dweller is dissatisfied because he feels that he cannot really influence the quality of his immediate environment: he cannot effectively influence the safety of his family, the education of his children, the behavior of his neighbors, the appearance of his physical surroundings, the cleanliness of his block, the purity of his air, the adequacy of local transportation (or even the quality and safety of the food he buys).

The problem arose basically when the industrial revolution brought about a greater separation between the place of residence and the place of work. As more and more people started working at sites fairly distant from where they lived, they could no longer pay attention to the immediate environment around their homes. They started paying people to educate their children, police their streets, pick up refuse, and so forth. In other words, we've been "contracting out" for our local needs, but we've done it without writing very good specifications for the work to be done and without establishing very good systems for measuring the performance of the contractor who is doing the work on our behalf.

A century ago the cities operated under the ward-heeler system (and it persists to this day in some cities.) The neighborhood politician made sure the potholes were repaired and other needs got taken care of; if you voted right on election day, somehow things got done for you in the neighborhood. We've gotten away from that system because of its many problems and inadequacies. Instead, we've gone to an increased professionalism, increased reliance on the merit system, and increased reliance on civil service. We have seen an increasing centralization of our municipal governments to the point where we are now encountering the diseconomies of scale.

In the case of New York City, we have more than 300,000 employees working for the City government. We have about 80,000 "social blocks" (a "social block" consists of two block faces which face each other across a street). With 300,000 employees and only 80,000 blocks, one can start speculating about the effect of having a block worker working on each block, monitoring services, organizing community groups for various pressure purposes, organizing block associations and generally producing a socio-political entity which can influence the quality of life in the area. Now, I'm not suggesting a Chinese commune, where everybody in the block shows up for calisthenics at 7 o'clock in the morning and afterwards marches into the community mess hall for breakfast, but one starts wondering what the effect would be on the quality of life in our cities if the government were structured geographically as well as functionally.

In New York City we are not attempting the kind of grand reordering of our Civil Service structure along the lines of speculation indicated above, but a few things are happening to strengthen the neighborhood government concept. You might say that we are now approaching the pupae stage. Some day we may emerge as a beautiful butterfly. The city has been divided into 62 neighborhoods, called community planning districts, with an average population of about 130,000. In many of these communities Neighborhood City Halls have been set up and a high-level city official assigned by Mayor Lindsay as his liaison for that community. His job is to make sure that the city agencies are responsive to local needs in that community, that the community has high-level access to city government, and that proper coordination is achieved between local departments where necessary.

Some significant pilot projects are being initiated now, including one in the uptown part of Manhattan, an area called Washington Heights. This is a community whose size in terms of population is about that of Jacksonville, Florida, or Salt Lake City: 190,000. Imagine! A city of 190,000 people that has never been considered as an entity. Who would even dream of running Jacksonville, or Salt Lake City as an anonymous appendix to something else!

One of the elements of this neighborhood government experiment is a discretionary community budget. A sum of money will be allocated and some kind of a community council (yet to be devised) will determine how to use that money. This discretionary budget amounts to only 1/100th of the city budget for that area, so the objective is to use that money in an attempt to establish an irreversible process, i.e., some kind of neighborhood government apparatus, so that the community can influence the "regular" city expenditures in that area.

The subject of neighborhood government is replete with problems. In the first place, should only residents be involved in setting of goals and determination of priorities? What about merchants who may not live in the area but work in the area? Secondly, how many people represent "a majority" in the community? Clearly, the ideal of a New England town meeting is out of the question in a practical-sized urban neighborhood of 100,000 to 200,000 population. But after all, the President of the United States was elected by the votes of only 16% of the population.

Perhaps there is a significant role to be played by polling techniques. They can be used to study the problems of a community, their relative priority, and the relative acceptability of alternative solutions. Political scientists can legitimately ask whether one is subverting democracy by relying on polling techniques rather than formal voting, but until we get to the point where voting can work effectively at a local level, polling may be the modern technological key that can provide communities with the kind of feedback guidance that is needed to keep the system moving toward improved quality of living.

Regardless of whether or not the lay public is good at planning, it is very good at monitoring the performance of services; and the public will complain when service is poor. We have had a city-wide complaint system in operation for some years but only recently have we started thinking about taking full advantage of all of the information it provides. In addition to handling the individual complaints made by the public, we are planning an information system which would pull together the information on complaints to see what kind of patterns emerge; whether a particular service is inadequate in a particular part of the city.

As part of the neighborhood government experiment in Washington Heights, a Citizens' Environmental Committee has been established with the backing of the Mayor. The purpose of this Committee is to monitor the service in the area, to receive information about the schedule of services and to report back problems and to change the goals and priorities of the agencies concerned with sanitation, water, sewage, air pollution, noise pollution, street and sidewalk conditions. This is a minor-loop feedback control system. But the system poses some problems.

For example, how do you measure the cleanliness of an area? The Department of Sanitation regularly reports the number of tons of refuse it picks up, but that's meaningless for control purposes: the important parameter is not how many tons are picked up but how many tons are left behind, and the latter is much harder to measure. One can always weigh the amount of refuse carried in the truck to the dump but one cannot readily measure how much has been left behind. One surrogate approach that we are trying in several places in the city is to get ordinary citizens to serve as monitors, reporting whether the truck reached them in their route today. The monitor could be a person that traditionally hangs out of the window just watching what goes on all day long. You talk to that person and say "look, let me know if the truck passed by today, okay?", and if you get someone like that near the beginning of the route and near the middle of the route and near the end of the route you have constructed a reporting system which in some ways is superior to the present quantitative reporting system on tonnages. This is one way that the systems approach, relying on basic cybernetic principles, can be applied to improving the city.

This same Citizens' Environmental Committee is expanding its goals: instead of being concerned only with the conventional things like refuse collection, street sweeping, litter baskets, and abandoned autos, it has started reporting buildings which emit too much smoke. By observing and reporting any regularity in the emission pattern, inspectors from the Department of Air Pollution can schedule their inspection visits and detect violations. This is infinitely better than the "smoke-chasing" approach of responding to a telephoned complaint of a current (but temporary) smoke condition. This is an example of the use of feedforward control instead of feedback control to guide the deployment of inspectors.

The concept of neighborhood government raises additional fundamental questions. Questions like what should be the budget allocated to a particular community? Well right now in New York City we just don't know how much money is being spent by the city in each community. We've never looked at our city in that way. We know down to the last penny how much money is allocated for pencils or for street cleaning but we don't know how the money is allocated geographically. That dimension has simply never been looked at before our current involvement and move toward neighborhood government. We are now examining our budget data and trying to estimate how much we spend in each neighborhood. It is not easy to do, in part because the conventional service districts are incongruent with the neighborhood boundaries. Over the years, each individual department has created its own administrative districts, and so police precincts have nothing whatsoever to do with fire battalion areas and those areas in turn have nothing to do with ambulance service areas, which in turn are unrelated to health areas, etc. Every city agency has drawn its own boundaries for its own purposes and therefore to translate budgets from conventional service districts to community districts in the neighborhoods, while not intellectually challenging, is difficult.

Budgets relate to the allocation of resources. The conventional political way to allocate resources is to equalize the complaint level. That's the "squeaky wheel" approach. If you get a lot of complaints from one neighborhood then you do something there: you take a cop from one place and put him in another, or you take a garbage truck and reallocate it from here to there. Actually, one does not equalize the complaint level, but rather the weighted complaint level. That is, some complaints, such as those by state senators or city councilmen, are very likely to get more prompt attention than the complaint of an ordinary citizen. So our urban structures allocate resources in such a way as to equalize the weighted complaint level, and this may be the best approach in a democracy after all. Remember through that the implicit weighting formula depends on the value judgements of the political leadership.

On the other hand, if you attack the allocation question from a technological point of view, you might conclude that you should equalize service levels throughout the city: for example, you should deploy policemen, firemen and ambulances in such a way that each citizen gets the benefit of equal protection even at unequal cost. Take the case of ambulances; perhaps one should deploy them in such a way that all citizens and all areas experience the same average response time. But further reflection raises some doubts about this plausible strategy. It probably costs much more to provide such service in a remote peninsula in the city than it does in an area close to the dense core of the city. This observation suggests that one should strive to allocate equal ambulance budgets to the two areas. But this tentative conclusion, if carried to its logical end, demands that all neighborhoods receive equal (per capita) budgets, and yet we know that the essence of the urban problem is the large concentration of people with greater-than-average needs for public services. Public policy has long accepted the notion that society as a whole will benefit if slum neighborhoods are on the receiving end of net income transfer from richer neighborhoods.

These are some of the questions that challenge the urban systems analyst, the challenge of looking at the neighborhood system and relying on basic concepts of cybernetics to improve the functioning of the city. There is an enormous amount of work that could be done at the neighborhood level, research that would challenge practically any department of a university. An economics department could be challenged by a study of the total revenue generated by a neighborhood, through real estate taxes, scale taxes, income taxes and various hidden taxes. Sociologists could be engaged in determining and ranking the various priorities in the community, the needs ranked by socio-economic groups, by age, by race, by sex, by income! What are the needs as perceived by members of these sub-groups within a geographic area? A great deal could be done in systematic polling at the neighborhood level in order to get answers to these kinds of questions. How do you evaluate the quality of life at the neighborhood level? How do you measure the adequacy of transportation service in a neighborhood? The cleanliness of a neighborhood?

The task of redesigning our urban systems and institutions is the challenge of the decade. It encompasses the need to reorient our national priorities and to preserve our environment. In the most fundamental sense, it challenges us to improve the quality of our lives.

REFERENCES

1. Savas, E.S., Cybernetics in City Hall, Science, 168:1066-1071 (1970).
2. Servan-Schreiber, J. J., The America Challenge, Atheneum, New York, 1968.

Chapter 6

A MULTILEVEL APPROACH AND THE CITY: A PROPOSED STRATEGY FOR RESEARCH

J. Richardson & T. Pelsoci

6-1 INTRODUCTION

In a recent issue of the IEEE Spectrum, Gabor Strasser suggests that "the solution of most of our domestic societal problems will primarily depend on the proper orchestration of a host of disciplines and other resources."

"Much of what will need orchestrating," he continues, " will be 'off the shelf' knowledge, procedures, or hardware, where the sources may be as varied and often as incommensurable as hard physical and natural sciences, technologies, political and social sciences, economics, management and institutional arrangements, alternative sources of funds, government practices, and so on. The engineer, and especially the systems engineer, may be as well suited as any to attempt this orchestrating role." (1971).

The authors are respectively, Assistant Professor of Political Science, Case Western Reserve University, Cleveland, Ohio, and a graduate student in the Department of Political Science, University of Minnesota, Minneapolis, Minnesota.

This research was partially supported by the National Science Foundation under Grant GS 2955 and by a Postdoctoral Research Training Fellowship from the Social Science Research Council. Institutional support and computer time were provided by the Systems Research Center, Case Western Reserve University and by the Center for Comparative Studies in Technological Development and Social Change, University of Minnesota.

The Systems Research Center at Case Western Reserve University has been concerned with the development of interdisciplinary approaches to the structuring and control of large scale systems for more than a decade. Concerns similar to those expressed by Strasser have provided both motivation and an integrative focus for these investigations. Of particular importance has been the belief that effective cooperation between disciplines would be greatly facilitated by the resolution of certain crucial theoretical issues regarding the representation and control of large scale systems.

The Theory of Hierarchical Multilevel Systems [Ref. 2], which emerged from these efforts provides a methodology for the analysis and synthesis of complex systems based on hierararchical concepts. Three types of hierarchical structures are used for analysis and explanation, levels of abstraction or stratification based on the different levels of aggregation or abstraction; levels of decision complexity, based on "layers" of decision making (pp. 9-10; 46-48) and levels of priority of action based on the hierarchical, echelon type arrangements which characterize complex organizations. For purposes of synthesis, specific techniques have been developed for decomposition and coordination of systems to harmonize their functioning toward an overall goal. This methodology provides a descriptive framework for formulating precise statements regarding complex or poorly defined systems which is vastly superior to verbal description. From such a starting point, one can first investigate the structural dimensions of the systems involved. From such investigations, it is often possible to progress to deeper mathematical theories. [Ibid.]

Ineffective, fragmented, maladaptive institutions have been identified by many observers as the root causes of contemporary urban ills. Multilevel systems theory seems particularly applicable to the task of designing new institutions for a coordinated attack on these ills.

Designing political institutions may be a new area of concern for the systems engineer, but systems engineers should be familiar with a general class of design problems which are quite similar. Indeed, Herbert Simon has suggested that such design problems constitute a major integrative focus for the engineering sciences, or, as he calls them, "The Sciences of the Artificial" [Ref. 3]. "Such sciences, "he argues," are concerned with the way in which adaptation of means to environments is brought about - and central to that is the process of design itself". The object of design is the synthesis of artificial systems whose purpose is to attain goals and to function (p. 5). "Fulfillment of purpose or adaptation to a goal involves a relation among three terms: the purpose or goal, the character of the artifact, and the environment in which the artifact performs" (p. 6).

The general paradigm suggested by Simon's analysis is depicted in figure 1. At this level of abstraction it is, of course, identical to the process-control paradigm of classical control theory. The outer environment corresponds to the process and the inner environment to the controller [Ref. 6]. The analogy to control theory is appropriate, moreover, since it is reasonable to view the task of designing more effective institutions as one of designing more effective controllers [Ref. 5].

If the problem were to design a controller for a physical process, the task of design could be readily simplified by disaggregation into well defined layers, or subproblems [Ref. 2]. A typical disaggregation has been suggested by Holt and Bailey [Ref. 4]:

(1) Determine the goal and the constraints.
(2) Construct a model of the outer environment (process).
(3) Construct a model of any existing inner environment (controller).
(4) Determine the control output which the controller must use to achieve the goal.
(5) Design (or modify) the controller so that it will generate the appropriate output.
(6) Modify the real-world controller in accordance with the design. In most instances, the designer would be aided in the completion of these subtasks by the availability of effective problem solving procedures (or at least well developed heuristics).

The designer of urban political institutions, of course, faces quite a different situation. In his field, there is little consensus about the way these problems should be approached. Goals are indeterminate and conflicting. Proponents of competing models and paradigms clamor for attention. Environments (i.e., urban systems) are of the nonlinear, high order type for which few or no analytic solutions have been obtained. Existing controllers (political and social institutions) are heavily constrained and seem to offer scant opportunities for modification. The need for a systems science which will integrate this field, and provide a basis for concrete solutions to pressing problems is immediate and apparant.

During the past year, the authors of this paper, along with others, have been concerned with the development of mathemat-

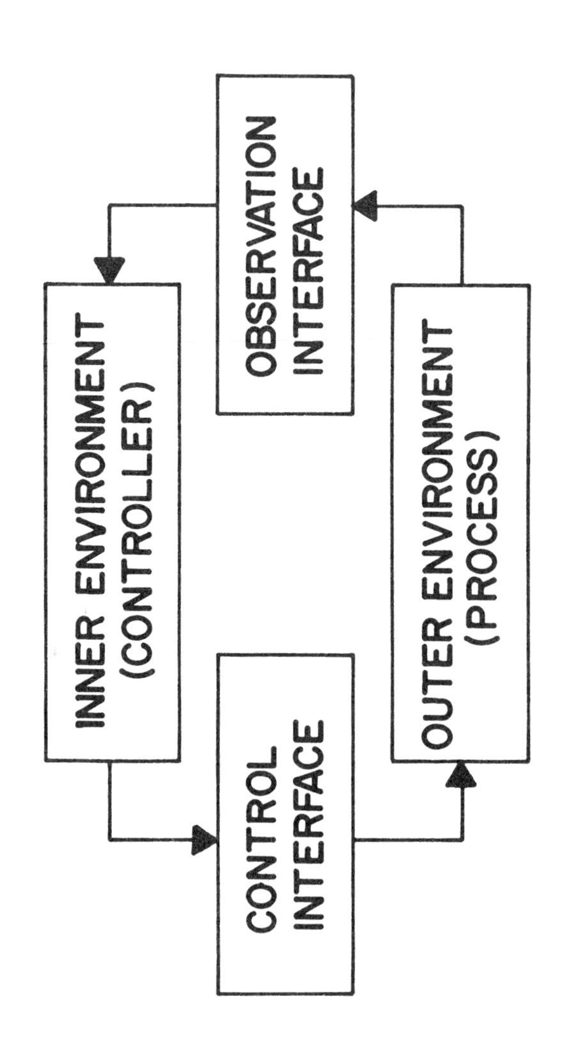

Figure 1
Basic Paradigm for an "Artificial Science"

ical and computer simulation models which would facilitate the investigation of specific urban institutional design problems. [Ref. 4, 5, 6, 7, 8] Much of this research has been strongly influenced by the multi-level approach. While there are, as yet, few concrete results to report, a number of interesting avenues for further investigation have been suggested. Below, we will present a formal mathematical description of a simplified urban political system using the multiechelon approach, and will discuss a computer simulation model which has been suggested by this description and by the control theoretic paradigm. We believe that these models represent a feasible solution to the problem of modeling the inner and outer environment and, raise some interesting issues regarding the operational definition of overall systems goals. In addition the models provide an excellent example of the broad applicability and fruitfulness of the multilevel approach.

6-2 MULTIECHELON SYSTEMS: A FORMAL MATHEMATICAL DESCRIPTION OF ORGANIZATIONAL STRUCTURE

Structural descriptions of political systems from a design oriented (normative) perspective normally fall within the purview of organization theory or management science. The multilevel approach cuts across the three approaches which are most generally used by organization theorists, namely, the classical (institutional) approach, the systems approach and the behavioral approach. [Ref. 2, p. 19] Its emphasis on hierarchical structure is quite similar to that of the classical approach. Hierarchical arrangement of decision making units is viewed as one of the primary characteristics of an organization. The model of the participant or decision unit as a goal seeking system is closer to the modern behavioral or, more specifically, motivational approaches. For example, levels of satisfaction and discrepancies between actual and operational goals [Ref. 9, 10] are accepted concepts. Finally, the multilevel approach recognizes explicitly that an organization invariably consists of an interconnection of decision making subsystems.

The multiechelon model of organizational decision making which incorporates these characteristics is diagrammed schematically in figure 2.

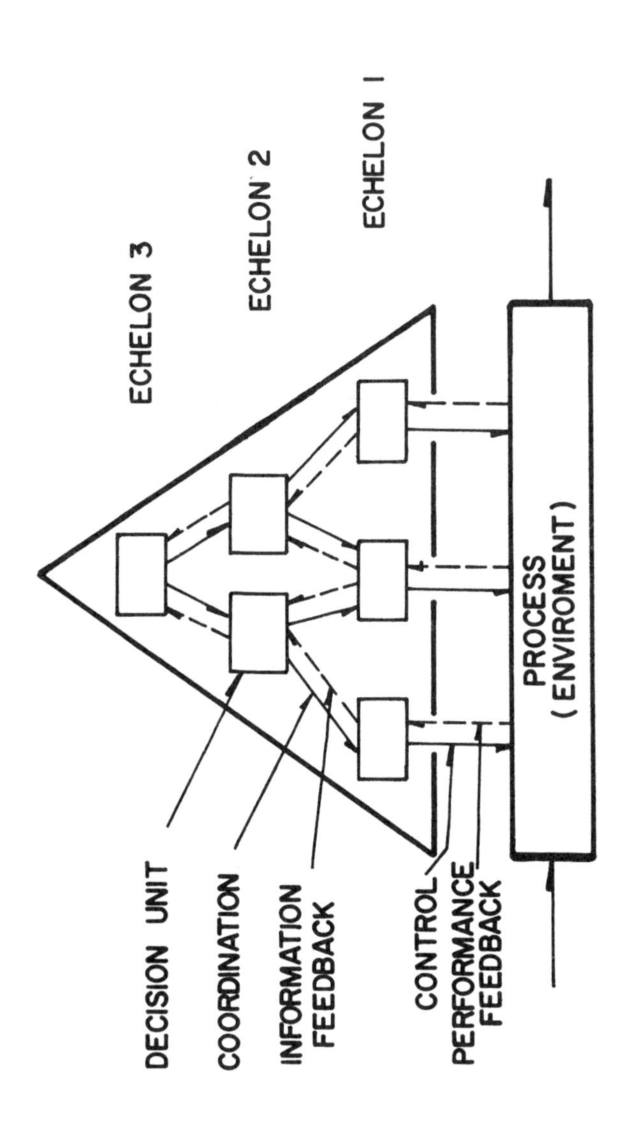

Figure 2
Multiechelon System

In the model, each unit is assumed to be a decision-making system[1] which solves either optimization problems[2] or satisfaction problems[3]. Higher level or supremal units control and coordinate the activities of lower level (infimal) units by manipulating the parameters of infimal decision problems. Infimal units convey information about the state of the environment and

---

[1] A decision making system may be represented schematically as follows:

More formally, a system $S \subseteq X \times Y$ is a decision making system if there is given a family of decision problems, $\mathcal{D}_x \varepsilon X$ with a solution set Z and a mapping $T: Z \rightarrow Y$ such that:

$$(\forall x)_X (\forall y)_Y [(x, y) \varepsilon S \leftrightarrow (\exists z)_Z : P(z, \mathcal{D}_x) \wedge T(z) = y]$$

where the predicate $P(z, \mathcal{D}_x)$ is true iff z is a solution of $\mathcal{D}_x$.

[2] A general optimization problem is defined as follows: Let $g: X \rightarrow V$, where:

X is a set of inputs, the decision set;

V is a set of values, linearly or partially ordered under $\leq$;

and g is the objective function.

Define X^f as the set of feasible decisions. A given $\hat{x}$ in X^f is a solution of the optimization problem iff the condition:

$$(\forall x)_{X^f} [g(\hat{x}) \; R \; g(x)]$$

is satisfied (where R is a specified relation, usually $\geq$ or $\leq$). Optimization problems may be represented by pairs of the form (g, X^f).

[3] A general satisfaction problem is defined as follows: Let $g: X \times \Omega \rightarrow V$, and $\tau: \Omega \rightarrow V$, where

τ is a tolerance function

Ω is the uncertainty set, and where

X, V, and g are defined as in note 30 above. A given x in X^f is a solution to the satisfaction problem iff the condition:

$$(\forall \omega)_\Omega \; [g(\hat{x}, \omega) R_\tau \tau(w)]$$

is satisfied. Satisfaction problems may be represented by quadruples of the form (g, τ, X^f, Ω).

about their own preformance through feedback inputs to the supremals.

The multiechelon structure incorporates the notion of specialization of functions both within and between levels. Moreover independent behavior on the part of infimal units is permitted. Effective coordination, therefore, is a crucial determinant of overall performance. Since lack of coordination is regarded as one of the major obstacles to effective urban management, the emphasis on coordination in Mesarovic's work is particularly appropriate.

In general, a system will be coordinable only if there is consistency between the infimal, supremal, and overall decision problems. The effective control of a process (to achieve an overall goal) depends, then, on the design of an appropriate structure (including the definition of decision problems at each level) and the selection of an appropriate coordination mode[4] and

---

[4] Five coordination modes are discussed by the authors. [Ref. 2, pp. 59-60]. The general characteristics of these modes are summarized as follows:

(i) Interaction Prediction Coordination: The supremal unit specifies the interface input and the infimal units proceed to solve their local decision problems on the assumption that the interface input will be exactly as predicted by the supremal unit.

(ii) Interaction Estimation Coordination: The supremal unit specifies a range of values for the interface inputs and the infimal units treat the interface inputs as disturbances which can assume any value in the given range.

(iii) Interaction Decoupling Coordination: The infimal units treat the interface input as an additional decision variable; they solve their decision problems as if the value of the interface input could be chosen at will.

(iv) Load Type Coordination: The infimal units recognize the existence of other decision units on the same level; the supremal unit specifies what kinds of communications are allowed between them. This leads to a coalition or a competitive (game theoretic type) relationship between the infimal units.

(v) Coalition type coordination: The infimal units recognize the existence of other decision units on the same level; the supremal unit specifies what kind of communications are allowed between them. This leads to a coalition or a competitive relationship between the infimal units.

coordination inputs. Coordinability also depends on the set of coordination inputs which will be accepted by the infimal units (a particularly important consideration when studying political systems). It will be useful to examine these aspects of the multi-echelon approach in a little more detail in order to suggest some specific applications.

6-3 CO-ORDINABILITY AND CONSISTENCY POSTULATES FOR A TWO LEVEL MULTIECHELON SYSTEM

In this section a formal definition of coordinability and consistency postulates for a two level system will be presented. A two level system is chosen for discussion because of its relative simplicity and because it can be used as the basic module for the synthesis of multiechelon systems in general. [Ref. 2, p. 85] A diagram of a two level system is presented in Figure 3. The general characteristics of this type of system are summarized by the authors as follows:

> [In the model diagrammed in Figure 3], the blocks represent the subsystems while their arrangement reflects the hierarchical structure of the overall system. The diagrammed system has n + 2 basic subsystems; the supremal control system C_o, the n infimal control systems $C_1, \ldots, C_n$ and the controlled process, P. Notice the two kinds of vertical interaction between the subsystems. One is a downward transmission of command signals; the signals from the infimal control systems to the process will be termed control inputs, while the signals from the supremal to the infimal control systems will be termed coordination inputs or interventions. The other kind of vertical interaction is an upward transmission of information or feedback signals to the various control systems of the hierarchy. These signal transmissions are represented by dashed lines in the diagram. The simplest way to describe the subsystems of a two level system is in terms of their terminal variables: inputs and outputs. It is convenient to describe the sybsystems as being functional in the sense that the inputs give unique outputs; one might view this as the situation in which the present state is given. Each of the blocks in Figure 3, therefore, represents a mapping. [p. 8].

Given this representation of a two level system, the general requirements for consistency (coordinability) may be defined. First, define a general optimization decision problem (See note 2,

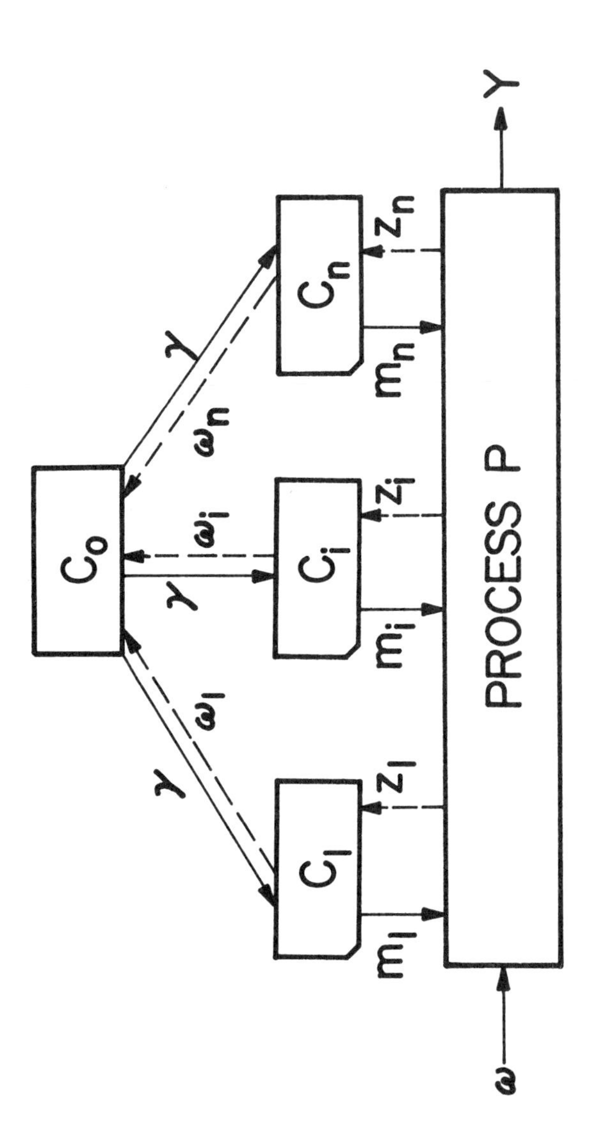

Figure 5

above.) for the system by the triplet (P, g, X^f) where P is the process to be controlled, g is an objective function, $g: x \rightarrow V$, X^f is a set of feasible alternatives and V is a set of values, linearly or partially ordered under $\leq$. Given a decision problem, $\mathcal{D}$, define a predicate P over the Cartesian product $((x \in X^f) \times \mathcal{D}$ such that $P(x, \mathcal{D})$ is true if and only if x is a solution of $\mathcal{D}$.

Let $\mathcal{D}_o$ be a single supremal decision problem and for each coordination input γ in the set of possible supremal coordination outputs, ξ, let $\mathcal{D}_i(\gamma)$ be the decision problem specified for the ith infimal decision unit so that $\overline{\mathcal{D}}(\gamma)$, the overall infimal decision problem, equals $\{\mathcal{D}_1(\gamma), \ldots, \mathcal{D}_h(\gamma)\}$, i.e., the solutions of $\overline{\mathcal{D}}(\gamma)$ are a set of n-tuples $(x_1, \ldots, x_n)$ such that for each i, $1 \leq i \leq n$, x_i is a solution of $\mathcal{D}_i(\gamma)$. Since with fixed feedback information (see note 5 above), the control inputs, M, depend wholly upon the decisions of the infimal units, we may define the relationship between infimal decisions and control inputs in terms of a mapping $\Pi_M: X \rightarrow M$ where M is a control input to P.

Given these definitions, the basic conditions of coordinability for any two-level system S, process P and overall decision problem $\mathcal{D}$ may be specified. In general the total system will be coordinable iff

(1) the infimal decision problems are coordinable relative to the overall decision problem,

(2) the infimal decision problems are coordinable relative to the supremal decision problem, and

(3) the solution sets for (1) and (2) are consistent.

These conditions are formalized by Mesarovic et. al. as follows:

(1) The infimal decision problems of S are coordinable relative to a given overall decision problem $\mathcal{D}$ iff:

$$(\exists \gamma)(\exists x)[P(x, \overline{\mathcal{D}}(\gamma)) \wedge P(\Pi_M(x), \mathcal{D})] \qquad (1)$$

(2) The infimal decision problems are coordinable relative to the supremal decision problem iff:

$$(\exists \gamma)(\exists x)[P(x, \overline{\mathcal{D}}(\gamma)) \wedge P(\gamma, \mathcal{D}_o)] \qquad (2)$$

But recall that the solution of $\mathcal{D}_o$ depends explicitly upon the outputs of the infimal units. Accordingly, a predicate $Q_o(\gamma, x)$ may be defined over the Cartesian product $G \times (x_1 \times \ldots \times x_n)$ such that Q_o is true if and only if x is the solution of $\mathcal{D}_o$. The dependence of the solution of the supremal problem may be expressed formally as

$$P(\gamma, \mathcal{D}_o) \leftrightarrow (\exists x)[Q_o(\gamma, x)] \qquad (2.1)$$

and if it is assumed that the x in (1) is equivalent to the x in (2) it follows that $\mathcal{D}(\gamma)$ is coordinable relative to $\mathcal{D}_o$ iff:

$$(\exists \gamma)(\exists x)[P(x, \overline{\mathcal{D}}(\gamma)) \wedge Q_o(\gamma, x)] \qquad (2.2)$$

(3) Since solutions to the overall decision problem depend on the choice of an x in X, supremal coordinability relative to the overall decision problem requires that all goals of the system be in harmony. This relationship is expressed by a consistency postulate for the decision problems of the system. The decision problems $\mathcal{D}$, $\mathcal{D}_o$ and $\overline{\mathcal{D}}$ are consistent if and only if:

$$(\forall y)(\forall x)[[P(x, \overline{\mathcal{D}}(\gamma)) \wedge Q_o(\gamma, x)] \rightarrow [P(x, \overline{\mathcal{D}}(\gamma)) \wedge P(\Pi_M(X), \mathcal{D})]] \qquad (3)$$

From the perspective of the engineer or designer, the consistency postulate, plus its application to specific coordination modes, provides a basis for determining; in particular instances, whether or not a given structure has the capability to attain a given goal or whether either the goal or the structure must be modified. The general notion that a system's overall goal must be in harmony with its subsystem goals if the system is to function effectively is as applicable to the problems of political organizations as to the somewhat less complex control mechanisms which Mesarovic and his associates have examined in detail [Ref. 16,17]. But urban governments in democratic systems differ significantly in at least one respect from the standard canonical form of the multiechelon structure. We are referring, of course to the presence of an electoral decision-mechanism as a direct, albeit infrequent, feedback input to the supremal unit. In addition, externally imposed constraints (for example, by state constitutions) upon the set of alternatives available to decision units must be taken into account.

The mathematical description of a simplified urban political system presented below utilizes the multiechelon framework. No claim is made that it resolves all of the problems involved in developing multiechelon models of political structures, but it does represent a preliminary attempt to deal with some of them. This preliminary work can, we believe, serve to focus and direct further analytic and empirical investigations of political coordination problems. In addition, the formalization has been extremely helpful in identifying the parameters which need to be incorporated in our simulation model.

6-4 AN ILLUSTRATIVE MULTIECHELON MODEL OF AN URBAN GOVERNMENT

The basic concept upon which this model is based is that the principal decisions of government are resource allocation decisions and that its principal function is to maximize (or at least increase) the level of "satisfaction" in the urban environment. This latter assumption is derived from the idea that the overall objective of government (the overall system goal) is to "promote the general welfare and happiness". This overall goal, however, is not the goal of any of the component decision units within the system. The goal of the supremal unit is to remain in power by winning elections. The goal of the infimal (administrative) units is to successfully administer the sectors of the urban environment for which they are responsible by maximizing specified performance functions. The model is diagrammed schematically in figure 4.

6-4.1 Definition of the Component Elements in the System

Define the urban governmental system (the overall system) as a mapping $G: R \rightarrow \bar{\bar{A}}$, where R is the amount of resources available to the system and $\bar{\bar{A}}$ is the set of possible allocations of those resources to the urban environment, U. Thus, $\bar{\bar{A}}$ may be defined formally as a set of matrices, $\bar{\bar{a}}$, with elements a^{ij}, where an element refers to the amount of resources allocated to the jth program of the ith sector of governmental activity.

The <u>supremal</u> <u>unit</u> of G (the elected executive) is a decision system, $C_o: R \rightarrow \bar{A}$, where $\bar{A} \varepsilon R \times R \times R \times R$. The infimal decision units for the four sectors of governmental activity are decision systems $C_i: A^i \rightarrow \bar{A}^i$ where A^i is the set of resource allocations to the ith sector, $\bar{A}^i \varepsilon A^{i1} \times \ldots \times A^{in}$, where $j=1, \ldots, n$ are indices for the <u>programs</u> in the ith sector to which resources may be allocated. For every program in G there exists a <u>performance function</u> ψ^{ij} and a corresponding performance value p^{ij} which indicates the level of performance for the jth program in the ith sector.

The urban environment, U, is defined as a mapping $U: Y \times \bar{A} \times \Omega \rightarrow Y$, $Y = S \times \bar{\bar{P}}$, where the elements of $\mathcal{S}$ are aggregate levels of satisfaction in U, $\bar{\bar{P}}$ is a set of matrices, P of the performance values p^{ij}, and Ω is the uncertainty set. The sub-

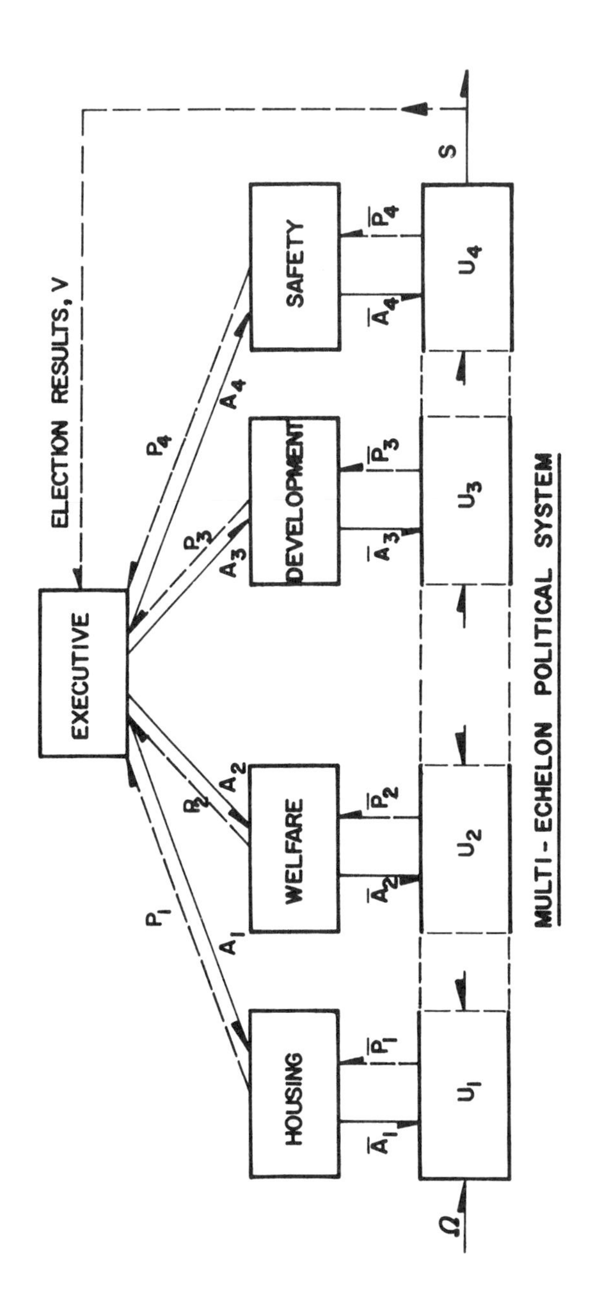

Figure 4
Multi-Echelon Political System

processes of the environment are defined as mappings U_i: $Y^i \times \bar{A}^i \times U^i \times \Omega \to Y^i$, $Y = S^i \times \bar{P}^i$, where an element $\sigma^i \varepsilon S^i$ is the net contribution to aggregate satisfaction of the ith subprocess, $\bar{\rho}^i$ is a vector of performance values, and $\mathcal{U}^i$ is the set of interface inputs to the ith subprocess, U_i.

Values of σ^i are defined by a mapping f: S^i $S^i \times p^i \times \Omega \to \mathcal{S}^i$ for any time, t, $\sigma_t = f^{\mathcal{S}(i)}(\sigma t_{-1}, p^i_t, \omega^i_t)$ and by the definition of σ^i, $\sigma_t = \sigma_i \Sigma_{i=1} \sigma^i_t$. Since $y_t = U(y_{t-1}, \bar{\bar{a}}_t, \omega_t)$ we may also define an aggregate satisfaction function, $f^{\mathcal{S}}$: $Y \times \bar{\bar{A}} \times \Omega \to \mathcal{S}$, so that for any time, t, $= f^{\mathcal{S}}(y_{t-1}, \bar{\bar{a}}_t, \omega_t)$.

6-4.2 Decision Problems and Feedback

Up to this point, the discussion has differed from the basic theory of Mesarovic and his associates only in the names of certain variables. But in attempting to formulate realistic decision problems for the different levels in the system, one must deal with two distinctly political aspects of organizational behavior which are not usually encountered in systems engineering.

First, the relationship between the supremal goal (winning elections) and the overall goal (maximizing satisfaction) is problematic. If we make the not unconventional assumption that the budgetary process (by which a specific $\bar{\bar{a}}_t$ in $\bar{\bar{A}}$ is chosen for some value R_t) is an annual one and the electoral decision process occurs less frequently, then the supremal decision unit must base its coordination inputs on very limited information. Moreover, the goal of the supremal unit can only be defined in terms of an expected probability function, i. e., the expected probability of maximizing the percentage of votes cast in the next election (or, perhaps, simply winning the next election). Additional complexity arises from the fact that the frequency of elections and the decision rule for elections (majority, two thirds majority, etc.) are also variable parameters whose properties can be investigated. Although we will assume that these parameters are fixed, for purposes of this discussion, the reader should be aware that this need not be the case. It is quite possible that the system may be coordinable for some sets of assumptions about the electoral process and not for others.

It follows from the discussion above, that the relationship between the supremal and infimal goals is also problematic. There are not necessarily any obvious performance criteria which could be chosen for the infimal problems. In the formalization below, only the case where $\bar{\psi}$ is fixed is considered. But the reader should recognize that this assumption glosses over an intriguing and important problem which must be investigated in much greater detail.

Finally, the political system is a mixed system in the sense that some decision problems for some units (especially lower level "administrative" units) are likely to be optimization problems, while the decision problems for higher level (executive) units are certainly going to be satisfaction problems. To the writer's knowledge no detailed investigation of such systems has yet been made.

It must be recognized, then, that the formalization below is only an exploratory venture into largely uncharted territory. Much more work must be done before the unique problems involved in modeling political systems and examining their coordinability characteristics are satisfactorily resolved.

In formalizing the structure of G, we assume that R_t, the resources available to the system at time t, and $\overline{p}$, the performance criteria for the infimals, are fixed. The overall goal of the system, as noted above, is to maximize satisfaction. The goal of the supremal unit is to maximize the probability of winning the next election. The goal of the infimals is to maximize the values of their respective performance functions, ψ^i. These general notions will now be developed more precisely below, so that the basic coordinability requirements for the system can be stated.

For fixed resources r_t, the overall decision problem, $\mathcal{D}(r_t)$ may be represented by the triple $(f^{\mathcal{S}}, \bar{\bar{A}}_t, \omega_t)$. An allocation, $\hat{\bar{\bar{a}}} \in \bar{\bar{A}}_t$ is a solution to $\mathcal{D}(r_t)$ iff:

$$(\forall \bar{\bar{a}})_{\bar{\bar{A}}_t} (\forall \omega t)_{\Omega} \; [f^{\mathcal{S}}(y_{t-1}, \hat{\bar{\bar{a}}}_t, \omega_t) \geq f^{\mathcal{S}}(y_{t-1}, \bar{\bar{a}}_t, \omega_t)] \tag{4}$$

Assuming fixed performance criteria, $\overline{P}$ and specified coordination inputs $\overline{a}_t$ to the infimal units, then the infimal decision problems $\mathcal{D}_i(a^i_t)$ may be represented by triple $(p^i, \overline{A}^i, \Omega^i)$. An allocation vector $\hat{\bar{a}}^i_t \in \overline{A}^i_t$, where $\overline{A}^i_t = \{ \bar{a}^i_t, \bar{a}^i_t$ is a possible allocation vector of $C_i\}$ is a solution to $\mathcal{D}_i(a^i_t)$ iff:

$$(\forall \bar{a}^i_t)_{\overline{A}^i_t} (\forall \omega^i_t)_{\overline{\Omega}^i_t} [\psi^i(\bar{p}^i_{t-1}, \hat{a}^i_t, \bar{u}^i_t, \omega^i_t) \geq \psi^i(\bar{\psi}^i(\bar{p}^i_{t-1}, \bar{a}^i_t, \bar{u}^i_t, \omega^i_t))] \tag{5}$$

Then the solutions to the overall infimal decision problem, $\mathcal{D}(\overline{a})$, are a set of four-tuples, $A^1 = (\bar{a}^1_t, \ldots, \bar{a}^4_t)$ such that for $i = 1, \ldots, 4$, $\bar{a}^i_t$ is a solution to $\mathcal{D}_i(a^i_t)$.

Formalization of the supremal decision problem is - as noted above - somewhat more complex. First, some additional definitions are needed. Let V = the set of electoral outcomes, where $v_t \varepsilon V$ is the percentage of votes received by the "administration" if an election is held at time t. The value of v_t is defined by a function $f^V: \mathcal{S} \times \Omega^o \rightarrow V$. The electoral decision rule, D is a pair (K, V^D) where K is the number of years between elections and V^D is the percentage of votes needed to win under decision rule D.

Finally, let $\xi^{D(x)}$ be defined as the expected probability that the "administration" will win the next election, assuming that decision rule D^x is in force. Given these definitions, the supremal decision problem $\mathcal{D}_o(\tau_t)$ may be formalized as $\mathcal{D}_o = (f^{\xi(D(x))}, \bar{A}_t, \Omega^o_t)$ where Ω^o_t is an uncertainty set and where $\xi^{D(x)} = f^{\xi(D(x))}(\bar{a}^i_t, \omega^o_t)$. In principle then, a solution to $\mathcal{D}_o(\tau_t)$ may be defined as an allocation vector $\hat{\bar{a}}_t$ such that

$$(\forall \bar{a}_t)_{\bar{A}_t} (\forall \omega^o_t)_{\Omega^o_t} [f^{\xi(D(x))}(\hat{\bar{a}}, \omega^o_t) \geq f^{\xi(D(x))}(\bar{a}_t, \omega^o_t)] \qquad (6)$$

But this formalization does not in itself provide a very useful or realistic basis for investigating the consistency of the system. The difficulty lies in the fact that the properties of the function $f^{\xi(D(x))}$ are extremely difficult to define (and certainly now known to the supremal unit.) It seems plausible to require that the supremal decision problem be defined in terms of feedback inputs to the supremal unit, i.e. to assume that the supremal bases its decisions on measurable quantities. Two definitions of the supremal decision problem which meet this requirement have been investigated [Ref. 5]. In the one to be discussed here, called the performance approach, the supremal bases its decisions on past performance of the system.

It is assumed that the supremal unit is a two term incumbent and uses the immediately preceeding election as a base point for its calculations. (However, less restrictive assumptions will not affect the development of the model presented here.) It has no direct information of any sort regarding the value of $\mathcal{S}$ between elections. Thus it can only base its decisions on feedback information regarding performance from the infimals. In order to resolve this dilemma, it assumes that $\xi^{D(x)}$ will be maintained at a satisfactory level if the overall level of governmental performance is maintained or improved, in comparison to the level in the year when the election was won. This assumption provides a basis for defining an operational goal, based on available information, for the supremal decision problem.

The goal-attainment function for the supremal unit is defined as follows: Let $g^{p^i}(p^i_t, p^i_{t+x}) = \Delta_t p^i_{t+x}$ is a metric for the net change in performance in the ith sector during the interval from t to t+x. It will be assumed that this metric is defined in such a way that changes in the respective sectors are measured in comparable units. (For example, $g^{p^i}(p_t^i, p_{t+x}^i) = (p_{t+x}^i - p_t^i)/p_t^i$. Let $\Delta_t p_{t+x} = (\Delta_t p_{t+x}^1, \ldots, \Delta_t p_{t+x}^4)$ and assume that the supremal unit aggregates the values of $\overline{\Delta_t p_{t+x}}$ into an overall measure of change in performance, $\Delta_t p_{t+x}$, by a function $f^p(\Delta_t p_{t+x}) = \Delta_t p_{t+x}$. This function defines the way in which the supremal unit assesses the relative significance of changes in each sector (with regard to its goal of winning elections). Thus the supremal image of the environment (on which it bases its decisions) may be expressed as $\Delta_t p_{t+x}) \geq \xi_{t+x}^{D(x)} \geq \xi_T^{D(x)}$.

The expression $f^{\Delta p}(\Delta_t p_{t+x})$ may be rewritten as

$$f^{\Delta p}(g^{p1}(p_t^1, \psi^1(\bar{\psi}^1(\bar{p}_{t+x-1}^1, \ddot{a}_{t+x}^1, \ddot{u}_{t+x}^1, \bar{\omega}_{t+x}^1))), \ldots, (g^{p4}(p_t^4, \psi^4((\bar{\psi}^4 (\bar{p}_{t+x-1}^4, \ddot{a}_{t+x}^4, \ddot{u}_{t+x}^4, \bar{\psi}_{t+x}^4))))$$

reflecting the dependence of $(\Delta_t p_{t+x})$ upon infimal performance. Then the supremal decision problem $\mathcal{D}_o^p(r_{t+x})$ may be defined by the pair $(f^{\Delta p}, \bar{A}_{t+x})$ where an allocation vector $\hat{\hat{a}}_{t+x}$ is a solution to $\mathcal{D}_o^p(r_{t+x})$ iff:

$$(\forall \ddot{a}_{t+x})_{\bar{A}_{t+x}} [f^{\Delta p}(g^{p^i}(p_t^i, \psi^i(\bar{\psi}^i(\bar{p}_{t+x-1}^i, \hat{\bar{a}}_{t+x}^i, \bar{u}_{t+x}^i, \omega_{t+x}^i)))) \geq f^{\Delta p}(g^{p^i}(p_t^i, \psi^i(\bar{\psi}^i(\bar{p}_{t+x-1}^i, \hat{a}_{t+x}^i, \omega_{t+x}^i))))] \qquad (7)$$

for $i=1,\ldots,4$, and where $\hat{\hat{a}}_{t+x}^i$ is a solution to $\mathcal{D}_i(\hat{\bar{a}}_{t+x})$ and $(\forall \ddot{a}_{t+x})_{\bar{A}_{t+x}}, \hat{\bar{a}}_{t+x}^i$ is a solution to $\mathcal{D}_i(\bar{a}_{t+x})$.

6-4.3 Coordinability and Consistency Postulates

Recall (from pp. 7-10) above that the total system will be coordinable iff:

(1) the infimal decision problems are coordinable relative to the overall decision problem,

(2) the infimal decision problems are coordinable relative to the supremal decision problem and

(3) the solution sets for (1) and (2) are consistent.

6-4.3.1 Infimal-overall Coordinability

Define a predicate $P(\bar{\bar{a}}, \mathcal{D}(r))$ such that $P(\bar{\bar{a}}, \mathcal{D}(r))$ is true if $\bar{\bar{a}}$ is a solution of $\mathcal{D}(a)$ and a predicate $P(\bar{\bar{a}}', \overline{\mathcal{D}(\bar{a})})$ such that

$P(\bar{\bar{a}}', \overline{\mathcal{D}(\bar{a})})$ is true iff a' is a solution of $\overline{\mathcal{D}(a)}$. Then the infimal decision problems of G are coordinable, relative to the overall problem, $\mathcal{D}(r)$ iff

$$(\forall \bar{a})_{\bar{A}}(\forall \bar{\bar{a}}')_{\bar{\bar{A}}}[(\forall \omega)_{\Omega} \; [P(\bar{\bar{a}}, \; \bar{\mathcal{D}}(\bar{a})) \wedge P(\bar{\bar{a}}', \; \mathcal{D}(r))]] \qquad (8)$$

6-4.3.2 Infimal-Supremal Coordinability

The definition of infimal supremal coordinability will assume that the performance approach is being used, i.e., $\mathcal{D}_o = (f^{\Delta p}, \bar{A})$. Given this assumption, the infimal decision problems are coordinable, relative to the supremal problem iff

$$(\exists \bar{a})_{\bar{A}}(\exists \bar{\bar{a}}')_{\bar{\bar{A}}}[(\forall \omega)_{\Omega} \; [P(\bar{\bar{a}}', \; \bar{\mathcal{D}}(\bar{a}) \wedge P(\bar{a}, \; \mathcal{D}_o^p(r))]] \qquad (9)$$

Moreover, since the solution of $\mathcal{D}_o^h$ depends on $\bar{\bar{a}}$, a predicate Q_o^p may be defined over $\bar{A} \times \bar{\bar{A}}$ such that $P(\bar{a}, \mathcal{D}_o^p(r)) \leftrightarrow (\; \bar{\bar{a}}')_A [Q_o^p(\bar{a}, \bar{\bar{a}}')]$, where a is a solution of $\overline{\mathcal{D}}(\bar{a})$. Then $\overline{\mathcal{D}}(\bar{\bar{a}})$ is coordinable, relative to $\mathcal{D}_o^p(r)$ iff:

$$(\forall \bar{a})_{\bar{A}}(\forall \bar{\bar{a}}')_{\bar{\bar{A}}}[(\forall \omega)_{\Omega} \; [P(\bar{\bar{a}}', \; \overline{\mathcal{D}}(\bar{a}) \wedge Q_o^p(\bar{a}, \bar{\bar{a}}')]] \qquad (9.1)$$

6-4.3.3 Consistency

The solution sets for (1) and (2) will be consistent iff:

$$(\forall \bar{a})_{\bar{A}}(\forall \bar{\bar{a}}')_{\bar{\bar{A}}}(\forall \omega)_{\Omega} \; [[P(\bar{a}'), \; \mathcal{D}(\bar{a})), \wedge \; Q_o^p(\bar{a}, \bar{\bar{a}}')] \rightarrow$$

$$[P(\bar{\bar{a}}', \; \overline{\mathcal{D}}(\bar{a}) \wedge P(\bar{\bar{a}}', \; \mathcal{D}(r))]] \qquad (10)$$

But note that consistency in this sense does not ensure that the "administration" (supremal unit) will achieve its goal of remaining in power. This will only be true if it is the case that the supremal decision problems $\mathcal{D}_o$ and $\mathcal{D}_o^p$ are also consistent. Thus systems which satisfy (1) may be highly unstable. To illustrate the point, suppose that for some value of $\omega \in \Omega$,

$$(\exists \bar{a})_{\bar{A}}(\exists \bar{\bar{a}}')_{\bar{\bar{A}}}[P(\bar{\bar{a}}', \; \mathcal{D}(\bar{a}) \wedge Q_o^p(\bar{a}, \bar{\bar{a}}') \rightarrow \sim \; P(\bar{a}, \; \mathcal{D}_o(r))]$$

If this condition persists, it is quite possible that an administration which has achieved a high level of performance with respect to the overall decision problem $\mathcal{D}(r)$ may still lose elections. Moreover, a successor administration may be even less successful in maintaining desirable levels of $\mathcal{S}$.

Thus, if long term effectiveness is the criterion, the consistency postulate for the type of political system under consideration must be redefined as:

$$(\forall \bar{a})_{\bar{A}}(\forall \bar{\bar{a}}')_{\bar{\bar{A}}}(\forall \omega)_{\Omega}\ [[P(\ddot{\bar{a}}', \bar{\mathcal{D}}(\bar{a})) \wedge Q_o^p(\ddot{a}, \bar{\bar{a}})] \rightarrow [P(\bar{a}, \mathcal{D}_o(r)) \wedge P(\bar{\bar{a}}', \bar{\mathcal{D}}(\bar{a})) \wedge P(\ddot{\bar{a}}', \mathcal{D}(r))]] \tag{11}$$

Further extension of the formalization, through the definition of specific coordination modes and (for example) the introduction of "political" constraints on the control and information interface between the supremal and infimal units can be accomplished in a relatively straightforward manner. However, in view of the complexity of the urban environment, it seems unlikely that the theorems developed in Part II of [Ref. 2] can be directly applied to the political case. Although a number of scholars are beginning to investigate analytic techniques which may be applicable to political type systems [Ref. 4, 6, 11, 12] such techniques are presently available only for the simplest, most idealized (and least interesting) cases [Ref. 15, 13]. Accordingly, like other scholars who have encountered gaps between the problems they wished to solve and the analytic mathematical techniques available to solve them, we propose to use computer simulation as an alternative "method of solution". [Ref. 13 , 14]

6-5 A SIMULATION APPROACH TO THE PROBLEM OF ORGANIZATIONAL DESIGN FOR URBAN SYSTEMS

6-5.1 Introduction

The model we are developing uses the <u>Forrester Urban Dynamics</u> Model [Ref. 15] (hereafter referred to as the FUDM) to simulate an <u>outer environment</u>. The FUDM is modified by deleting rudementary governmental functions and by adding an "electoral" and a "satisfaction" mechanism. These mechanisms provide a basis for operationally defining overall and supremal decision problems. Electoral outcome and satisfaction performance criteria, then, constitute two major components of the <u>information interface</u>. The third component of the information interface is a set of infimal performance criteria. The elements of this set are, for the most part, already existing FUDM variables.

The existing <u>control interface</u> of the FUDM, comprising ten urban development programs [Ref. 15, pp. 51-106] is incorporated into our model without significant modification. However unit costs for the respective programs must be defined, so that they can be taken into account in governmental resource allocation decisions.

The inner environment comprises a supremal "executive" and five administrative "infimal" decision units. All of the decision units have similar behavioral characteristics. These characteristics are based upon principles of decision making which are most frequently used to describe the behavior of administrative and political systems, namely (a) crisis management, (b) incrementalism, and (c) satisficing.

Space does not permit a detailed technical description of the model; however some of the more significant aspects and some preliminary results will be examined briefly. Since our work is still in an exploratory and developmental stage, it is hoped that this discussion will elicit critical comments and suggestions which can be incorporated in future revisions and modifications.

6-5.2 The FUDM as Outer Environment

Since our long term objective is the development of more effective institutions, it seemed desirable to utilize a previously developed model of the outer environment so that major attention could be directed to the governmental and interface components. Many of the available models have been critically examined by Lowry [1967]. In general, they have been developed by and for planners and tend to be oriented towards the investigations of land use or location patterns. Although the majority tend to be static and to place a heavy reliance on linear assumptions, some non linear, time dependent models have also been developed. [Ref. 19, 20] The FUDM is more speculative, and has been severely criticized for its failure to take empirical data into account, [Ref. 21] for its neglect of factors outside of the city [Ref. 22], for its conceptual treatment of the issue of parameter sensitivity [Ref. 23] and for various other shortcomings [Ref. 24]. However Forrester's superb documentation, concern with long term policy issues and explicitly defined control interface rendered the model particularly attractive for our purposes.

The possible "lack of realism" in the FUDM does not particularly concern us, since principles for designing effective institutional control mechanisms for this outer environment should be readily transferable to more realistic ones.

6-5.2.1 The FUDM: Basic Characteristics

Essentially, the FUDM is an attempt to demonstrate that Forrester's industrial dynamics techniques [Ref. 25, 26] are applicable to non-industrial settings. The city is conceived as an essentially closed system in a "limitless environment" [Ref. 15, pp. 17, passim]. The parameters of the environment are not affected by developments within the city, and remain stable, relative to the city, over time.

The basic components of Forrester's city are three classes of state variables: housing, industry and population. Within these classes, state variables of premium, worker and underemployed housing; new, mature and declining industry and managerial, worker and underemployed persons are defined. The housing and industry sectors are essentially simple "filtration" models. New industry and premium housing are constructed, gradually become obsolescent and are eventually demolished. The population sector is a relatively standard population dynamic model which includes migration, net birth rate and mobility (both upward and downward) components. There is also a rudamentary government which (a) assesses basic tax needs, (b) adjusts these needs to the political influence of the respective classes (c) computes a tax rate based on needs and available resources (d) adjusts this rate for public awareness, political support and feasibility and (e) makes "public expenditures". [Ref. 15, pp. 201-206]. Intra-sectoral flows and flows between the city and the "limitless environment" are controlled by complex feedback loops which are influenced both by the state variables and by auxiliary variables which incorporate various social, political and economic factors.

6-5.2.2 The FUDM: Control Interface

Like the authors of this paper, Forrester is concerned with discovering policies which, when implemented will produce better cities. In the language of our paradigm, the ten "urban development programs" [p. 211, ff.] which he defines, for purposes of experimentation, comprise the control interface of the model.

6-5.3 Modifying the FUDM

Two major, and several minor modifications to the FUDM have been necessary to make it compatible with our objectives. The major modifications involve the introduction of variables and equations which define electoral outcomes and states of "satisfaction" for the urban environment. Neither of these dimensions are considered explicitly by Forrester, although he makes an implicit distinction between more and less desirable states of the city [Ref. 27]. In addition, the rudamentary government which is integral to the FUDM, has been deleted. In making these modifications considerable attention has been devoted to preserving the basic character of Forrester's model. A brief discussion of the election and satisfaction mechanisms will serve to illustrate the way in which this has been accomplished.

6-5.3.1 The Satisfaction Mechanism

Satisfaction is measured by a three component vector. The components measure levels of satisfaction in the manager, worker and underemployed classes. As noted above, satisfaction levels are used to provide an indicator of overall system performance. The rationale for such an indicator is the assumption that the overall function of government in a democratic society is to provide for the general welfare and happiness. The equations which define satisfaction levels are based on the following specific assumptions:

1. It is assumed that the level of satisfaction for any individual in the urban environment is determined by his housing and employment situation, by "symbolic" effects of governmental programs, and by governmental expenditures within the public sector. This assumption is dictated in large degree by the commitment to the Urban Dynamics model as a useful representation of the urban environment and by a desire to avoid modifications of the internal structure of the model.

2. It is assumed that the components of satisfaction may be aggregated within classes, but not across classes. For example, it is believed that the concept "level of satisfaction for the underemployed sector", comprising components of (a) housing satisfaction, (b) employment satisfaction and (c) public expenditure satisfaction, is conceptually meaningful, but that the concept "level of housing satisfaction", comprising components of managerial housing satisfaction, labor housing satisfaction and underemployed housing satisfaction, is not.

3. It is assumed that each component of within-class satisfaction comprises three subcomponents. These subcomponents are (a) the aggregate effects of specific individual experiences with actual system states, (b) the effects of popularly perceived system states and (c) the symbolic effects of governmental programs.

4. Finally it is assumed, that aggregate class satisfaction levels cannot be further aggregated to define a simple coefficient

of overall system satisfaction. Accordingly, as noted above, the overall level of satisfaction in the system for any year is defined as a vector having managerial, labor and underemployed components.

Maximizing the above defined measure of satisfaction is the overall system goal for a multi-echelon governmental decision making structure. However information about overall system performance with respect to this goal is never directly available to governmental decision-makers. Feedback to the governmental decision units takes the form of measurable system parameters, demands and electoral results. The major function of the satisfaction level variables, then is to provide an outside observer with an objective measure to evaluate the effects of alternative governmental policies, aimed at reviving the stagnant urban environment.

A number of computer simulation runs of the Forrester Urban Dynamics model, with satisfaction calculations incorporated, have been completed. The primary purpose of these experiments has been to examine the degree to which satisfaction coefficient values are consistent with Forrester's implied normative assumptions and conclusions.

Results for the 250 year long urban growth period are presented in Figure 5 a-d. The general impression of declining satisfaction during the stagnation period is clearly conveyed. Of course, the reader might legitimately question the degree to which any model of satisfaction could remain valid for a period of 250 years. A discussion of that issue, however, would not be germane to our purposes here.

Additional results, depicting the effects of remedial programs on satisfaction are presented in figures 5b, 5c and 5d. Once again, the satisfaction values are more or less consistent with Forrester's implicit normative assumptions. Thus, the "urban revival policy", SHDR .05 and WHCN = .015 produces improved satisfaction levels for the labor and underemployed classes while other policies are less successful.

The somewhat irregular effects of the underemployed training program, UTR = .05 are of interest because they illustrate a political problem commonly encountered in urban systems. Note that this program produces short term beneficial effects for the underemployed class, relatively long term beneficial effects for the managerial class (because of indirect influences on the premium housing adequacy multiplier, (PHAM) and no significant effects for the labor class Because of the short term beneficial effects, evidently this policy would be more politically attractive

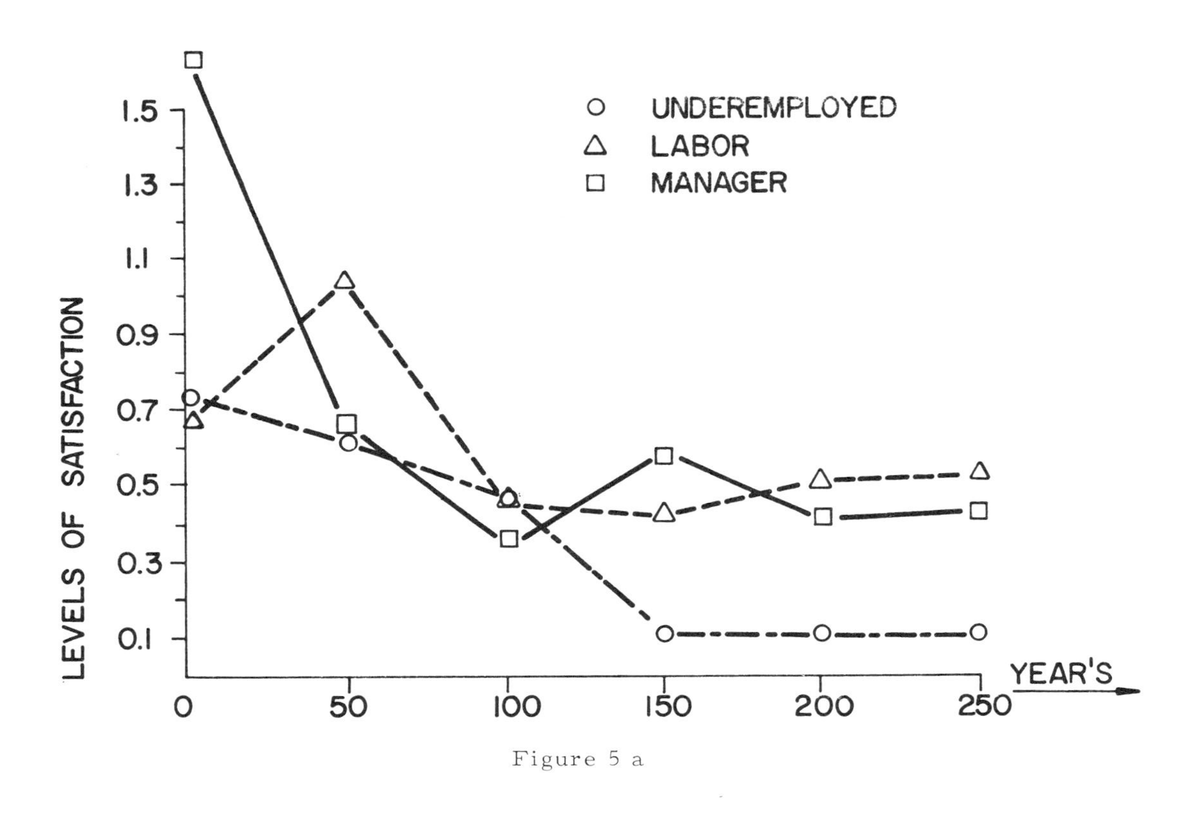
UNDEREMPLOYED
LABOR
MANAGER
LEVELS OF SATISFACTION
1.5
1.3
1.1
0.9
0.7
0.5
0.3
0.1
0
50
100
150
200
250
YEAR'S

Figure 5 a

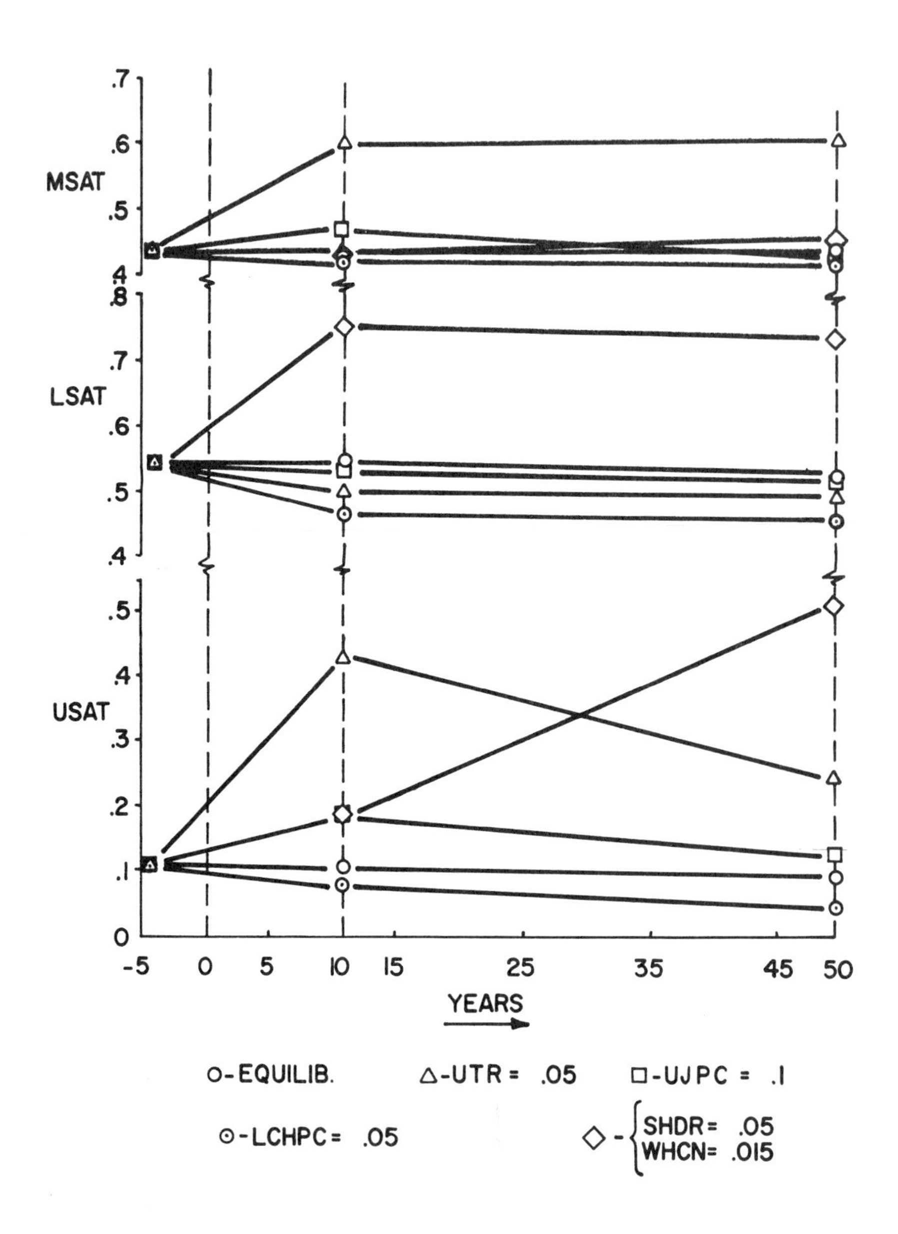

MSAT
LSAT
USAT
.7
.6
.5
.4
.8
.7
.6
.5
.4
.5
.4
.3
.2
.1
0
-5
0
5
10
15
25
35
45
50
YEARS
○-EQUILIB.
△-UTR = .05
□-UJPC = .1
⊙-LCHPC = .05
◇ - SHDR = .05
WHCN = .015

Figure 5 (b, c, d)

than the "revival" policy. This would be especially true if we make the reasonable assumption that preponderant political influence is exercised by the managerial class. These results are, of course, consistent with the observation of Forrester (and many others) that politically feasible policies may not be the most effective in the long run. Dealing with such situations represents, in our judgement, one of the more significant design problems for the prospective urban organizational engineer. When our model has been more fully developed, the search for solutions to this type of problem will be a major objective.

6-5.3.2 The Electoral Mechanism

The work of the Survey Research Center at the University of Michigan, [Ref. 28] has provided the underlying rationale for the mechanism through which the modified FUDM generates electoral returns. To achieve compatibility with the FUDM plus a degree of realism, the following specific assumptions have been made:

(1) Urban political arrangements are non-machine, i.e. the effects of party identifications are deemed insignificant.

(2) Institutional roadblocks to voting, such as property or literacy qualifications, are absent.

(3) The effects of age, sex, region, history, and religion are ignored.

(4) Forrester's class divisions into underemployed, laborer, and manager types are taken to reflect socio-economic-status (SES) characteristics, such as education, income, etc.

(5) The underemployed class is disaggregated into white and non-white subgroups. Labor and manager types, however, are assumed to act as groups without significant racial cleavages. The mechanism generates hypothetical election decisions at two levels: first the decision to vote or not to vote is made; then if the first decision has been for voting participation, the direction of the vote is decided.

The structure of the voting model consists of three successive filtering processes, through which two contending mayoral candidates are evaluated by the electorate. These filters should be viewed as attributes of social classes, not of individuals; i.e. the evaluation process occurs conceptually at aggregate group levels for the underemployed-non-whites, the underemployed-whites, the laborers, and managers. Hence the contending candidates A and B, one of whom is an incumbent, are evaluated as indicated in Figure 5 e. Each filtering process outputs a level of intensity (hostile or supportive), generated for each

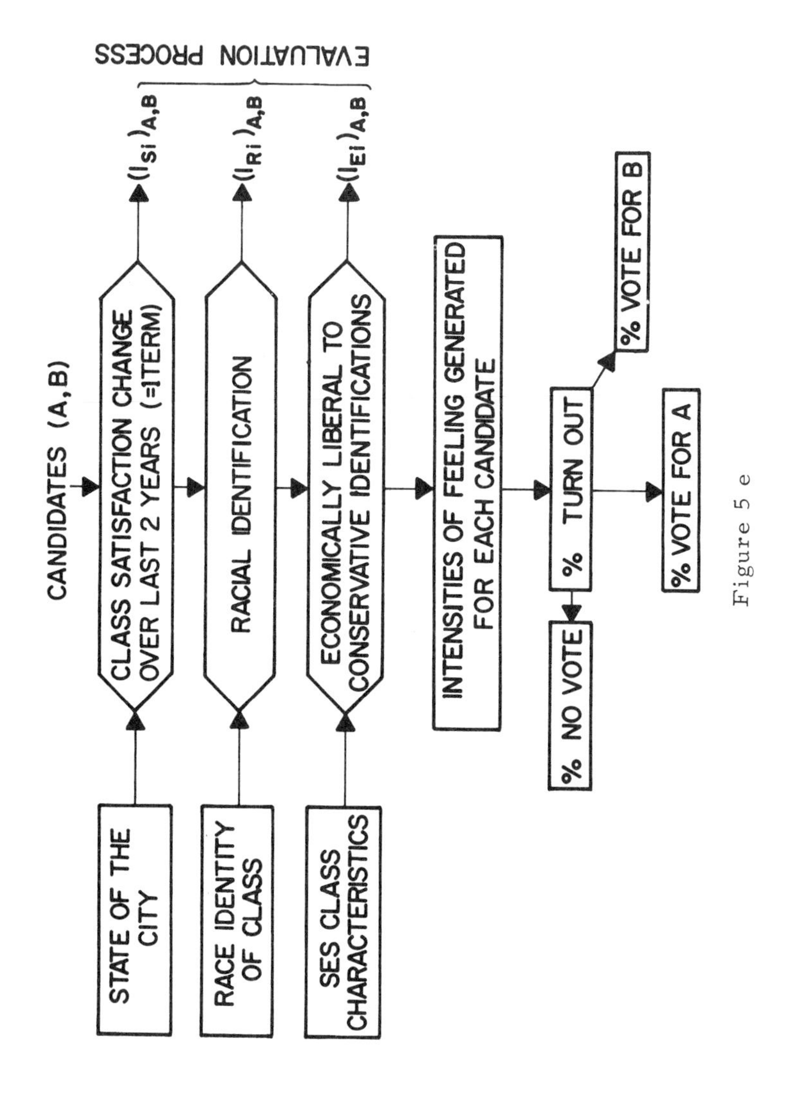
CANDIDATES (A,B)
CLASS SATISFACTION CHANGE OVER LAST 2 YEARS (=1TERM)
RACIAL IDENTIFICATION
ECONOMICALLY LIBERAL TO CONSERVATIVE IDENTIFICATIONS
STATE OF THE CITY
RACE IDENTITY OF CLASS
SES CLASS CHARACTERISTICS
$(I_{Si})_{A,B}$
$(I_{Ri})_{A,B}$
$(I_{Ei})_{A,B}$
EVALUATION PROCESS
INTENSITIES OF FEELING GENERATED FOR EACH CANDIDATE
% TURN OUT
% NO VOTE
% VOTE FOR A
% VOTE FOR B

Figure 5 e

candidate. Specifically the generic term $(I_{ji})k$ refers to the level of intensity generated through the evaluation process j, of locial class i, by candidate k. The percent turnout from social class i, generated by the contending mayoral candidates A and B, is then computed in the following fashion.

The shape of functions f_1, f_2, and f_3 are assumed such that $(I_{satisfaction,\,i})_k = (I_{s,\,i})_k = f_1(\%\ \Delta$ of class i satisfaction over last mayoral term, incumbency of k)
$(I_{race,\,i})_k = (I_{r,\,i})_k = f_2$(k's racial image, class i's racial perceptions)
$(I_{economics,\,i})_k = (I_{e,\,i})_k = f_3$(k's economic image, class i's economic perceptions)
The quantities α_i, β_i, γ_i are then computed for all i so that

$$\alpha_i = |\,(I_{si})_A + (I_{si})_B\,|$$

$$B_i = |\,|(I_{ri})_A| \pm |(I_{ri})_B|\,|$$

$$\gamma_i = |\,|(I_{ei})_A| \pm |(I_{ei})_B|\,|$$

If $(I_{ji})_A$ and $(I_{ji})_B$ of the same sign then minus otherwise plus.

α_i, β_i, and γ_i are then ordered for all i so that

$$z_i = \max\,(\alpha_i, \beta_i, \gamma_i)$$

$$v_i = \max\,(\{\alpha_i, \beta_i, \gamma_i) - (z_i)\}$$

$$w_i = \{\alpha_i, \beta_i, \gamma_i\} - \{z_i, v_i\}$$

Finally the percent turnout for class i is calculated by $g_i(z_i, v_i, w_i)$ for all i such that

$$g_i(z_i, v_i, w_i) = [(100 - \tau_i)/100][25 z_i = 15\ v_i + 10\ w_i]$$

where τ_i represents the normal class i turnout due to feelings of civic duty. Thus decision one, the issue of voting or not voting, is decided for individuals in each class. To determine the direction of the vote of class i (i.e. decision two), we calculate

$$A_i = [(I_{s,\,i})_A + (I_{r,\,i})_A + (I_{e,\,i})A]$$

$$B_i = [(I_{s,\,i})_B + (I_{r,\,i})_B + (I_{e,\,i})_B] \text{ for all i}$$

Then if $A_i > B_i$, candidate A receives all the votes from social class i. However if $A_i < B_i$, then candidate B receives all the votes from social class i.

Initial manual operation of this model has yielded results which appear intuitively acceptable.

To cite an example, assume that the following are given: 1. a minus 5 percent change in underemployed class satisfaction over the last mayoral term and 2. a candidate A who is non-incumbent, black, and economically moderate, and 3. a candidate B who is incumbent, white conservative, and economically conservative. The model then generates a 53 percent underemployed non-white turnout, all voting for A; and a 42 percent underemployed-white turnout, all voting for A as well. These results show up the tendency of underemployed persons to support progressive and out of office candidates. Moreover, the smaller underemployed white response to A reflects the negative effect of feelings of racial hostility induced by candidate A's racial image.

Further experimentation, with the electoral mechanism and the FUDM fully integrated will be undertaken in the near future.

6-5.4 Towards a Model of Inner Environment

The problem of designing more effective political organizations is viewed as a problem of reliable design and unreliable components, under conditions of uncertainty and imperfect information. Although we have not, as yet obtained results from a fully developed governmental model, considerable attention has been devoted to developing the basic components or decision units. We assume that the decision units of the government regardless of their position in the organizational hierarchy and specific function will have essentially similar behavioral characteristics. The characteristics of a typical infimal decision unit will be described briefly.

6-5.4.1 A Typical Infimal Decision Unit

As noted above, the behavior of the decision unit is based upon three principles of decision making which are most frequently used to describe the behavior of administrative and political systems namely (1) crisis management, (2) incrementalism and (3) satisficing [Ref. 29, 30, 31, 32]. Each of these principles describes a method by which fallable decision makers with limited competence cope with a highly complex interdependent outer environment. The crisis management approach to decision making emphasizes the fact that decision units only search for and implement new policies when certain crucial thresholds in the outer environment are crossed. The incremental approach emphasizes that decisions, at any given time, are heavily constrained by previous decisions. Policy changes in a given year tend to be incremental modifications of previous policies. The satisficing approach emphasizes that a decision unit will not seek out the best alternative. Instead it will content itself with one which is "good enough". [Ref. 32]

A flow chart for the model which incorporates these characteristics is presented in figure 5 f. The model has five basic components, namely:

1) a set of performance criteria, which are taken into account in decision making processes;

2) for each performance criterion, a defined threshold value which, if exceeded, causes the decision unit to search for and possibly implement a new policy or ordering of program priorities:

3) a set of programs (control inputs) through which the decision units attempt to modify performance criterion values in desired directions;

4) a set of policy-making strategies, based on alternative priority orderings of programs; and

5) a memory in which the decision unit remembers the successes and failures of previous policy-making strategies.

The inputs from the supremal unit are annual inputs of resources, changes in threshold values and (occasionally) specific directives regarding policies. From the outer environment the decision unit receives inputs regarding environmental states relevant to the defined performance criteria. Outputs of the decision unit are performance feedback inputs to the supremal unit and control inputs (programs) to the outer environment.

Since results for governmental model are not available at this time, we have chosen to discuss other more fully tested aspects in greater detail.[5] However the close correspondence between the simulation model and the formal mathematical model presented above should be noted. Each of the components in the formal model is given a concrete representation in terms of FUDM or newly defined variables. Thus further development of the formalization and experimentation using simulation will be interrelated and mutually reinforcing.

---

5 For a more detailed technical discussion of the governmental model, see Richardson and Pelsoci, 1971c.

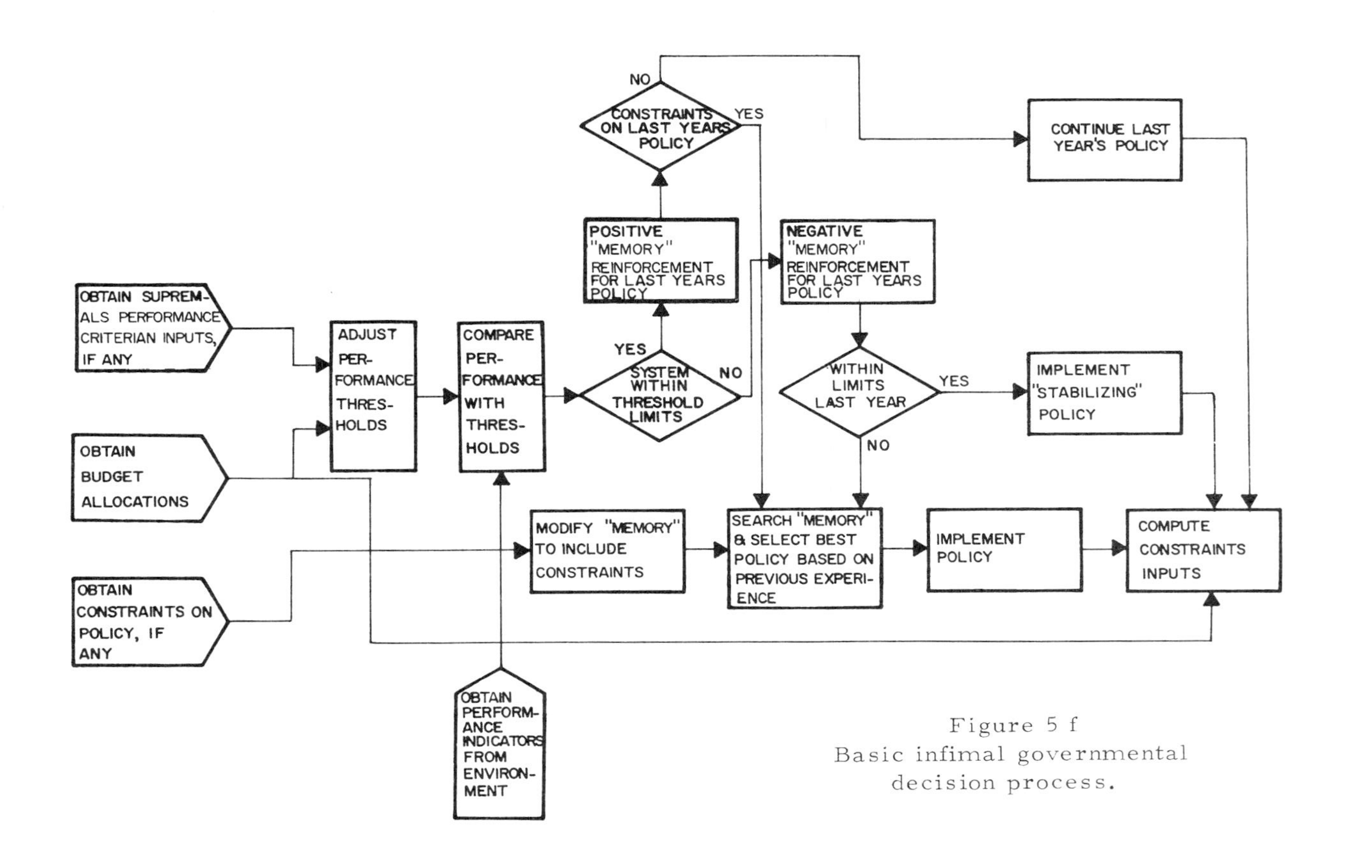

Figure 5 f
Basic infimal governmental decision process.

6-6 CONCLUSION

In this paper, the underlying philosophy of a program of systems oriented urban research has been discussed and some preliminary models and results have been presented. This research has been guided by a commitment to the organization-environment paradigm and to the multilevel approach. We believe that these two elements can provide a solid foundation for a systems oriented science of the artificial, focusing on urban problems. The development of this science will involve, as Strasser has suggested, a substantial task of "orchestration and coordination". For progress to be made, much greater consensus must be achieved regarding the kinds of theoretical problems which are most crucial and the way in which they should be approached. A major objective of our research is to provide a concrete basis for discussing these issues.

REFERENCES

1. Gabor Strasser, "Impediments to Social Problem Solving", IEEE Spectrum, July, 1971.
2. M.D. Mesarovic, D. Macko and Y. Takahara, Theory of Hierarchical, Multilevel Systems (New York and London: Academic Press, 1970).
3. Simon, Herbert A., The Sciences of the Artificial (Cambridge; MIT Press, 1969).
4. Bailey, F.N. and Robert T. Holt, Towards a Science of Complex Systems (Center for Comparative Studies in Technological Development and Social Change, University of Minnesota, March, 1971a).
5. Richardson, J.M. Jr., Notes on a Science of Urban Organizational Engineering (Systems Research Center Report, CWRU, 1970).
6. Bailey, F.N. and Robert T. Holt, "The Analysis of Governmental Structures in Urban Systems" Proc. Joint National Conference on Major Systems, Anaheim, California, October, 1971b.
7. Richardson, J.M. Jr., and Thomas Pelsoci, "Measuring Satisfaction in Urban Systems, "Preliminary Report, Urban Organizational Design and Control Project, Systems Research Center, CWRU, 1971a.

8. Richardson, J.M. Jr. and Thomas Pelsoci, "Simulating Governmental and Electoral Behavior in Urban Systems". Second Semiannual Report, Urban Organizational Design and Control Project, Systems Research Center, CWRU, 1971b.
9. Simon, Herbert A., Models of Man Social and Rational (New York: Wiley, 1957).
10. March, James G. and Herbert A. Simon, Organizations (New York: Wiley, 1958).
11. Bailey, F.N. and H.K. Ramaprian, "Decentralized Control of Weakly Coupled Systems", Department of Electrical Engineering, University of Minnesota, 1971. Mimeo.
12. Hurwicz, Leonid, "Centralization and Decentralization in Economic Processes," 1970, mimeo.
13. Moore, James A., "Simulation and International Interaction Analysis", Support Study no. 6, University of Southern California, 1970, mimeo.
14. Ashby, W. Ross, "Simulation of a Brain". In Harold Borko, ed., Computer Applications in the Behavioral Sciences (Englewood Cliffs, Prentice Hall, 1970).
15. Forrester, Jay, Urban Dynamics (Boston: MIT Press, 1969).
16. Richardson, John M. Jr., Partners in Development (East Lansing: MSU Press, 1969).
17. Holt, Robert T. and John E. Turner, The Political Basis of Economic Development (Princeton: Van Nostrand, 1967).
18. Ira S. Lowry, Seven Models of Urban Development: A Structural Comparison (Santa Monica: RAND Corporation, 1968).
19. Crecine, John P., "A Time-Oriented Metropolitan Model for Spatial Location", CRP Technical Bulletin No. 6, Department of City Planning: Pittsburgh, 1964.
20. Seidman, David R., "A Decision-Oriented Model of Urban Growth". Paper presented to the Fourth Annual Conference of the International Federation of Operations Research Societies (Boston, 29 August, 2 September, 1966).
21. Garn, Harvey A. and Robert H. Wilson, "A Look at Urban Dynamics: The Forrester Model and Public Policy", Proc. Second Annual Conference on Modeling and Simulation, University of Pittsburgh, March 29-30, 1971.
22. Rochberg, Richard H., "Some Questionable Assumptions in the Model and Method of Urban Dynamics". Proc. Second Annual Conference on Modeling and Simulation, University of Pittsburgh, March 29-30, 1971.

23. Averich, Harvey A. and Robert A. Levine, "Two Models of the Urban Crisis: An Analytical Essay on Banfield and Forrester". Rand Report, RM6366 - RC. Santa Monica: Rand Corporation, 1970.
24. Chen, Kan, Urban Dynamics: <u>Extensions and Reflections</u> (San Francisco: The San Francisco Press, Forthcoming in Jan. 1972).
25. Forrester, Jay, <u>Industrial Dynamics</u> (Cambridge: MIT Press, 1961).
26. Forrester, Jay, Principles of Systems, Preliminary Edition, Privately Printed, Cambridge, 1968.
27. Barney, Gerald O., "Understanding Urban Dynamics". 1970, mimeo.
28. Campbell, Angus, Phillip Converse, Warren Miller and Conald Stokes, <u>The American Voter</u> (New York: Wiley, 1960).
29. Crecine, John P., Governmental Problem Solving: <u>A Computer Simulation of Municipal Budgeting</u> (Chicago: Rand McNally, 1967).
30. Cyert, Richard M. and James G. March, <u>A Behavioral Theory of the Firm</u> (Englewood Cliffs: Prentice Hall, 1963).
31. Braybrooke, David and Charles E. Lindbloom, <u>A Strategy of Decision</u> (New York: MacMillan, 1963).
32. Simon, Herbert A., <u>Administrative Behavior</u> (New York, Free Press, 1945).

Chapter 7

AFTERTHOUGHTS ON FOUR URBAN SYSTEMS STUDIES

E. Cushen

In November 1967, the Department of Housing and Urban Development sponsored a one-year experiment to determine how the systems approach might be productively used in solving some city problems[1]. This chapter provides a short narrative

The author is Chief of Technical Analysis Division, (TAD), National Bureau of Standards, U. S. Department of Commerce, Washington D. C.

[1] An extremely large debt of gratitude is due to many people whose efforts caused these studies to have the results they did. Particular gratitude should go to John M. Patriarche, City Manager of East Lansing and then President of the International City Managers Association; G. Michael Conlisk, East Lansing's Planning Director and project leader; William J. Veeder, City Manager of Charlotte; Graham W. Watt, then City Manager of Dayton; Theodore Maurer, then City Manager of Poughkeepsie, Harold Weber, then City Planner of Poughkeepsie; Louis C. Santone, Deputy Chief of TAD; Geoffrey Berlin, who with Santone provided the TAD analysis for the East Lansing project; June Cornog, who designed the original TAD questionnaire for the Charlotte project; Marcia Maltese, who adapted the fire station locator to the Montgomery County school example; Harold R.

Millie, who assisted in starting the project; William Besuden, Assistant Director of ICMA; Robert Havlick, ICMA Project Manager; Leopold A. Goldschmidt, Research Director of ASPO; and John Parker, then with Fels Institute. The value judgments made about the studies should not be attributed to any of them, but to myself. Comments that seem to be negative, should not be regarded as critical of individuals or of cities, but simply an attempt to summarize a situation as I saw it existing. Throughout the studies, I was deeply impressed by the professional skills of all we met, by their devotion to improving their cities, and by the cities themselves. Finally, I wish to thank Arnold Reisman and Al Drake for their sacrifices in transcribing, editing, and improving an earlier manuscript for this Chapter.

The four city project was funded under the Demonstration Cities and Metropolitan Development Act of 1966 (Public Law 89-754). Supplemental work on the school districtor, the management game, and intergovernmental relations was supported by funds appropriated to the National Bureau of Standards.

---

description of those studies, the circumstances of their operation, a subjective evaluation of their results, and a set of "lessons learned" for the future[2].

The concept of the experiment was to conduct two training courses in systems analysis for city managers and planners, one in November 1967, and the second a year later. In the intervening year, three to five cities would be helped in carrying out short systems studies which could be reported and evaluated at the second training session.

Participants in the exercise came from five groups:

(a) The International City Management Association, prime contractor, and the professional society for city managers,

(b) The American Society of Planning Officials, a professional society for planners,

(c) The Fels Institute of State and Local Government, degree-granting graduate college in the University of Pennsylvania complex, from whom many city managers had received their Master's degree,

(d) The Technical Analysis Division of the National Bureau of Standards, a two-year old group of operations research, economics, human factors, and engineering specialists whose purpose is to assist agencies of government to use the systems approach, and

(e) City Managers, City Planners, and city staff people from the cities forming the basis of the studies.

2

Other reports describing the studies in greater detail, or from the points of view of city personnel involved have appeared:

William E. Bseuden, "Systems Analysis for Urban Problem Solving," Public Management, 51, February, 1969, pp. 14-15; G. Michael Conlisk, "Systems Analysis--How It Works in East Lansing," Public Management, 51, February, 1969, pp. 7-8; Louis C. Santone, "Application of Systems Analysis to Urban Problems," Municipal Yearbook, 1969, International City Managers Association, pp. 225-232; Louis C. Santone and Geoffrey Berlin, "Location of Fire Stations," in Systems Analysis for Social Problems, edited by Alfred Blumstein, Murray Kamrass, and Armand B. Weiss, Washington, D.C., Washington Operations Research Council, 1970, pp. 80-91; Alan J. Goldman, "Discussion" Op. Cit., 92-103.

The one-week training courses in systems analysis were designed by John Parker and others associated with Fels Institute, with some modifying suggestions from ICMA and TAD. Thirty attendees at the two courses were selected by ICMA and ASPO, with some emphasis given to cities whose managers and/or planners might have conditions more favorable to a successful experiment, and with some small midifications in achieving a balance geographically and by city size. Four manager-planner teams from the same city were invited.

The courses included play of a city game (Richard Duke's Metropolis); lectures on OR techniques, systems concepts, data processing, and the planning-programming-budgeting (PPB) concept that was then just being introduced to civilian agencies; workshops in setting up a project; and a critique in search of ways to match up city needs with the analytic smorgasbord that had been displayed. Lectures were given by Fels and TAD people, under the encouraging supervision of ICMA and ASPO.

At the conclusion of the first training session, a number of city managers volunteered their cities to serve as a locale for the conduct of one of the systems studies. The plan called for a conference to decide which study to undertake in each of the cities, and to develop a plan of action if it was decided to proceed. Typically, the study was to be a joint undertaking of several institutions. The city manager and/or planner was to serve as the primary user of the results of the study, and was expected to participate personally as much as his schedule would permit. TAD was to serve as technical coach to the city, helping them design the project, monitoring progress, and advising whenever problem areas arose. A university or college in, or near to, the city was to be involved; and some commercial organization was to participate in the project. When the project was completed, an evaluation was to be made by all participants, and a final report provided to HUD. At the completion of the study, the city was to be in a position to initiate and use systems studies with relatively little continuing advice from those who had conducted the first study.

Within a month after the completion of the training course, at least five cities had indicated a serious willingness to be one of the experimental locales. A visitation team consisting of ICMA, Fels, ASPO, and TAD travelled to the first four cities that volunteered--East Lansing (Michigan), Dayton (Ohio), Charlotte (North Carolina), and Poughkeepsie (New York). In each city, some preliminary thinking on the nature of their preferred systems study had been done, and discussed by phone and letter with ICMA, ASPO, Fels, and TAD prior to the visit.

The first visit to each city resulted in a rather explicit study design for each location, and this was accomplished without exception in the course of an evening's exploration, followed by a five or six hour planning session the following day. Each city had prepared well for the sessions: they had roughly identified what problem they wanted solved, they had selected a project that was reasonable in size for a 9-month study, and they assembled a large city representation to help in the discussions. One or two hour tours of each city were made at the beginning or midpoint of the project discussions.

East Lansing knew it would have to replace a central city fire station, and suggested that the team undertake to determine how many fire stations the city needed, and where they should be located. Dayton suggested one of several resource allocation studies that could mesh with another experiment being conducted there--Dayton was one of the 5-5-5 cities[3] in which the PPB system was being explored. Charlotte requested help in setting up a management information system. Poughkeepsie suggested an analytic comparison of ways to upgrade housing, concentrating on building code compliance.

Discussions of the city's proposed projects resulted in some modifications, usually as a result of urging by the visitors.

In East Lansing, the presence of the city planner permitted the study to be expanded to include several alternative future city environments, rather than a single time slice analysis. The existence of rather extensive data on fire alarms in the city over the preceding ten years provided a historical basis for comparison. Subsequent to the initial visit, Michigan State University was invited to participate, and their computer was used for production runs of the calculations.

In Dayton, resource allocation strategies for various programs were explored, largely in search of that set of programs that would have immediate visible results, and which might help Dayton to bring about reinforcing benefits from the various federal assistance programs they participated in. Dayton was especially interested in working on a problem that would have immediate relevance in improving the lot of the inhabitants of West Dayton, an area suffering from high un- and under-employment, adverse health experience, and relatively high crime rates. The Dayton planning discussions ranged more widely than those in

---

3 Five cities, five countries, and five states.

East Lansing, and occasionally lapsed into discussions of linear programming techniques. Dayton had hired an OR analyst for inhouse staff, and had been participating in the 5-5-5 program for a number of months, and this prompted earlier discussions of whether the linear programming approach to program selection might be regarded as productive. At the end of the planning session, it was decided to run a linear programming computation on the selection among various job training programs available to the city, but with some residual feeling that the project definition might require further modification. The city proposed that both university and commercial resources could be readily built into the study itself.

The Charlotte discussions were particularly significant, in that representatives of neighborhood groups were present at the project discussion, in addition to members of the city staff. Charlotte had already contracted with a commercial firm to prepare a management information system, and the TAD analysts urged that some other project than the one they proposed be undertaken. After about a half-hour's discussion, it was determined that a facility location problem be undertaken in the Model Cities area outside the city center. That suggestion was initially less appealing than some other alternative, since facility location was already apparently a good choice for East Lansing, and the team desired to have a broader set of examples of the systems approach to illustrate. The insistence of the neighborhood groups that this neighborhood facility location was important to them rapidly led to a new visualization of the problem that catapulted this problem formulation into a higher level of discourse. Basically, the plant location notion for East Lansing was a center of gravity calculation, with weightings at the nodes for countable entities--number of people whose lives were placed in jeopardy, frequency of fires, dollar damage potentials, and fire spread probabilities. But in Charlotte, there emerged a contrast between the perception of services rendered by the administration, and a perception of services received and acceptable to the target population. A problem statement whose purpose was to close a communication gap occurred nearly simultaneously to all persons present:

> Calculate the optimal number and location of neighborhood facilities using traditional time, distance, and population measures. Calculate the optimal number and location of neighborhood facilities using perceived distances and perceived value of services received by the neighborhood populations. Compare the two answers to discover the extent

to which improved usefulness of assistance programs might result if perceptions of the citizens formed the basis for choice of assistance programs.

From that point forward, a near-missionary zeal gripped most of the participants. The principal of the predominantly black high school offered to have a Model Cities Day, in which students and teachers could cover the neighborhood to check the adequacy of a two-year old housing inventory to serve as a basis for sampling inhabitants, who then would be asked to participate in an in-depth interview in search of the needed "perceived distances and utility." The Community College offered to help. The local newspapers built the project into the total Model Cities plan that was announced in the local newspapers shortly afterward[4].

Among the perceptions we hoped to measure were the following (all to be measured "actual" and "perceived"):

(a) Accessibility time
(b) Round trip time, including the service time
(c) Availability of the service
(d) Eligibility for service
(e) Need for the service
(f) Volume of service provided
(g) Timeliness of service
(h) Value of the service
(i) Contribution to expanded opportunity
(j) Threshold levels for noticeable value
(k) Availability of a subsequent job

The study team itself had some misgivings about whether the project as stated was researchable, since there was little basis for believing that perceived distances could be treated analytically, much less perceived values. But the ardor of the Charlotte participants and my own stubbornness prevailed, and a study plan was created.

The model neighborhood area in Charlotte is really a collection of six neighborhoods whose collective geographical boundary resembles the shape of a butterfly. The area adjoins the city center, and is served by two bus lines moving toward the city center. It is traversed by a connecting link to Interstate

---

4

"400 Families to be Interviewed in Pilot HUD Project," The Charlotte Observer, April 24, 1968, followed by "$8 Million in U.S. Funds for Slums Headed Here," The Charlotte News, April 25, 1968.

Route 81. The neighborhoods include Greenville, Belmont, Villa Heights, and Wards 1, 3, and 4, and their racial and economic compositions are sufficiently distinct to make the boundaries a little more pronounced than might otherwise be expected. Neighborhood recreation facilities, located more or less in accord with traditional plant location theory, have not been patronized as had originally been hoped. This observation led to proposing the idea that plant location theory needed more realistic behavioral/psychological data inputs if the final facility locations were to be those that would be used. If locations are not used, the intended service is not delivered.

The full range of possible locations then could be

(1) none--let Model neighborhood residents use facilities already in existence, largely in the center city area,
(2) one somewhere in or near the Model neighborhoods, possibly at the edge of the city center,
(3) two, one for each half of the butterfly-shaped area,
(4) three, four, or five, representing some compromise mixes to achieve better neighborhood accessibility,
(5) six, one for each of the model neighborhoods.

If perceptual distances from a projected user to the possible facility location could be obtained, then the calculations leading to the optimal location could be weighted by number of users, age of users, relative need of users, ability to pay for travel, etc., so that a variety of choices could be offered.

The range of possible services could include

(1) medical, including health education and medical examinations
(2) educational, including training in homemaking, budgeting, household maintenance, and (where appropriate) African studies
(3) counseling--marriage, family, social
(4) legal aid and advice, help in filling out forms
(5) employment assistance
(6) housing referrals
(7) recreational facilities and training
(8) financial advice--loans, budgets, expenditures
(9) day care and pre-school training for youngsters
(10) services intended to bring out the latent talents among individuals and groups in the neighborhood
(11) creation of a unifying cultural point, e.g., an African culture center, an experimental neighborhood-owned tufting mill, etc.
(12) provision of facilities--typing, sewing, shops

(13) a social facility, e.g., for holding family reunions
(14) a chapel

A short unstructured conference developed a list of 26 service possibilities, and more might have been generated.

Each of the above services might be provided at a number of possible levels, ranging from a mere directory of available services in the city area, through a modest advisory service, to a full-fledged staffed and equipped service center.

Furthermore, it would be necessary to determine how much "clustering" of services would be required before the center might take on a real attraction to the neighborhood. A one-stop general service facility would possibly be better patronized, and deliver services more efficiently and usefully, than a scattered set of locations.

This, of course, meant that not only should accessibility as perceived by the user be measured, but also his perception of the value (utility) of the services offered.

In Poughkeepsie, the project selected was the creation of a benefit-cost comparison of various strategies the city might use in bringing about building code compliance in the designated Model Cities neighborhood. A number of possible strategies appeared possible-strict code compliance with consequent shutdowns, low cost loans to owners or occupants to improve the property, tax incentives, etc. Furthermore, there was some feeling that a "Jones effect" might be capitalized on. For example, if several houses were to be significantly improved, giving visible tokens of neighborhood improvement, the remainder of owners and occupants might be motivated to build a new neighborhood sprit and a revitalized neighborhood. The neighborhood building inspectors agreed to help with data collection by providing some information beyond that which they needed to accomplish their task. The city social welfare workers agreed to estimate probable social effects of improved housing. City staff members agreed to estimate probable economic consequences of different improvement strategies. Thus a project was designed in which the unit of calculation was the individual house, and in which the analysis required comparing each of the different improvement strategies as it might be applied to each housing unit.

Each of the study designs was reported to HUD in order to obtain their inputs and approvals. Useful suggestions were received for all cases except the Dayton one, in which the study design fell outside the definition of legally permissible ways to use the appropriation from which the study was funded. The question could have been resolved favorably if facilities had been involved; in the Dayton study plan, only services were in question. ICMA requested a reconsideration of the Dayton exclusion, and the studies in the other three cities were started.

The East Lansing study got off to an easy start. City personnel, with the city planner acting as a vigorous project leader, gathered data, refined the study plan, visualized new alternatives to be compared, speculated on ways to combine services, estimated the extent to which future demands for fire services might deviate from a mean value, and watched milestone accomplishment carefully. TAD was able to adapt a computer subroutine for least-path calculation that had been developed in its portion of the Northeast Corridor transportation study with relative ease and at little cost. TAD had project coached in depth, and visited the city regularly. The city manager was an active participant in project definition and the early stages of the study, and was absent from the project only while computer program debugging and exploratory runs were in progress.

The Charlotte study was slower in making explicit that which had been agreed to in the project design stage. Although the tone and enthusiasm for the study continued, the necessity to develop a good questionnaire for a relatively new purpose required a large amount of design work by the TAD staff. Charlotte hired a professional sociologist for its staff, an event which was welcomed as a significant asset to the study. Independent university advice on the interview plan and purpose was obtained by Charlotte, and the pioneering nature of the intended work became still more apparent. Pressures for immediate action in the many Model Cities program competed severely with the time needed to conduct the systems study, although the interview training program that had been scheduled was completed only slightly later than planned. Trial runs on the interviews were conducted, and the questionnaire design itself went through several iterations to replace academic 1-type phraseology with language that conveyed meaning to those interviewed. By the conclusion of the study, however, the questionnaires were modified to provide a badly-needed socio-economic profile, and the intended "perception" questions fell a victim to the modified analytic plan.

The Poughkeepsie study recycled through the design stage approximately a half dozen times. City personnel in charge of, and those assigned to, the study were in constant demand for the solution of daily problems. The systems study was repeatedly placed in a "hold" status, and the city valiantly tried to keep the study moving by assigning replacement personnel, each of whom needed to be briefed on philosophy, concept, and status. Each new team revised the data collection sheets and the matrix format for analysis, at the same time that the neighborhood inspectors found the additional data items to be more cumbersome than anyone had expected.

The request for reconsideration of the Dayton study was decided in the negative. The advanced stage of progress on the East Lansing and Charlotte studies, as well as the difficulties with the Charlotte concept and the Poughkeepsie staffing, prompted a decision to hold the number of city studies to three, rather than adding a city half way through the experiment. Dayton, meantime, had solved its manpower program selections in a fulltime capacity, giving an increasing amount of attention to the 5-5-5 study that was nearing completion.

At this stage of the analysis, a number of hypotheses about "lessons learned" were beginning to emerge. They tended to concentrate in one of the following seven observations, most of which had been learned in World War II by the early operations analysts:

1. A nuts and bolts example is easy to understand, solve, and use. It has high chances of leading to tangible improvements, and will lead to still broader and more complicated analyses, which will also be easy to understand, solve, and use.

2. It is necessary to involve representatives of the user organization in the study, and in a depth sufficient to guarantee constant attention from them.

3. Perhaps more than in a typical federal organization, city staffs have less time to do analytic work than might be expected.

4. Problems related to social consequences, and intended to improve the inner city quality of life, have a degree of urgency that will not tolerate lengthy research investments. Conversely, a good systems solution to such problems appears to need just that research investment.

5. Communications gaps can be bridged if people with different objectives will jointly design a systems study. In those cases, there is less concern with the answers to the study that has been designed than there is a desire to capitalize on the communications opened up, so that other issues can be discussed.

6. There is little patience in cities with a long, large, expensive, comprehensive systems study so common in the defense-aerospace community.

7. The team approach can work, but there are at least two hazards--the lack of an explicitly designated team leader with responsibility to deliver a result, and the arrival of too many experts with conflicting advice.

The East Lansing study proved, by nearly any criterion, to be the most successful of the studies undertaken. Although the technical details have been described elsewhere (see Footnote 2), it will serve a useful purpose to summarize the concept of the analysis here.

East Lansing is a charming and quiet town with a population of about 37,000. When the students from Michigan State University are present, the population doubles. The City Manager provides fire protection service to the city and, under contract, he also provides this service to the university. In the lower part of the idealized city map in Figure 1, one of the fire stations is located on the school's campus. At the time of our project, one problem confronting East Lansing was the fact that the City Center station was obsolete and would have to be replaced sooner or later. The situation gave rise to the question: How many fire stations should East Lansing have and where should they be?

For the analysis, the city was represented as a series of nodes and connecting links, where a city block represented the area serviced by one of the nodes, and the streets that connected the intersections then formed the connecting links. The nodes which were possible candidates for fire station location were used as points of departure in finding the fastest path from that location to any other point in the city. Thus we have a network layout in which the elapsed travel time from intersection to intersection represents what we shall call "impedance". If we next assume that the fire engines will always take the fastest path, we can then draw a frequency plot indicating the number of nodes which can be reached by a fire engine in 30, 45, 60, etc. seconds, assuming that the engines are available when needed. It must be recognized that if any point is more than two minutes distant from a fire station, the fire protection is not considered adequate and the fire insurance rates on properties go up. In this study, we assume fire stations must be so situated that all locations can be serviced within two minutes travel time. No matter how we juggle the variables, we could not reach all nodes in the city in two minutes from a single fire station. We then tried two station locations. Two fire stations wouldn't do it either. A satisfactory solution was obtained as long as we were willing to allow for three fire stations in East Lansing.

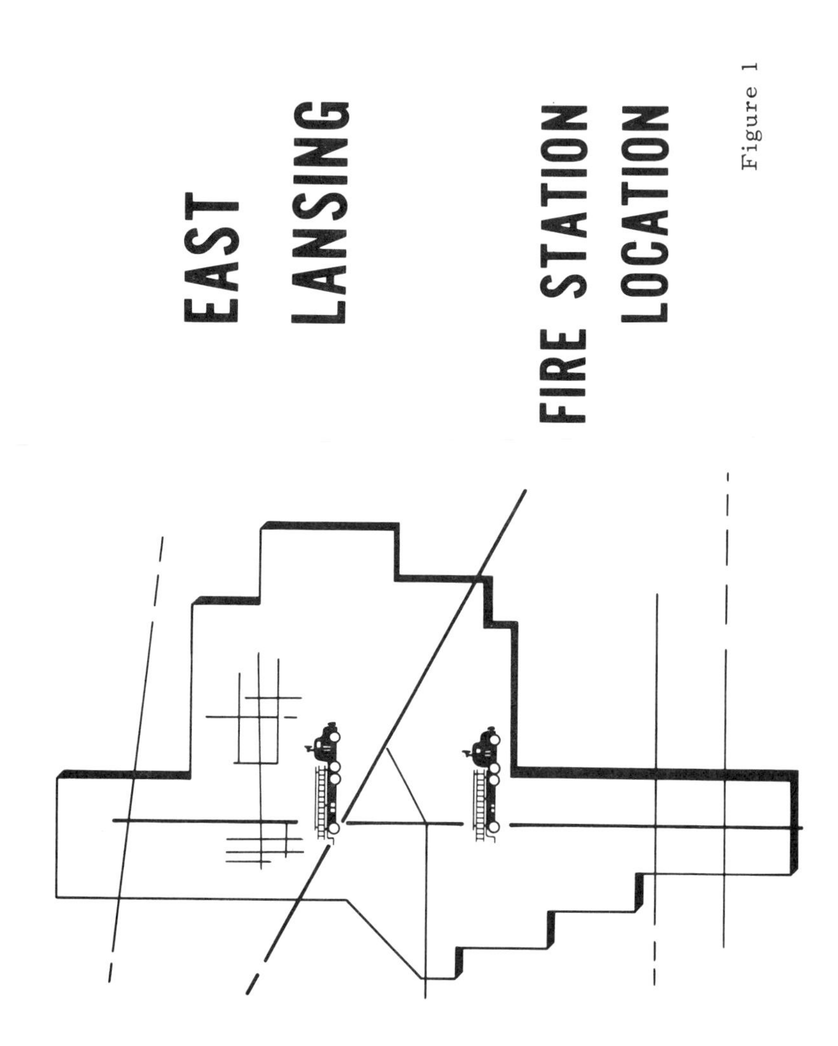

Figure 1

Of course the problem did not end at this point. In this exercise, the City Manager, Planner, and Fire Commissioner assisted in the design of the project. At no stage of the development were they at a loss as to what problem was being solved, or how it was being solved. They were involved in the design of the project, in the acquisition of data, and the process of obtaining solutions with the aid of the Michigan State University computer. They understood the basic logic and the formulation of the analysis. This made it possible for them to think more creatively using a numerically-described base case and take into consideration several dimensions that were left out in the first solution. For example, a densely populated area has a higher probability of fire than one which is sparsely populated. There is no point in treating a vacant lot similarly to a high-rise apartment. In a densely populated area more lives are in jeopardy and dollar damage caused by fire is also greater.

Success in obtaining good, plausible solutions from the analysis prompted the city officials to consider the long-range point of view. A time slice twelve years in the future was selected to symbolize possible ways the city might look. They considered probable population shifts, a possible deterioration of the downtown area with increased probability of fire, likely growth of high-rise apartments in the north of the city, etc. Given this new information, a three-station solution was the one that was again found to satisfy the required constraints.

Optimizing on the basis of the two minute travel time was not enough. Every additional fire station involves land acquisition, building, staffing, and equipment. Cost comparisons of different ways of providing adequate levels of fire service, then led to speculation about whether helicopters might replace some of the fire stations, whether fire prevention would be a better investment than fire fighting, or whether a better alarm system would be preferable.

When the nodes were weighted by population at those nodes, the preferred location for the center city fire station shifted five blocks away from the location implied by a flat map. When the nodes were weighted by frequency of fires, the shift was only two blocks. At this stage of the project, several research people associated with the fire underwriters became interested in the possibility of using the fire station locator to compare possible effects of changing criteria for rating cities.

By the time the study was completed and reported at the November 1968 training course, interest among city managers was relatively high. A number of cities requested a run of the model under their circumstances. ICMA had, for some months, felt that models of this sort might be generalized for wide use, and designed what has come to be called a "Technology Applications Program." One of the features of such a program would be the development of models, their test, and ultimate distribution to cities, along with technical advice on the use of the models. The fire station locator developed under this contract would probably be the first of a long series of such models. However, funding for adaptations to other cities was not available, and only occasional consulting could take place.

In the meantime, the Congress legislated the Fire Research and Safety Act of 1968 (PL 90-259). Conduct of research under this act is the responsibility of the Fire Research program of the Institute for Applied Technology at the NBS. The program director decided that a major contribution to fire service in the United States might be made if studies of an applied sort, similar to the fire station locator, were undertaken. Building on the East Lansing example, that program is now funding the development, application, and teaching of fire station locators, fire equipment calculators, shared service calculations, and community relations studies.

One of the impediments to rapid computations in the East Lansing case was the fact that districting of the city had to be done by hand, and the computation of optimal locations was therefore a series of volleys between the computer and the desk. When the East Lansing example was completed, TAD searched for a problem that could use the fire station locator, but would allow the development of a computer-based redistricting program. Such an example arose early in 1969, when the Montgomery County School Planning Board agreed to work with TAD to develop revised school district boundaries in an area now served by eight elementary schools in the Takoma Park vicinity of Maryland. The revised problem statement is, how many schools are needed in the immediate future, and where should the boundaries be? The components were the capacity of the schools, the number of teachers, the number of students, and the road network. Among the criteria used were the time it would take for the children to walk to school, how many children would have to walk three-quarters of a mile or more (or be bussed) in order to get to school, and also how to district in order to achieve better racial balance.

TAD was operating on very limited funds, so maximum advantage was taken of the offers to provide data by the County school staff, and the analytic work was simply adapted from work done earlier by Sidney Hess on political districting.

Of some interest to the OR community will be the cost estimates associated with the work described above. The HUD contract was awarded in the amount of $80,000. The TAD portion of the contract totalled $29,500 for analytic work and coaching. The fire research and safety efforts are still under way under separate funding, and the school districting work represented an additional expenditure of approximately $10,000.

The school districting problem also served as a student project stimulus to a group of five Western Maryland College mathematics majors in a January term class in operations research. Those students reworked the computer algorithms and tested the sensitivity of the answer to various exponents used in weighting locations of students in the network. When the number of students at each node was squared, the resulting solution was preferable to linear or cubic weightings by two criteria--compactness of districts, and fewest students residing farther than 3/4 mile from the school. Cubic weightings gave still greater compactness, but resulted in more isolated "islands" associated with a district.

Our exposure to these city problems was followed very quickly by the appearance of a study conducted by the Federal Executive Board in the city of Oakland, California. From an OR point of view, the problem statement might be stylized as follows:

> What can be done to maximize the collective effectiveness of Federal aid programs converging on a city?

Categorical grants and assistance programs from a variety of agencies (HUD, HEW, OEO, Labor, EDA, and others) had targeted on Oakland for a number of years. Still the indicators of city health were downward-bound. It was apparent in the study that programs frequently conflicted with one another, or their timing was mismatched, or the State and local priorities precluded taking maximum value from the programs, or State and local matching resources were insufficient. Each of our simpler city studies had brushed against other programs, all intended to improve the inner city condition. The year was one in which civil disorders were prominent national concerns. Each of our city studies was a suboptimization, targeted on a city service. We were troubled by the need to include a larger horizon, but were at a loss as to how to handle it. The Oakland example served an admirable purpose: it provided raw materials for creating the scenario for a management game. Because there was no funding for developing the game, the ensuing several pages simply suggest an outline of what might be done.

Any game requires three kinds of entities--the players, the things they exchange or use, and a set of rules.

In the Oakland game, players have been somewhat artificially limited to five:

(a) a spokesman for the ghetto community,
(b) an industrial player,
(c) the city administration,
(d) the county, and
(e) Federal agencies with programs to improve the community.

These five players are abstractions. For example, there is probably no single representative of the ghetto community; there may be no single-minded industrial representative. The city administration is simplified to represent a unified image of mayor, council, city manager, and the administrative departments. County and State represent elected officials and members of legislative, executive, and judicial branches of government. The Federal players represent the logical sum of Departments and Agencies such as HUD, OEO, Labor, Commerce, HEW, Transportation, etc. Although the simplification is somewhat artificial, there is some implicit grouping of purposes in this representation. One of the lessons of the game could be that much could be gained from cooperation of the separate entities in achieving a common purpose.

One thing stands out from gathering the many decision-making entities into these 5 groups. Each has a different perception of urban need, a different perception of what is possible, and a partially frustrating awareness of the limitations on his flexibility.

The pieces in this game include a map of the city, a profile of its people and services, a budget, and a set of resources to be used. Since only the concept of the game has been developed, the specifics relating to these pieces are only suggested in the set of instructions given to the players.

The game rules are also undeveloped. Instructions to start the game have been prepared for each of the "players," who have been identified as "Decision Loci," since each player represents the position developed from the rebuttal and debate of the institutions gathered together. An Umpire Manual would be needed to allow for calculation of results of decisions made, and to advance the scenario of the game.

Several initial plays of this game in classrooms, using simplified Umpire calculations, have shown that there is a built-in deterioration of the urban situation in early time frames, and that there is a rapid student response in either taking unilateral drastic corrective action or a move to cooperative action.

7-1 DECISION LOCUS #1 POVERTY PROGRAM RECIPIENTS

You represent the intended receivers of the poverty programs in Oakland, California.

Oakland, a city of 387,000 people, is largely a diversified manufacturing city, a kind of second-class cousin of its richer neighbors--Berkeley, San Francisco, etc. You represent 25,000 unemployed men, 38,000 unemployed women, and 114,000 people whose employment levels are at less than $4,000 per year. You represent 23,000 families who are now on welfare, and another 16,000 who are eligible, but are not receiving those benefits. About one-third of the citizens of Oakland are non-white, largely Negroes. Whites are moving out at the rate of 11,000 per year, and Negroes are moving in at the rate of 8,000 per year. There are probably no more than 2,000 jobs available in the city, and these are largely in the wage rate range of $1.60 to $2.25 per hour. Welfare recipients can get a maximum of $2,300 per year. You need roughly $3 per hour to make ends meet, and $4 per hour to receive an income that would be almost respectable.

Although many poverty programs have converged on Oakland, you see some bad effects, and not very many good ones. 28% of the people still live in substandard houses, and the two largest poverty programs have succeeded in removing 2,500 housing units, and 5,000 more have been lost to a new freeway. There are 1500 people on the waiting list for public housing, for which only 100 units have been built. Landlords will not rehabilitate the houses you live in because improvements are losing propositions to them.

Your city is listed as one of the more likely places to have a riot, and police relations are bad. There are 3% Negroes on the police force.

There has been extensive discussion about a rapid transit system to work areas, but bond issues have always been defeated at the polls. Your decision problem: How to break the cycle for your people?

7-2 DECISION LOCUS #2 INDUSTRY

You represent the industries now in Oakland, California, and those who might locate there.

Oakland is basically a city for diversified manufacturing, and uses 37% of its total employment in that way. It is part of the San Francisco-Berkeley complex, which is the area most of your professionals will want to live in. A freeway system is such as to make access easy. Oakland itself has large numbers of unemployed people (25,000 men and 38,000 women) and a larger number of people who are underemployed (about 114,000). Most of these people are unskilled or semiskilled. About 80% of them are Negroes. Productivity and job permanence of these workers are bad bets.

There have been riots in Oakland, and there is difficulty in securing risk, equity, and debt capital for building in the ghetto areas. Your taxes are among the highest in the state of California, and land is expensive compared with other possible locations. A recent urban renewal effort has cleared approximately 60 acres of land, about half of which is to be made into an industrial park. There has been a steady exodus of business firms into the suburbs; ghetto residents do not have cheap public transportation to get there.

Some job training programs have been undertaken by components of the Department of Labor, but the test program was in training cooks and pantrymen in a 54-week training program, with a large dropout rate.

Workers can be expected to report for duty and hold their jobs if you give them $3-$4 per hour; you are not competitive with the welfare program for wage rates much below that.
Your problem: Move in, stay, or relocate?

7-3 DECISION LOCUS #3 CITY ADMINISTRATION

You represent the City Manager, Mayor, Council, and Administrative structure of the city of Oakland.

Your city is a second-class neighbor to San Francisco and Berkeley. Industry is moving out; the racial balance is changing rapidly--Whites are moving out at the rate of 11,000 per year, and non-Whites (largely Negroes) are moving in at a rate of 8,000 per year. Your city has a high rate of unemployment (6.4% city-wide, and about three times that much in the ghetto target areas). 25,000 unemployed men, 38,000 unemployed women, and 114,000 underemployed persons are restive, and you have had a number of riots. Your city is a port of entry for low-income households.

Your current city income for operating expenses is $42,000,000, and your expenditures are $43,000,000. Your capital budget is approximately $11,000,000, not nearly enough to make the necessary maintenance improvements, let alone build a newer capital plant. Your tax rates are about as high as they can go for both industry and property bases. You have a 1% sales tax on most things except groceries, and 6% on tobacco. Your city debt is $43.3 million, and rising slowly. Your city receives $3,000,000 per year from the Federal government, and $3,000,000 per year from the State of California. Because of matching fund requirements, you need to deflect some of your priorities to take advantage of Federal antipoverty and economic development programs. Federal agencies believe you are receiving five times the amount of support your budget shows. Your pay rates for city employees is high ($8,200 - 8,800 for entering firemen and police), and you are an attractive city employer.

Your city shows a very apparent air of general deterioration; the exodus of industry and white middle class families is reducing your tax base, and the demand for services is increasing. Three quarters of the housing stock in the city was built before 1939, and 20% of the housing has been officially declared substandard. The city's capital plant is aging, and the expected subway was stalled because of an unanticipated increase in construction costs. Many streets are in poor condition, and a large part of the sewer system needs replacing. Many schools are inadequate; a recent estimate suggested that $75 million would be required to update the capital plant. You have repeatedly asked for a new school bond issue for the last ten years, and this has been repeatedly defeated at the polls. You do not have enough money to provide mainstream medical care for the poor, who underutilize the existing facilities. You have high disease rates among the poor. The welfare case load of 23,000 covers only about 2/3 of the families known to be eligible. Furthermore, about 80% of the 65,000 older people are eligible for welfare, and only one-fourth of them are covered.

One of the ghetto pressure points is the police force, which is only 3% Negro in staffing. The ghetto residents are calling for removal of the police from ghetto areas.

The Federal government wants to help, but you find it difficult to accept all their offers because of the matching fund requirements. Common sense tells you previous investments have only begun to address the problem. Your problem: How to improve the city and provide the needed services?

7-4 DECISION LOCUS #4 COUNTY BUDGET

Alameda County counts approximately one-third of its population in Oakland, which is in a difficult financial and social condition.

Your county income is $473 million, and your operating expenses are $466 million. There is an increasing demand for welfare payments in Oakland; the Federal government will pay for about 55% of those transfer payments. Your current welfare budget is $70 million; this would have to double to cover all known eligibles in the city of Oakland alone.

Your educational budget is $214 million, and Oakland's schools are in a state of disrepair. It was recently estimated that it would take $75,000,000 to upgrade the capital plant of the Oakland school system.
Your problem: How to manage the county and upgrade Oakland?

7-5 DECISION LOCUS #5 FEDERAL AGENCIES

You represent the panorama of Executive Branch Agencies charged with the conduct of the various programs in the War on Poverty. Your concern is nationwide, and your charter comes from the Legislative Branch. You are attempting to alleviate the national poverty situation, using the forces of State and local governments to the maximum. You are attempting to get them to exercise greater leadership in identifying their needs, and in making a coordinated effort to solve their local problems.

In the city of Oakland, California, you have recently conducted the following programs connected with improving the rather bad poverty situation in that city.

7-5.1 Housing and Urban Development

Over the past eight years, you have spent about $50 million in Oakland (counting your predecessor agency's investments). Two major projects are of recent vintage: an urban renewal project in the Acorn area, and a public housing project in Tassaforanga. The Acorn project was intended to remove a blighted area, move affected families, and turn 60 acres of land partly to residential and partly to industrial purposes. The Acorn project cost you $10 million, and Oakland put up about $5*, part of which was non-cash. The public housing project cost you $1.5 million. In Acorn, 1600 housing units were demolished, and about half of the residents moved from what had been substandard housing to standard housing. About 14% of the former residents moved from the city, and you don't know where the remainder are, nor does Oakland. You presume that the backlog of 1600 families needing public housing come in part from these missing people. You intended that Acorn have a long-term effect in raising local land values and in making the land availability more attractive to industry to relocate in Oakland and use the large volume of unemployed people there. The housing project was occupied in 1966, and created 105 dwelling units, in which there is now a 20% annual household turnover. The area still needs access to a shopping center and some recreational facilities, which Oakland seems in no position to create. Of all your and Oakland's investments, most of the money went for land acquisition costs, and only $1.5 million found its way into Oakland payrolls, where it created 72 man-years worth of work for Oakland residents. Their expenditure patterns led to a tax revenue to Oakland of $11,000.

You are now working on a Neighborhood Centers Pilot Project in Oakland.

7-5.2 Economic Development Administration

You and your predecessor agency have spent at least $25 million in Oakland over the past eight years to improve its regional economic development posture. Oakland is a redevelopment area. Recently you undertook, with Oakland, a Block B Public Works project, costing you $1.7 million. The purpose of this project was to develop an industrial park adjacent to the airport; Oakland matched your investment, 50-50. Your investment resulted in a temporary payroll in Oakland of 9 man-years' effort, approximately $300,000 of which was spent in Oakland, and resulted in city tax revenues of $1800. You paid your share after the project was completed; Oakland had to borrow the mon-

---

*$5 million.

ey to cover the costs of the project until it passed your standards.

7-5.3 Office of Economic Opportunity

You have had a rather small dollar investment in Oakland, totalling only $3 million since your agency came into existence. You currently have two projects there, one a legal service project for ghetto inhabitants, costing you $200,000, and a family planning services project in the ghetto costing you $40,000. The legal services project has turned out to be a fantastic success; you expected to handle a case load of 4800 per year, and are actually handling about 8000. There are some anomalies in that the four offices that have been set up show a case load from unexpected areas, including the Mexican-Americans in the city. Most of the legal service money ends up outside of Oakland, since the professional people do not live in Oakland. However, there has been work for a little over 2 Oakland residents. You are footing only two-thirds of the bill for the family planning activity, but the whole bill for the legal services, except for the office space.

7-5.4 Health Education and Welfare

You have a rather large series of programs in Oakland, and in the past two years have spent approximately $122 million in the city, all of which has been handled by way of the California State Department of Education. There is still a tremendous need, both for capital plan in education, upgraded educational services, upgraded health services, and a welfare demand that is easily twice what you are now spending. Two rather interesting, although relatively small, projects are now underway. The first of these is a Compensatory Education Program (ESEA Title 1), for about $2.5 million. This program goes largely to language development, enrichment, basic skill development, and remedial training for youngsters in the ghetto areas. The second program, Mental Retardation, turned out to require medical and social services, in addition to the educational investments. It is a five year program, now in its third year. The programs have resulted in the creation of 36 man years of work for Oakland residents, although the bulk of the professionals involved live outside of Oakland. Oakland has received the benefits of approximately $1 million of the expenditures, and this resulted in Oakland tax revenues of $6400.

7-5.5 Department of Labor

Over the past eight years you have spent approximately $12.5 million in Oakland. You now have going a Concentrated Employment Program there, plus two other projects that can show some positive results. The first of these was a training program for culinary workers; it cost you $600,000 to run a 54-week program for 48 people. About half dropped out. The second project is the Neighborhood Youth Corps, started in 1965. You have had about 1600 participants in this program, and nearly 40% of them have gone on to other productive employment or further training. Four city agencies do most of the job training, and most of that training is in maintenance, gardening, custodial, and office aide work. Your students are about 85% Negro; they drop out at a rate of about 13% per month. Figuring on the average, you have done well in placing money into immediate circulation in Oakland--half the investment has been spent there, and there is continuing employment at a rate of about 22 man-years per year.

You wish you could do more, and hope that the CEP program will find jobs for about 25,000 people.

CONCLUDING REMARKS

Finally, we rediscovered something we had all learned in high school in classes that were then called Problems of Democracy. The American government is, in reality, a system of governments, some 100,000 in all: State, local, national, and special units. If one wanted to do grand strategic operations research in the delivery of services and freedoms to the American people, he would need to describe ways to improve the rules related to the delivery of those services as they pass through the various units of government. We experimented with a simple descriptive narrative and accompanying diagram to show how one type of project is delivered to the public, and illustrating the steps taken and by whom.

To develop these ideas further would require more than my allotted pages in this volume; they will appear in greater detail in a future issue of Operations Research.

Chapter 8

APPLICATION OF MULTILEVEL SYSTEMS THEORY TO THE DESIGN OF A FREEWAY CONTROL SYSTEM

D. Drew

8-1 INTRODUCTION

The urban problem is not limited to any single country, society, or historical era. Symptoms of the problem are slums, unemployment, an inadequate water supply, ill-health, accidents, pollution, social violence, ugliness, and congestion. Traffic congestion is among the most serious problems; its solution could play a giant part in alleviating the urban crisis.

The world appears to be on the threshhold of the new revolution in transportation technology. For example, "Guideways" -- automated highways -- are technically feasible today. The basic components are a wire embedded in the road; pickup coils installed on the car to sense its position in relation to the wire; a computer to keep track of vehicle positions, headways between successive vehicles, and relative speeds; a control system with a steering subsystem to keep the vehicle automatically on course; and a speed-change subsystem to accelerate, slow-down, or stop the vehicle. Off the guideway, vehicles would be driven individually. Other high-density public systems also appear to hold great promise for the future.

The author is Chairman, Systems Engineering and Management Center, Asian Institute of Technology, Bangkok, Thailand.

But, the urban transportation problem can't wait twenty years. Cities right now are practically choked from traffic volumes such that the average urbanite spends 15 percent of his working day in transit. There is a danger that the transportation experts of the 1970's, following the pattern of the past five years will continue to deplore the inefficient ways in which high-density automobile traffic is handled on existing systems on the one hand while painting scenarios of futuristic solutions on the other, as if the two were unrelated. Unless there is a change in thinking just as innovative as anticipated developments in technology, the only means of rationing traffic flow on existing highway facilities -- many of which were built at costs of millions of dollars per mile -- will continue to be congestion and delay.

By making minor, inexpensive additions to thousands of miles of urban freeways, a new era could be opened to the commuter now as well as providing an initial step in the transition to the guideways of the future. This paper describes a strategy for making urban freeways congestion-free and therefore safer and more efficient through the design and installation of multilevel freeway control systems.

8-2 DECOMPOSITION OF THE FREEWAY CONTROL FUNCTION

If urban freeways are to operate at acceptable levels of service during periods of peak traffic demand, these facilities must be controlled. In early attempts at freeway control, the control action depended upon prior calibration using historical data on traffic demand. For the purpose of classification, this will be referred to as "zero" level control because of the absence of feedback or memory. Examples of this form of control are freeway entrance ramp closure and fixed ramp metering. The use of visual surveillance -- aerial or closed-circuit television -- represents a means of replacing crude open-loop control with "first" level control -- one with a higher sensitivity to input changes.

A freeway control system may be viewed as an array of surveillance, communication, and control components with the control law being split into degrees of sophistication or levels. This multilevel approach is directed towards establishing a hierarchy of control that results not only in an efficient system but one that can be implemented in stages. The hierarchy of control is established such that the zero- and first-level systems are directed towards recognizing the influence of short term factors whereas the higher levels are reserved for factors that influence performance on a long term basis. There is also a certain ordering of hierarchical levels based on the degree of complexity of computation, the degree of uncertainty, and the required speed of reaction to a change in operating conditions. The central idea is to share the effort of solution among two or more levels, each of which communicates both with the level directly above and that directly below. In the general case, the (n+1)st level influences or even directs the decisions of level n.

Figure 1 illustrates the conceptual form of a four level freeway control configuration. These levels are, in ascending order of sophistication, the regulating, the optimizing, the adaptive, and the self-organizing control functions. The fundamental variables are the input vector r (traffic demand, desired speeds, etc.) and the output vector c (volume or rate of traffic flow, density, etc.). Still referring to Fig. 1, d is a disturbance vector which represents unintended inputs to the freeway system which cannot be adjusted such as environmental factors, weather conditions, accidents, and incidents; m represents signals supplied to the controller regarding those disturbance vector components and output vector components which are instrumentable; n stands for signals from the controller to the control devices in the actuator subsystem; and u is a manipulatable vector which represents those freeway inputs which can be influenced by control. The vector w represents the broad set of operating specifications, restrictions, and hypotheses pertinent to the control problem.

The basic control activities associated with each of the four functions shown in Fig. 1 differ. The Regulating Function Controller accomplishes what might be called the basic "subgoal" of the control system. Although the goal of the control system is to provide the best possible level of service on the freeway, its components, and its interfaces; various subgoals have been advanced upon which the regulating control subsystem may be based. Implicit is the assumption that optimizing the subgoals will optimize the primary performance criterion. The optimal use of available gaps in the freeway merging process is such a subgoal and it is accomplished by the regulating controller in the block diagram in Fig. 1. This controller translates the directions of the higher level controllers into direct actions in the operation (the timely release of ramp vehicles by a standard traffic control signal located on the entrance ramp).

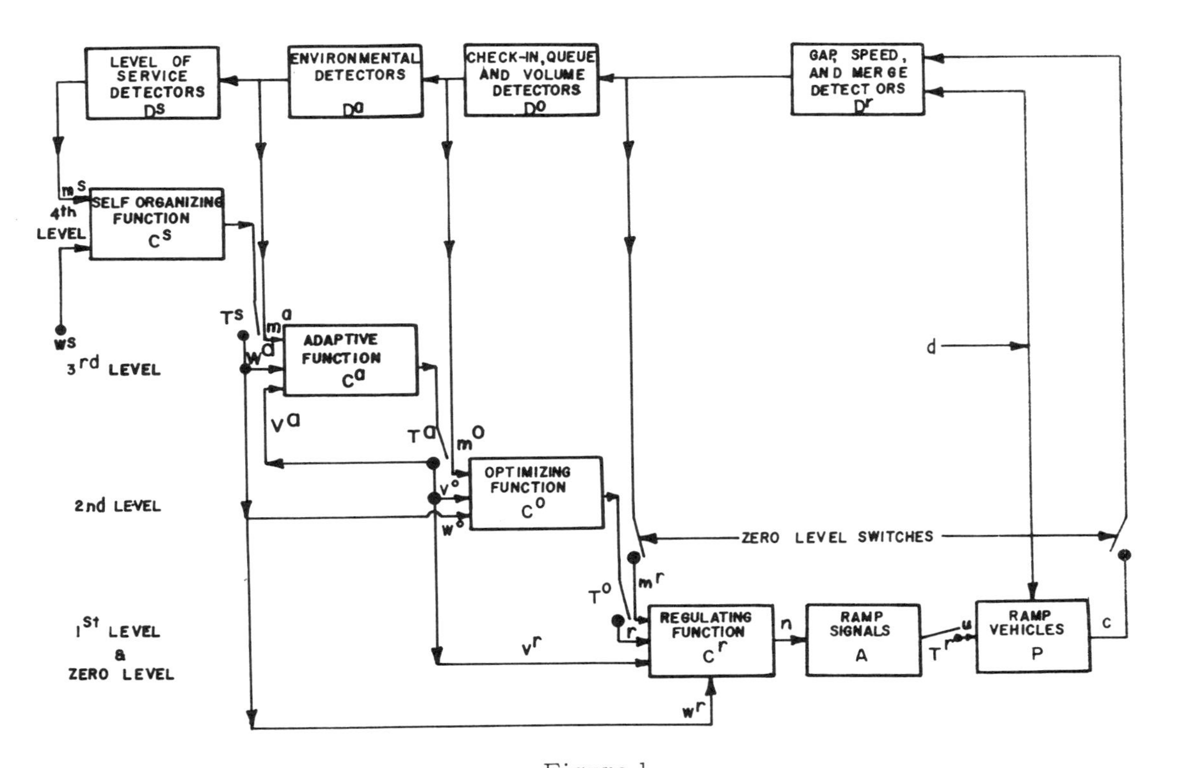

Figure 1
Decomposition of Freeway Control Function

The objective of the Optimizing Function Controller (Fig. 1) is the determination of optimum operating conditions based on the appropriate performance criterion and mathematical model of the process. For example, if the setting on the regulating controller is too high, many marginal gaps are left unfilled; on the other hand if the setting is too low many metered vehicles will reject the gaps and be forced to stop in the merging area where their presence, as detected by a loop detector, pre-empt metering. Obviously then, the optimum gap setting is somewhere between "too high" and "too low". The form of this optimizing function for a freeway control model will be discussed later in the paper.

While the two lowest levels of the control hierarchy are developed on mathematical models approximating the real system, the adaptive controller's function in Fig. 1 is to compensate for the errors introduced by the models by adjusting the parameter values, v^o and v^r. Note that a parameter vector v^a is supplied to the adaptive controller so that, in effect, it can see what it has been doing. The parameter vectors v^r, v^o, and v^a alter the coefficients of the control laws that are applicable at the lower control levels but do not change the laws themselves.

The Self-organizing Function Controller determines what the "worth" of "decision" vectors w^r, w^o, and w^a should be on the basis of those measurable freeway characteristics m^s and the intervention of humans in the system as represented by w^s. The worth or decision vectors generated by the self organizing function act to control the overall system to achieve the best total system performance. These decisions are based on the accumulated experience and understanding of the system. They are subject to the specifications, goals, and constraints embodied in the worth vector w^s. These decisions w^r, w^o, and w^a alter the control laws that are applicable at each level in the hierarchy.

In conclusion, it is apparent that the higher up in the hierarchy one goes, the less rapidly do the environmental and operational conditions pertinent to a given level vary. For this reason, the outputs of the higher levels are considered to be discrete. This is represented in Fig. 1 by the samples ("switches" in the logic circuits) which operate with different periods T in which $T^s > T^a > T^o > T^r$.

8-3 DESIGN SPECIFICATION

How does one go about the design of a freeway control system? The problem is, of course, one of system design. A design specification consists of four parts: (1) a description of the location (the plant) and the environment, (2) the description of the inputs (traffic demand, speeds, and other variables), (3) specification of outputs, and (4) measures of effectiveness. Since freeways are built to well accepted standards, they vary very little making part one unnecessary.

Regarding (2), in the past it has been assumed that a complete description of a freeway's operating characteristics was needed before the control system could be designed and installed. The implication was that all bottlenecks and thus capacities must be determined; that trip origins and destinations for the freeway and its ramps must be known; and that gap availability and gap acceptance characteristics must be established -- all before the control system could be designed. The procedure has been: (1) to perform manual system input-output studies requiring as many as 30 people, (2) use time-lapse aerial photography, (3) conduct questionnaire studies to obtain trip information, and (4) employ moving-vehicle study techniques. As it turns out, this part of the specification for the actual design of a freeway control system is unnecessary because it affects neither the choice nor the location of the control system components -- only their calibration, a function that must be performed after installation anyway.

One does not just install a control system and then "see what he can do with it". A specification of the system outputs is needed. However, a problem often exists in finding a suitable analytic description of what is desired. While it is generally accepted that the function of the freeway is to present an environment which permits a driver to operate his vehicle economically, safely and with a minimum amount of anxiety, it is easier to give a qualitative description than to find an analytic one. The output specification that is recommended is based on two objectives: (1) the optimal use of acceptable freeway gaps by merging ramp vehicles, and (2) the prevention of congestion. The underlying philosophy of the first specification is that minimizing intervehicular interference at entrance ramps (1) reduces the probability of rear-end collissions in merging areas due to false starts, (2) reduces the tension on a merging driver, and (3) prevents shock waves from developing on the freeway in the vicinity of entrance ramps. The theory behind this first specification is based on the utilization of gap availability and gap acceptance models. Behind the second specification is the idea that the prevention of congestion ultimately results in moving more traffic faster since, theoretically, congestion is prevented if traffic demand never exceeds the freeway capacity.

The fourth, and in many ways the most difficult, component part of a system specification is a set of measures by which the success of a system can be evaluated. Various figures of merit have been proposed to evaluate freeway operations. To some extent they can be categorized according to (1) whether they are macroscopic or microscopic in nature, (2) rational or empirical, (3) designer or user-oriented, (4) their sensitivity, and (5) capability of automatic measurement. Because of the complexity of the freeway phenomenon and the relevancy of most common measures of effectiveness to the two output specifications discussed above, not one, but several measures should be employed in the design and evaluation of the control system. Those recommended are speed, travel time, kinetic energy of the traffic stream (the product of speed and traffic volume), delay to a ramp vehicle, queue lengths, and number of accidents.

8-4 FIRST LEVEL CONTROL

To meet the merging control specifications one must be able to (1) detect acceptable gaps on the outside freeway lane, (2) be able to predict when these acceptable gaps will reach the merging point and (3) arrange for the speed adjustment of the merging ramp vehicle so it hits the gap at the merge point. Referring to Fig. 2, a sensor is required to measure the time interval between successive freeway vehicles (gap detection) and vehicular speed (gap projection). A standard traffic signal installation on the ramp offers a conventional means of communicating with the ramp driver to standardize the required speed adjustment in the merging maneuver. The effect is to stop all ramp vehicles at some point on the ramp far enough from the merge point to allow their vehicles to accelerate to the speed of the freeway traffic system. A call for the green signal is made when the projected gap reaches the position in time, designated as the decision point, at which the travel time of the gap to the

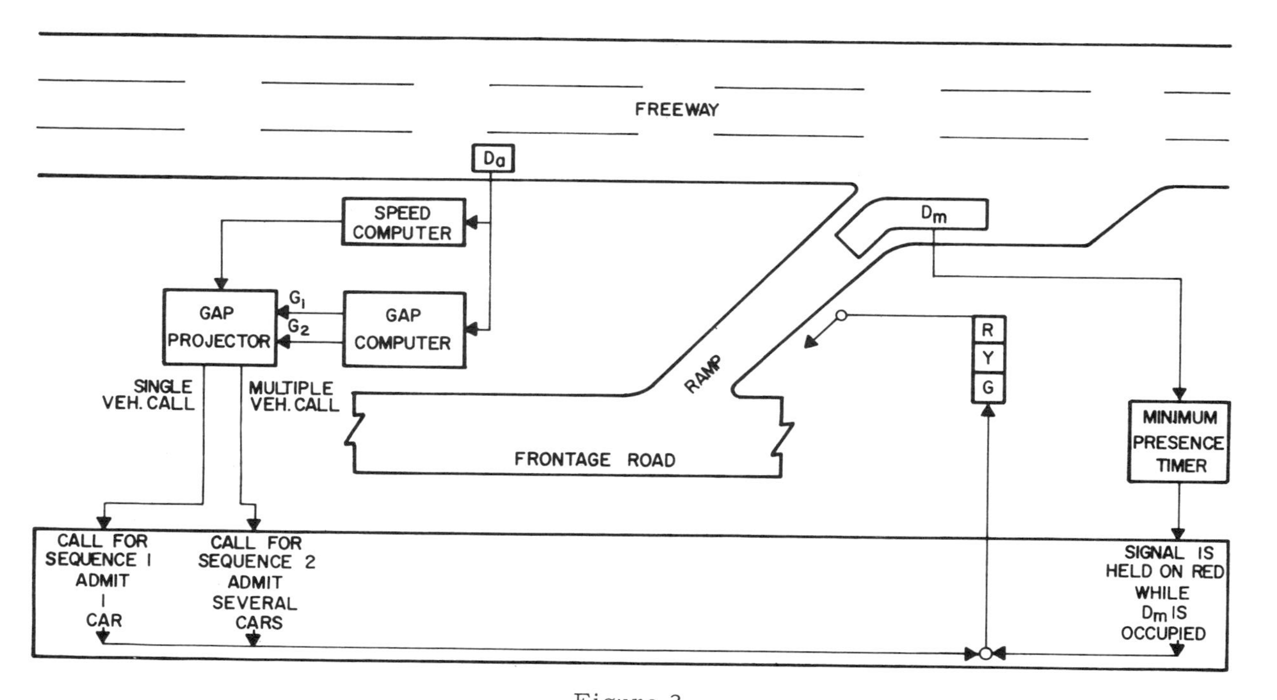

Figure 2
First Level Freeway Control Showing Regulating Function Components

merge area is the same as the travel time of the ramp vehicle from the signal to the merge area. If the gap is equal to or greater than the designated acceptable gap size for more than one vehicle, the controller holds the green signal until the gap passes the decision point. A loop detector is placed in the pavement of the ramp just upstream of the merge area. All vehicles entering the freeway from the ramp will actuate the detector. If a vehicle stops on the ramp in this area, blocking the entrance to the freeway, the detector will "time out" and the signal controller will hold on red until the detector is cleared.

8-5 SECOND LEVEL CONTROL

The block diagrams in Figs. 2 and 3 serve to illustrate the functions and components of the first-level and second-level control systems. As shown in Fig. 2 the control of the ramp signal is accomplished at the first-level basically through the detection and projection of gaps. As a provision for keeping the ramp area from the ramp signal to the freeway clear, the presence of a vehicle in the merge area precludes a green signal indication thus preventing a queue forming at the merge point, reducing driver anxiety and the potentiality of accidents.

Whereas the first-level regulating controller compares measured gaps in the outside lane of the freeway to arbitrary gap setting and then meters ramp vehicles into acceptable gaps; there is no assurance that the resulting operation has been optimized with respect to any criteria. The objective of the second-level optimizing controller is to adjust the gap settings on the first-level regulating controller in response to the outside lane freeway operation (volume and speed) so as to maximize the ramp service volume. This is accomplished by the gap selector computer component shown in the second-level freeway control block diagram in Fig. 3. Optimization is based on the family of curves plotted in Fig. 4 which were developed from a queueing model [Ref. 5]. Reference to high, intermediate, and low-type operation is based on the criteria of relative speed of merging vehicles with respect to the freeway traffic stream under stable flow conditions.

Two additional functions of the second-level controller are depicted by sequences 2 and 3 in Fig. 3. The ramp may be located such that an excessively long queue at the ramp signal will back into an intersection of the frontage road with a cross-street,

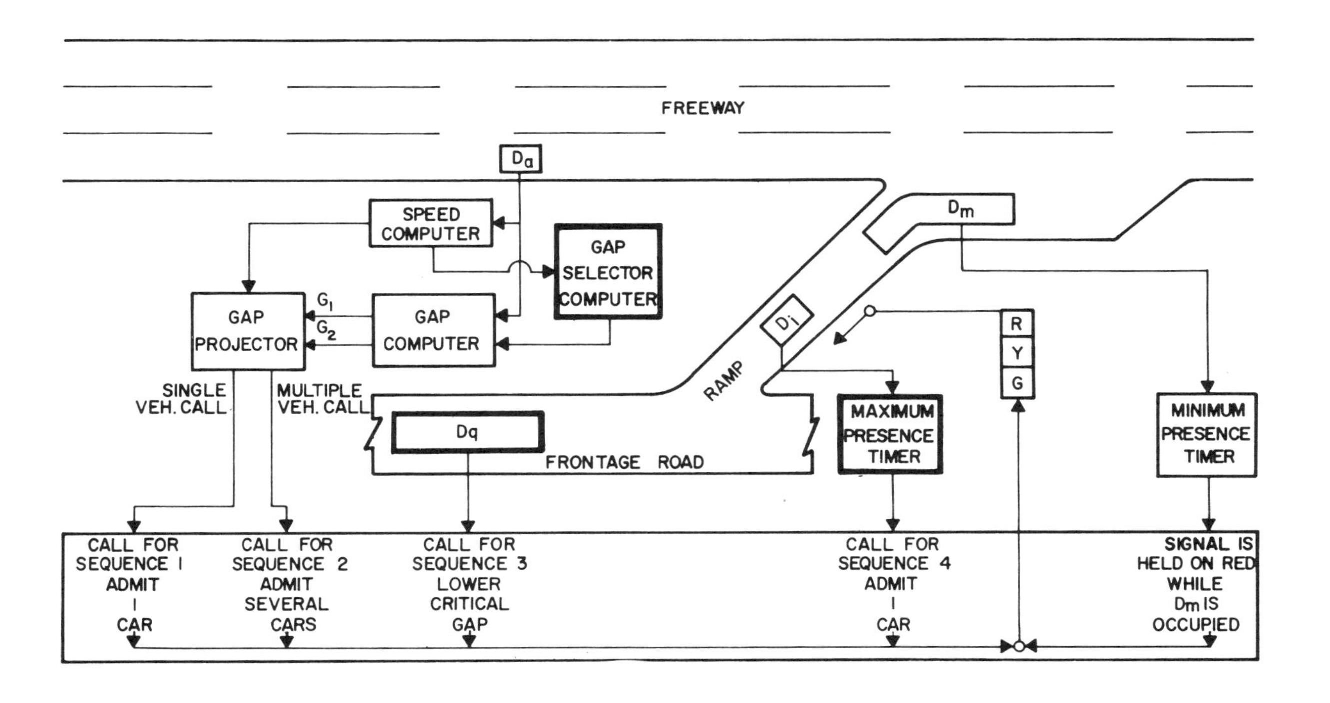

Figure 3
Second Level Freeway Control With Optimizing Function Components Identified

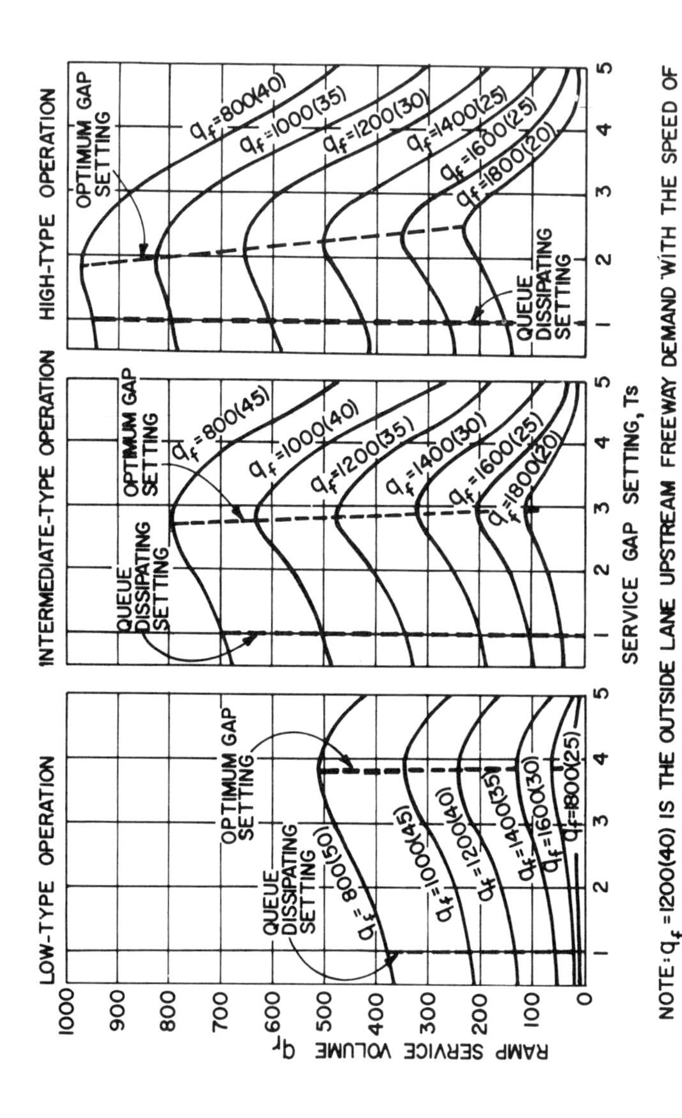

Figure 4
Service Volume Control Curves

thus adversely influencing an adjacent traffic system. To minimize this interference with off-freeway traffic systems, it is necessary to detect such an occurrence with a suitably placed presence detector D_q which, if continuously occupied for longer than a certain period, will cause vehicles to be metered at a faster rate by reducing the service gap to a minimum gap setting (see Fig. 4).

Another loop detector D_i is placed in the vicinity of the ramp signal. If a vehicle is delayed at the signal for longer than a certain period, chances are that the driver will assume the signal to be out of order and proceed past the signal anyway, thus violating the control. This period varies among drivers of course, but is considerably shorter than at a traffic signal on a regular surface street intersection probably because of the somewhat unconventional location of the signal and the absence of any immediate danger in violating the signal. The violating driver, more often than not is then forced to stop in the merge area. It is therefore necessary to have a maximum red phase, insuring that the signal will turn green every so often. A 20 second maximum waiting time is suggested.

8-6 THIRD LEVEL CONTROL

In designing a freeway control system the automatic detection and location of a reduction in capacity must be given a high priority. This reduction may result from either a bottleneck caused by a deficiency in design or from an accident blocking one or more lanes.

In the vicinity of each entrance ramp, six detectors are installed to monitor the accumulation of traffic in an entrance ramp subsystem (see Fig. 5). The three detectors located from 1000 to 1500 feet upstream of the entrance ramp nose are used to determine the freeway demand. The speeds at the location of these detectors are also monitored. Three more detectors are located from 500 to 1000 feet downstream of the ramp nose (past the end of the acceleration lane) and are used primarily to detect reduced capacity operation. Low speeds in this area indicate congestion from a downstream bottleneck with volume counts at this location used to estimate the capacity of the critical bottleneck.

A downstream capacity reduction, other than a geometric bottleneck, may be caused by either the prevailing ambient con-

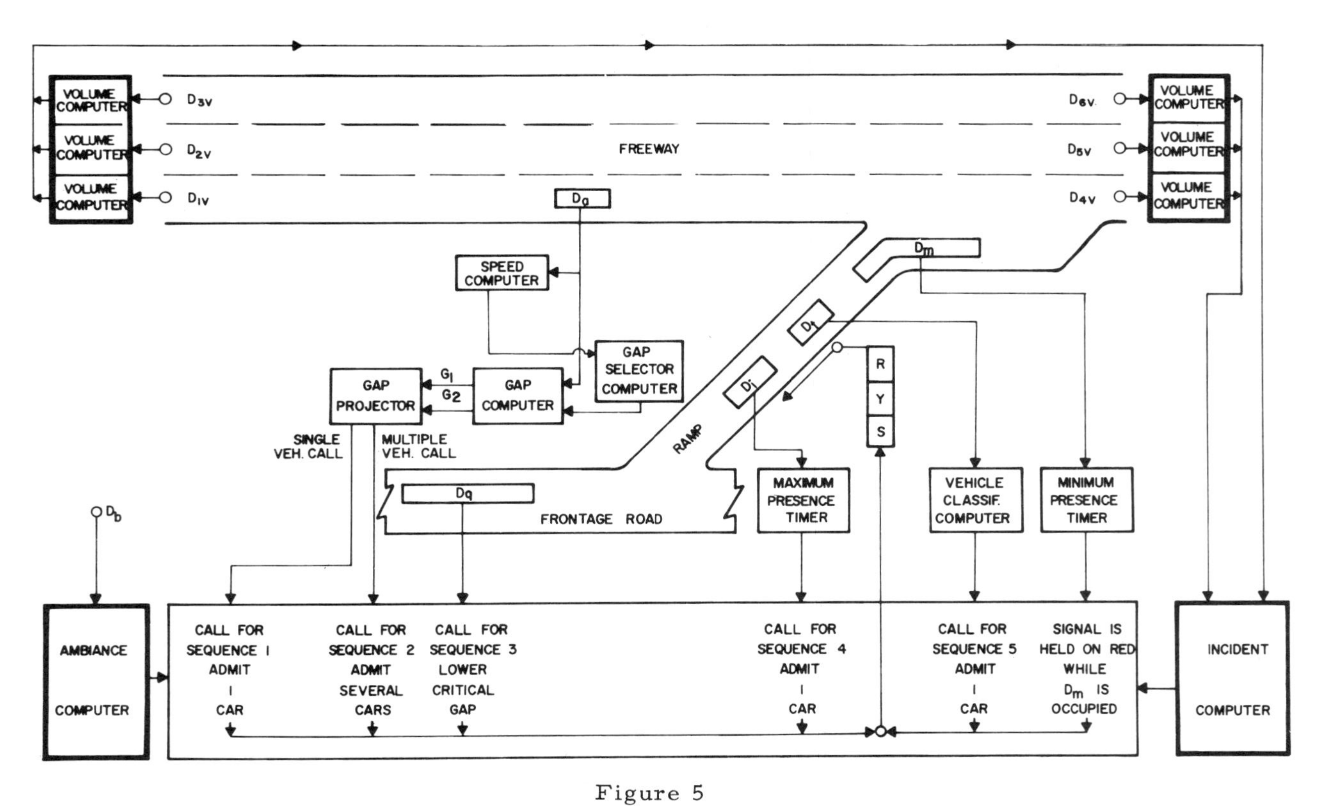

Figure 5
Third Level Freeway Control With Adaptive Function Components Identified

ditions or by an accident on the freeway. When this effect is detected and the degree of capacity reduction is measured, using the third-level adaptive function components illustrated in the block diagram of Fig. 5, these new parameters must be fed to the optimizing controller which in turn will modify the critical gap setting on the regulating controller. It should be pointed out that a capacity reduction due to the ambient conditions is predictable whereas an accident can only be ascertained after the fact. The ambiance components are envisioned ultimately as containing a light-meter, thermometer, and rain gauge so as to describe driving conditions as evidenced by visibility and the condition of the pavement.

It should be borne in mind that the adaptive controller's function is to adjust the parameters of the lower controllers to compensate for deviations from the assumptions inherent in the mathematical models governing the lower control functions. Another way of looking at this is that the third-level or adaptive controller handles the unexpected inputs, often referred to as environmental factors. Ambient conditions and accidents represent two such environmental factors;trucks on the entrance ramp may be interpreted as a third.

The two detectors, D_i and D_t, are used to classify vehicles. Classification, though normally thought of as distinguishing between trucks and normal passenger vehicles, is not that simple. The significant difference between the two classes of vehicles as inputs to a control system is not their size, shape, weight, etc. but rather their accelerating characteristics. A fast accelerating, empty truck may well be placed in the same category as an ordinary passenger vehicle. On the other hand, a slow accelerating passenger vehicle may have the same effect as a truck. Consequently, the purpose of the classification really is to distinguish between fast and slow accelerating vehicles and, once slow accelerating vehicles have been detected, revise the operation of the metering equipment accordingly.

8-7 FOURTH LEVEL CONTROL

In automatic control technology, a number of expressions have been coined to designate the various control systems. In this paper, we list regulating, optimizing, adaptive, and self-organizing. The fundamental property of a self-organizing, or learning, control system as it is often called is its ability to perform better as time progresses. Using the notation of Fig. 1, learning might be implemented as follows: Suppose the optimum performance with respect to a given output specification is accomplished for the parameter settings v_i^j (j = r, o, a) when the input is r_i. Corresponding to a given input, r_1, for example, the optimizing system and adaptive systems previously discussed would ultimately settle on the vectors v_1^r, v_1^o, & v_1^a with the search procedure carried out by C^a always being the same. However, in the fourth level, self-organizing control system, the results of previous computations are stored in memory which makes it unnecessary that the same lengthy process of attaining the optimum settings v_1^r, v_1^o, and v_1^a be repeated each time the command input r_1, is observed by D^a (see Fig. 1). The memory of this simple self-organizing level would ultimately consist of a table such that for each r_1 there would be a corresponding value of v_1^r, v_1^o, and v_1^a. Let us see how this concept may be used in the optimization of a freeway control system.

To begin with, the fourth-level computer can be programmed to automatically update the parameters used in the lower three levels of control. Capacity reductions offer an example of its application to third-level control. The capacities of geometric bottlenecks, an icy pavement, a section of freeway being paint-striped by a maintenance crew, etc. can be "learned". The curves of Fig. 4 afford an example of the utility of the self-organizing controller to second-level control. Since the classification of the merging operation at a given ramp as to high, intermediate and low is based on relative speed, one function of fourth-level computer is to measure these relative speeds so as to be sure that the assumption as to a given ramp's classification is, in fact, correct.

Secondly, complex control algorithms may be devised. One such approach [Ref. 5] utilizes a linear programming model. Briefly, this model maximizes the output of the freeway system subject to constraints which keep the demand less than the capacity through each subsystem and which maintain the feasibility of the solution. Additional constraints applicable to the freeway control system are the control of ramp queues and maintaining the sum of the merging volume in the outside freeway lane and the entrance ramp less than or equal to a specified merging capacity. Then these constraints may be translated into local controller gap-settings for the freeway control system. One special subroutine in this algorithm particularly dependent on this fourth-level concept, is the procedure for learning freeway trip origin and destinations as inputs into the linear programming model.

8-8 SYSTEM HARDWARE

The hardware required to implement a multilevel freeway control system can be categorized into six basic subsystems: sensors, controllers, traffic signals, transmission subsystem, digital computer, and displays. Sensors, of course, are devices embedded in or placed above the roadway to detect vehicles. They may be of one of the following types: pressure sensitive, inductive loop, ultrasonic, radar, and magnetic. Controllers transform the computer commands into controls for the signals on the entrance ramps. The traffic signals simply present the traffic control indications to the ramp drivers. The transmission or communication subsystem provides the means of transferring information from the sensors and controllers to the computer and transferring command information from the computer to the ramp controllers (if local controllers are used). The digital computer accumulates the incoming data, analyzes it, makes decisions, and sends commands up and down the four-level hierarchy. Displays are incorporated so that the operator and other observers can monitor the status of the computer, the individual ramps, and the overall freeway traffic operation. A schematic of the hardware configuration and components for a prototype multilevel control system which is in operation on the Gulf Freeway in Houston appears in Fig. 6.

The Gulf Freeway Surveillance and Control Center, located on the frontage road south of the Wayside Interchange, is shown on the schematic in Fig. 6. Analog controllers, built to the first and second level functional requirements described in this paper have been installed for the control of six inbound ramps. In addition, and IBM 1800 digital computer has been installed in the Control Center. This computer can be used to either perform the third and fourth-level functions in conjunction with the local analog controllers, or to perform all four levels as a central computer controller. With the analog and digital computers now installed and operating in the Surveillance and Control Center, a wide range of control measures can be effected, from the simplest to the most sophisticated. This redundancy is not recommended for future operational projects; but in a research project, this type of flexibility is needed to compare various control system configurations, to establish control warrants, and to perform cost-effectiveness analyses.

Figure 6
Schematic of Gulf Freeway prototype surveillance and control system (inbound).

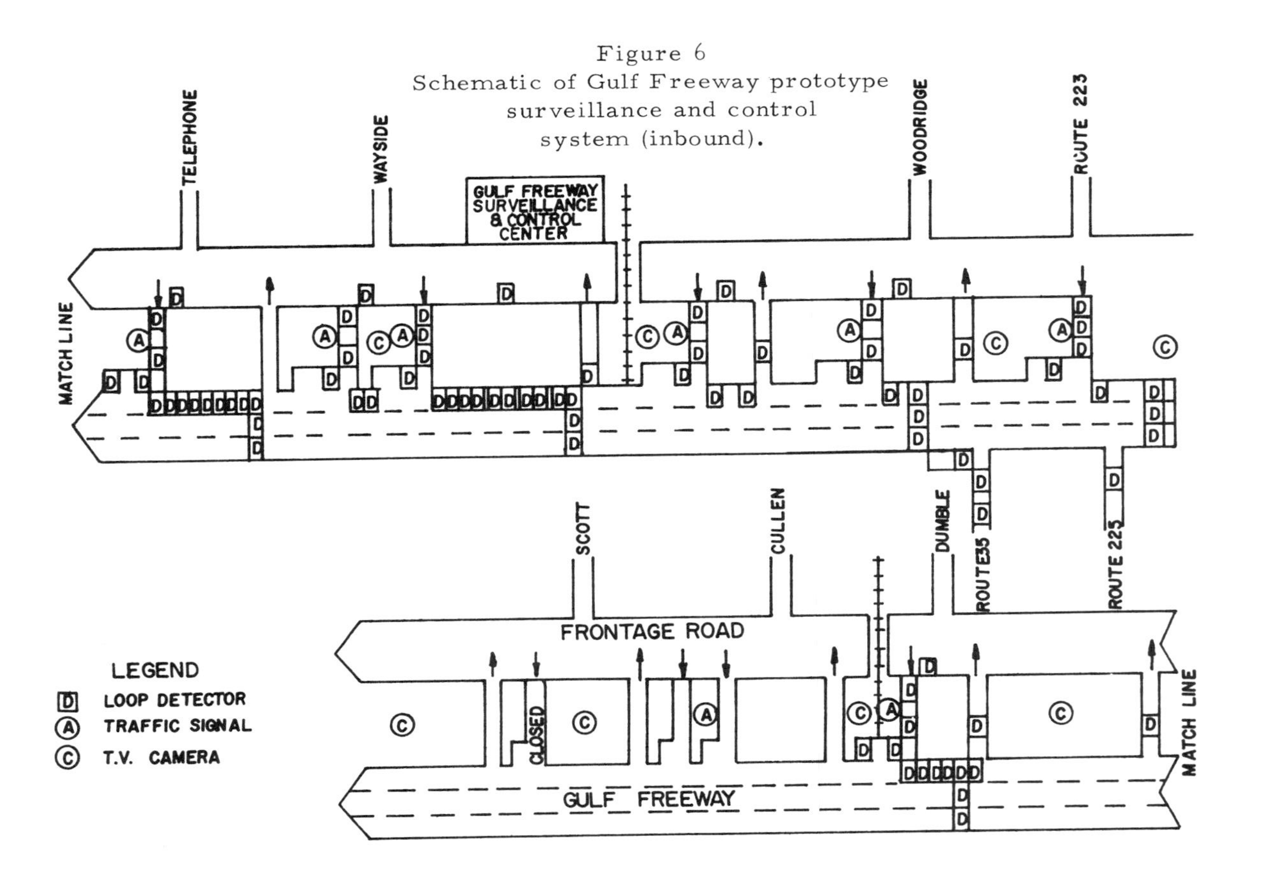

As the name implies, the Center contains surveillance equipment as well as control components. The location of a portion of the TV camera stations for the closed-circuit television surveillance system is shown in Fig. 6. It should be pointed out that closed circuit television is not a part of a freeway control system in the context of the multilevel approach advanced in this paper, it is merely a useful display device to complement the control system. Strategically placed cameras give the observer a view of critical merging areas and bottlenecks for observing the results of decisions changing system operational characteristics. There are other display devices in the control center such as the digital computer peripheral equipment (plotter, keyboard, etc.), meters in conjunction with the analog controllers, and various event recorders.

8-9 SUMMARY AND CONCLUDING REMARKS

The purpose of this paper has been to provide some rationale for the design and implementation of a freeway control system based on the multilevel systems approach used in process control. Basically, the components in the proposed control system consist of one traffic signal per entrance ramp, one merge detector, one gap-speed detector on the outside lane of the freeway, and a regulating controller (see Fig. 2). These are the minimum requirements for the control of a single entrance ramp. For the control of the entire freeway, one check-in detector, one vehicle classification detector, and one queue detector can be added per entrance ramp; one detector per exit ramp; one detector per freeway lane between entrance ramps; and preferably a real time central computer controller (see Fig. 5).

The implementation of a control system on a freeway should not be any more difficult than on any other street. Consider a major arterial, intersections become signalized one-by-one as the control is warranted. In the beginning the traffic engineer responsible for the operation does not have to know anything about network theory or control, he merely installs a detector on each intersection approach, the traffic signals, and the controller, and worries about calibration of the intersection's operation after the system is working. Eventually as more intersections are signalized, it may evolve into a complex network problem necessitating a central computer controller to coördinate the local controllers on the arterial and neighboring streets;

but even then the detection system, the signal system, and the local controller requirements do not change. Only the problem of calibration and coordination changes and if this has been anticipated at the time of the signalization of the first intersection (first-level control), the network problem becomes much more tractable.

Freeway control systems can evolve in the same way, entrance ramp by entrance ramp. Eventually, a collection of controlled entrance ramps would be regulated by some higher order of control just as in the case of a collection of interactions. Yet, the components employed in the control of the individual entrance ramps would be the same ones needed for the control of entire freeway. The control functions can be built-up level by level as more and more entrance ramps and freeways are brought under control. While this is being accomplished, those responsible for the operation of the freeway facilities could be gaining expertise. Although the multilevel systems approach has its basis in the theory of controlled processes, it is compatible with the practicalities of the stage implementation of freeway traffic control.

The benefits of such systems include cutting the number of on-ramp accidents in half and reducing a freeway motorist's travel time approximately two-minutes per mile during rush hours [Ref. 6]. Such controls on the 5,000 miles of urban freeways and 25,000 entrance ramps that will need them by 1975 could eliminate 100,000 merging accidents a year at an average cost of $300 an accident. As many as 500 lives a year could be saved and the average increase in capacity of 15% would be equivalent to adding another 1500 miles of urban freeways at an average cost of 2 million dollars a mile. The total cost would be about $10,000 a ramp, a fraction of the total benefits.

REFERENCES

1. Sanders, J. L., "Multilevel Control," Trans. IEEE, Applications and Industry, Nov. 1964, pp. 473-479.
2. Lefkowitz, I., "Multilevel Approach Applied to Control System Design," Journal Automatic Control Prymois, 1965, pp. 100-109.
3. Rekoff, M. G., "Multilevel Systems," 1966, Unpublished.

4. Mesarovic, M.D., "Self Organizing Control Systems, Discrete Adaptive Processes", Trans. IEEE, Sept. 1964, pp. 265-269.
5. Brewer, K.A., Buhr, J.H.; Drew, D.R.; Messer, C.J.; "Ramp Capacity and Service Volumes Under Merging Control"; Research Report RF 504-5, Texas Transportation Institute; 1968.
6. Drew, D.R., Traffic Flow Theory and Control, McGraw-Hill Book Co., New York, N.Y. 1968.
7. Moskowitz, Karl, "Analysis and Projection of Research on Traffic Surveillance, Communication, and Control", Roy Jorgensen Associates, Inc., Gaithersburg, Maryland, Dec. 1967.

Chapter 9

EVOLUTION OF MINIBUS SYSTEMS

R. L. Meier

Almost unobtrusively a new mode for passenger transport in urban regions has been coming into existence. Now that the minibus has established itself as a species we can look back and see that it had a number of intermediate steps in its line of evolution over the past ten or twenty years. By now its niche in metropolitan transport has developed sufficiently in a few locales so that it is possible to anticipate the forms and the systems that may be expected to invade most other metropolises. However, adaptation of the minibus networks, even in the locales where it has made the greatest penetration, is still far from complete, so it is useful to identify the factors that will improve the service in the future by integrating it even more closely into the life of the city.

The author is Professor of Environmental Design, Department of Architecture, University of California, Berkeley.

In Hong Kong the minibus already accounts for 20-25% of all vehicular trips and at that level operates with a superior profitability. Its share could rise to as much as 30% of all trips before a new equilibrium with other transport services is reached. Without any special planning Hong Kong has become the pacemaker in innovating this kind of transport.

Its origins can be traced to the various forms of jitney (also publico, jeepney, peresa, porpuesto, piratika, baht-bus, pak-pai and a multitude of other names) that appeared in every developing country as a substitute for bus and tram transport. In the 1950's American and European motor car companies, seeing the uses to which their passenger vehicles were being put, designed a vehicle with low operating cost and high reliability that could be used for journey to work in developed countries as well as for recreation. It was, in effect, a light van fitted with windows and seats. During the 1960's increasing proportions of jitney-type services were utilizing such vehicles.

Thus the fringe areas of the growing metropolises were connected with the inner core and the industrial estates as rapidly as the roads were extended and improved. The service was both more respectable and protected from the elements than either the bus or the bicycle, and the cost was intermediate between these and a taxi. Later densely travelled routes were invaded by the jitneys and had to be fought off by the licensed bus companies. The growing body of white collar workers and factory-employed women usually found themselves on the side of the intruders because they appreciated the convenience. The resolution of these conflicts has differed from metropolis to metropolis, but thus far the jitney-type service has rarely been allowed to provide as many as 10% of the trips. New conveniences and economies will cause the new borderline transport enterprises to appear, but prior franchises and agreements are likely to keep them minuscule as long as normalcy exists. Only some kind of cataclysmic event that required reconstruction of the transport system is likely to break up this premature rigidification of networks and systems.

The sequence of events leading up to the present minibus service in Hong Kong has therefore been quite typical. A need for dual purpose vehicles was demonstrated in the Colony in the 1950's, and a number of private operators attempted to serve it. They started adapting light vans about 1958-9. In general they connected market towns and roadside villages in the New Territories with the metropolis, picking up anyone and his baskets of produce who hailed him from the curbside. The passenger component grew far more rapidly than the goods movement. Operators of DPV's (dual purpose vehicles) filled their cars by poaching passengers from the queues awaiting the arrival of the bus. Later they were joined by private cars for hire.

All of this was, of course, outside of the Law and subject to fine, but the profits were more than adequate to make up for the extra risks. One law suit decided that the passenger need only carry a newspaper under his arm to qualify under the dual purpose clause. Protected by the ambiguity in the law, a standard shape of vehicle, well established routes, standard fares and convenient terminal areas came into being. Driver training costs were evaded by bidding for bus drivers, thus forcing the company to train replacements, because the alternative of matching wage offers was too expensive.

By 1967 perhaps as many as 2,000 "pak-pais" operated these routes and at least 80% of them resembled the minibus of today. Hundreds more served firms and hotels. The crisis came early in the year and lasted until September. An outgrowth of the Red Guard movement in China built up disturbances, first in Macao and then in Hong Kong, that were most intense in the left wing trade unions, one of which had organized the bus drivers. After early failures the Communist strategy for bringing the city to its knees was to call a strike by bus workers and organize street mobs to burn any equipment that kept on operating.[1]

Hong Kong was believed to be extraordinarily vulnerable to such tactics because, due to the huge refugee influx, its housing shortage and resettlement policies forced 90% of the workers to use public transport of some kind to get to their jobs. On the first day of the strike the bus services had been reduced by half and the cab drivers had been briefed to stay home; nevertheless near normalcy was restored within a few weeks. A factor that helped tremendously was the DPV that had been operating as an illegal "light bus". These small entrepreneur-drivers encountered fewer traffic jams and operated without police harassment, so their capacity was markedly increased. Key people, who might otherwise use cars or taxis, were able to get around, thus discovering the potential convenience of light buses for the first time.

---

1 John Cooper, Colony in Conflict, (Hong Kong: Swindon, 1970).

Cessation of the disturbances at the end of the year left the light bus system with an expanded network of routes and a number of newly converted vehicles, but it was still on the wrong side of the law. Lack of insurance also caused some complications. The reasonable thing to do was to license light buses as public carriers, regulating them in such a way that the controls were consistent with other transport services.[2]

Simple administrative procedures sufficed. Exploratory surveys of the working of the system were conducted on a confidential basis. Police made recommendations that would minimize accidents and unnecessary congestion. Impoundment of the vehicle became a standard penalty that really induced compliance. Altogether 3,500 official "public light buses" (see Fig. 1) were cleared in September, 1969. Neither fare schedules nor routes were regulated. Then everyone waited to see what would happen. Government in Hong Kong has a very pragmatic, laissez faire style until some kind of an intolerable situation develops that requires intervention of authority.

9-1 HOW THE SYSTEM WORKS NOW

Currently about 3,000 light buses speed through Kowloon and the New Territories, and another 800 work on Hong Kong Island (Fig. 1). Almost all move through the corridors that are most highly built up, but many drift through one-way streets at the beginning of the run whenever a high density of trip origins is known to exist.

The vehicles now coming into almost universal use are 14-passenger, Japanese-built (Nissan and Isuzu), diesel-powered minibuses, weighing close to two tons. Their cost, fully equipped and ready to go, is $4,000 (HK$25,000), which is amortized over three years. These vehicles normally run in two shifts, totalling eighteen hours a day, and cover 230-300 miles. Maintenance and repair is handled in an overnight shift.

---

2 Hong Kong Government, Road Traffic Regulations, 1969, (Amendments to "Registration and Licensing" and "Construction and Use"), Legal Supplement No. 2, Government Gazette (Extra), July 15, 1969.

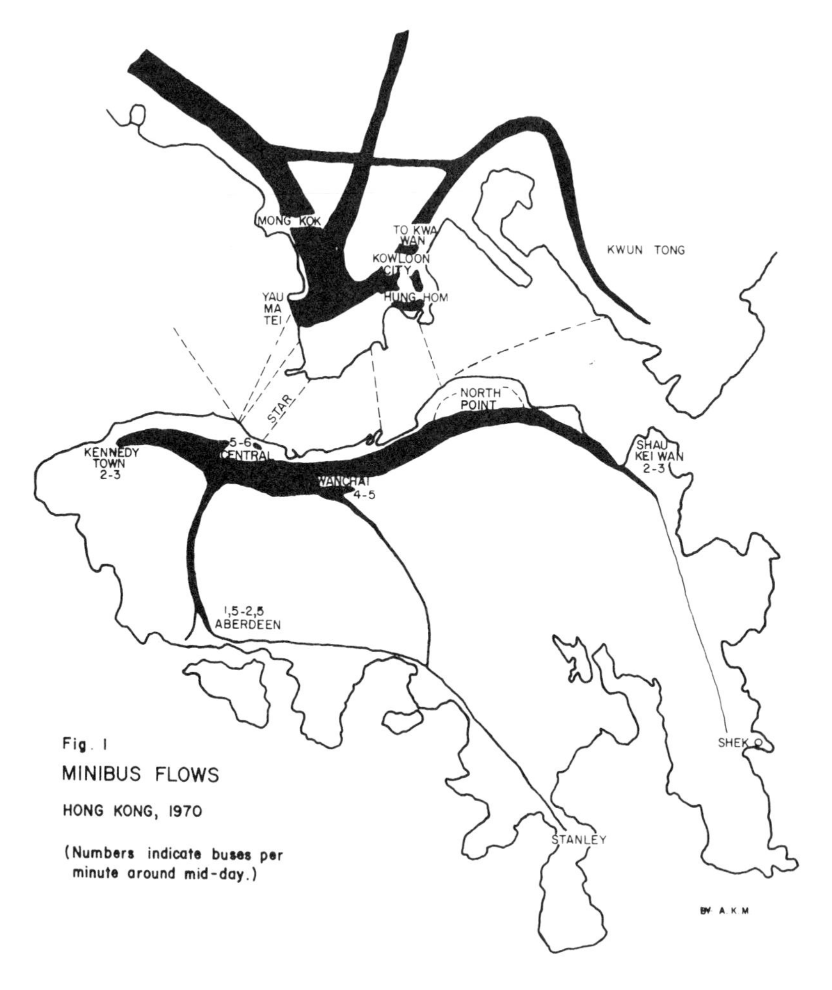

Fig. 1
MINIBUS FLOWS
HONG KONG, 1970
(Numbers indicate buses per minute around mid-day.)

Drivers are generally in their twenties and thirties and a substantial proportion are bilingual (an indication of secondary school education in Hong Kong). It is expected that the owner will drive the shorter, easier morning shift and that he will contract out the second shift for an amount that should yield the second man (usually a close relative) a net HK$45 (U.S. $7.50) per day, if the bus is one of the smooth riding, roomier new Nissans, which puts him into a relatively high income bracket for Hong Kong. The driver of an old minibus will make HK$10 less per day.

Passengers flag down a light bus as if it were a taxi. If this is done from an illegal zone, such as within 50 ft. of a bus stop or where a barrier is at the curb, the passenger is signalled to move to a legal point and is picked up by a subsequent bus. Thus within months more than 90% of the customers have been conditioned into hailing buses from legal positions. Passengers notify the driver of their destination when convenient and disembark at the nearest legal zone. Payment is HK$.50 (U.S. $0.083) for trips up to five to seven miles. All classes, ages and sexes ride the buses, but few school children use them (prestigious schools, hotels, factories and services now operate various sizes of private light buses totalling 1,300). Already a substantial fraction of the population (say 15%) has regulated its life around the availability of this light bus service. For them the imminent improvement of bus equipment is irrelevant.

The best indicator of success as a small business is the market value of the franchise. The cost of the franchise built up rapidly after legalization to HK$20,000 (U.S. $1,650) and varied between HK$15,000 and HK$25,000 until early in May 1970. Then the rumor that the bus companies would seek a fare rise accompanying the introduction of new equipment caused the franchise quotation to rise to HK$33,000 (U. S. $5,500).

A minibus entrepreneur needs at least enough capital to buy the franchise. The bus itself requires only a token down payment with three years to pay out of profits. The most important run on Hong Kong Island - Western Market to Shaukiwan - yields accounts approximately as follows:

	Daily Average (26 days/mo.)
Depreciation	HK$ 15
Diesel Fuel (250 miles)	30
Insurance, Maintenance, Repairs, Fines	15
Labor	80
Interest (15% of HK$33,000)	16
License (HK$3,000 per year)	9
Total Expense	HK$165
Total Fares Collected	$170-180
Net Profit before Income Tax	$ 5-15

The margin of profit is based upon sixteen fares per run - conditions under which 60-65% of the seat-miles of capacity are used - and is very sensitive to changes in the environment. For example, a sudden 10% increase in the supply of light buses is expected to erase the full value of the franchise (because no new traffic is likely to be generated when two to three light buses go by per minute already, and therefore the average take would be reduced to 14.5 fares per trip and interest could not be paid on franchise value).

Significantly, no payments for "protection" and trip promotion are listed in this accounting, although they existed before legalization. It is said that in parts of Kowloon a symbiotic relationship exists between young hoodlum gangs and the light bus operators on the most profitable runs, but the payoff is still small in proportion to other expenses. It is also notable that cleanliness of appearance must make a significant difference in obtaining fares because a bus cleaning service evolved at each major terminus.

Light bus owners have made careful calculations regarding the profitability of fleets or syndicates. They find that even under the most favorable circumstances the expected return on investment does not exceed 2% per month, and is therefore not worth the risk. A one-one business that also provides a job with good pay remains attractive, however. Thus economies of scale for the enterprise no longer exist in Hong Kong.

The economics of the light bus system are more understandable when treated in conjunction with those of the taxis and the bus companies. From the customers' point of view a typical three mile would cost H.K.$0.20-0.30 on the bus, H.K.$0.50

on the light bus, H. K. $1. 30-1. 75 by sharing a taxi. Travel time is rarely more than a minute's difference, but the traveller spends less time waiting for the light bus.

Although published very recently, the Hong Kong metropolitan transportation study[3] has been rendered obsolete in many respects by recent growth in the light bus system. The "pak-pai" category of the report has already advanced beyond 1986 projections in number of transports produced.

9-2 SYSTEM CHANGES IN THE FUTURE

A series of shocks will affect the Hong Kong transport system over the next decade, some of which have already been programmed. The cross-harbor tunnel will connect the centers of Victoria and Kowloon about 1972, thereby enhancing the value of a taxi and adding significantly to the opportunities presented to light buses. Crucial decisions regarding tolls to be collected and extensions of franchise have not been announced as yet. The designers assumed that three new bus lines would be created by the tunnel, but a good share of that demand may be taken up by light buses.

Mass transit is almost certain to come into being, starting about 1976, with final plans submitted this year. After that the minibus routes should properly function as feeders to the mass transit stations, but the chances are great that the time required for getting to and from trains would be sufficient to create a strong demand for surface transit along the corridors in the trip range of 1-5 miles. Although one's preliminary impression is that the first subway stage would not affect the present bus and minibus systems very much (because it will act primarily as a ferry substitute), the interim adaptation to the Cross Harbour Tunnel will have created a network that requires substantial reconstruction. Subsequent stages of subway development would require lesser route re-allocation.

3

Freeman Fox, Wilbur Smith and Associates. Hong Kong Mass Transportation Study (Hong Kong, Government Printer, 1967). It was revised to take into account new population trends in 1969.

Simultaneously an accelerated program of toll-financed expressway building is programmed, with locations for flyovers, causeways, bridges and tunnels already specified.[4] This new highway network will provide an opportunity for the minibuses to seize a large share of the express movements from the bus companies because they could offer a net time saving to the passenger of 10-15 minutes a trip. The 40-plus passenger buses will be at a disadvantage in this contest for riders.

Superimposed upon all of this road building is the necessary rationalization of automobile ownership. All densely settled countries at the stage of development of Hong Kong should have an excise tax on automobile imports of 100% or more, with traffic control as a prime objective. However the Hong Kong Government pays all of its obligations and still accumulates a steady surplus, thus making the tax increase politically difficult. Such an excise also selectively hits at the pocketbooks of the business and professional elite. Parking fees (much too low now for the value of land) may also be used to discourage excessive automobile ownership and operation.

Such matters are worrisome because Hong Kong is presently experiencing unprecedented rates of economic growth, and the boom is still accelerating. Increases in G. N. P. of 10-15% per year must now be considered likely over the next several years and the birth rate is still falling. If continued this means that per capita income will reach the level of present-day Tokyo by 1980. Consumerism may well become rampant. Already the rate of growth of private cars is 18% per year. However Hong Kong does not have room for Tokyo's 600 autos per thousand households or for Tokyo's 70 sq. ft. of living space per capita. What is the alternative? Very possibly a minibus (and hydrofoil?) system with higher fares and greater flexibility might serve a more deconcentrated Hong Kong than is presently envisaged.

Other implications of the current economic growth are not being faced. For example, little opportunity exists for increasing labor productivity on buses (particularly Hong Kong's, since output-per-man-day has reached European levels) so a trebling of wages will force the companies to raise bus fares, much as we see occurring all over Japan today. Due to the riots over

---

4 Freeman Fox, Wilbur Smith and Associates, Hong Kong Long Term Road Study, 1968.

the "Star" Ferry fare increase in 1966, companies are reluctant to face the issue. Thus far the minibuses have also operated on a fixed fare structure, partly due to convenience in handling coins while driving. Minibuses will also need some innovations in pricing within a few years.

9-3 PROSPECTIVE DEVELOPMENTS

With such substantial reorganization of transport in the offing it is apparent that the transportation system as a whole must be in a position that allows it to find a new optimum quickly. This means that there must be a quick feedback of information to the driver-owners and managements as well as to Government, even if units of the transport system do not extend their responsibilities. Buses, ferries and trams already have systems for data collection which permit calculations to be made regarding services to be provided on specific routes at specific times, but the light buses lack such information. Given the opportunities for keeping two sets of books, it will not be easy to produce reliable economic data at reasonable cost.

One of the easiest and least corruptible sources of evidence at hand is the mileage metered during the operating day. Recognizing that each light bus must make H. K. $0. 60 per mile to break even, one of the tasks of a small body of inspectors would be to compile mileages on specific runs, assuming that drivers will be optimizing with all the information available to them, and will therefore soon cease operating below the break-even point. The mileage operated is an index of the service provided on that route.

Passengers are not yet optimizing, because the only information they get is obtained by observation, explanations from friends, and personal inquiry. Their knowledge is therefore fragmentary and often incorrect. A guide to the use of the services offered by the public light bus system should be assembled and published. At H. K. $2. 00 or so it should be a best seller. The six associations of owners should also combine to take a page in the telephone directory, since the bulk of the users of the light buses in Hong Kong will soon have access to a telephone. The mere compilation of such a guide and the designation of a telephone that would be answered should lead to a further rationalization of the system. Feedback from <u>potential</u> customers produces an even closer fit to the demand for transport than responding to queue length alone.

Some indicator of overall system health is also needed; it would be especially useful for the regulation of the supply of franchises. The Hong Kong style of "minimum government" suggests that the market mechanism itself be used. For example, the Government may increase the supply of light buses whenever the value of a franchise exceeded a certain "tipping price", a policy it applies already in the leasing of Crown land. Perhaps that price should be set at the cost of a new bus (now H.K. $25,000) in order to minimize the effects of inflation. Conversely, the supply is diminished whenever operators are unwilling to pay the H.K.$3,000 annual license. Formal arrangements should be made for setting up bid and asked quotations, each with an indicated zone of operations, in order to stabilize the market. At present franchises turn over at a rate of 80 per week, so the market could become quite active and would support full time brokers. The funds so obtained may be used for the benefit of the light bus system - safety engineering, fare collection schemes, pricing experiments, etc.

A related task will be the regulation of traffic through the Cross Harbour Tunnel. In that instance there is clear gain if the peak flows can be redistributed. Why not set up a flow-dependent toll? When a given proportion of capacity is exceeded a high toll is charged, at intermediate flows a low toll, and the remainder of the time no toll. The result would be a more efficient utilization of facilities and equipment and the modal split between buses, light buses and cars less arbitrary. That policy could bring about some major readjustments in human activity over space and over time, and greater realism in land violation. Light buses would take on a greater share of the cross-harbor movement.

If economic boom still produces congestion in the central areas one could even charge a toll for the number of seats <u>not</u> used, whether on autos or minibuses! (All cars in Hong Kong are already marked with their official capacity.) The Chinese drive close bargains, so an arrangement like that would lead to some profound recalculations, very likely opening up some new capacity for the tunnel and its approaches. It could postpone a much more expensive second tunnel or bridge for a number of years.

Meanwhile one discovers a continual probing for new routes and new formulas. We found light trucks in the New Territories sprouting retractable steps so as to fill the dual purpose vehicle need that initiated this whole phenomenon. Hired cars also develop new route possibilities, particularly for Sundays and holidays. Express services are being attempted between more distant points.

Thus far the telephone system has not visibly affected the light bus system, but it must very soon. The number of working lines has been increasing by 19% per year, reaching 430,000 by mid-1970. Telephoned requests could provide valuable off-peak contract work.

Radio-dispatched cabs came to Hong Kong early in 1970. Within a few years their special uses in this metropolis will be understood and extension to minibuses rather obvious. Marginal routes could be developed at double the normal fares, because they would still be cheaper than the taxi alternative. The same provisions might apply to late hours when the ferries and the subway shut down.

Hong Kong settled prematurely upon one standard for a light bus that it now regrets - the two ton maximum weight when empty provision. Originally intended to keep the light bus flowing smoothly with automobile traffic, it pushed the designers into accepting a somewhat underpowered diesel engine (particularly for the mountain sides of Hong Kong Island) and it prevented the use of double rear wheels. One of the best engineered European-built minibuses could not be made to fit the public service, although it was used privately to advantage by many schools and hotels. A 2500 kg. limit may be more reasonable.

These rather obvious next stages in adaptation are mentioned because they take the Asian city much closer to conditions found in Japanese, European and North American cities. In such circumstances, the minibus system, with strong support from modern subway, bus and taxi systems, could serve as an automobile substitute. In this case, however, it would replace the first car in the household. Air pollution conditions in most Asian metropolises will require a restriction on the amount of fuel consumed by vehicles. Thus the planners' task is to design and introduce transport systems that offer at least as great convenience and freedom to move as would be provided by the possession of a private automobile but require only a small fraction of the fuel, as well as less capital investment. Integration of the minibus mode as a substitute for cars hold great promise for densely settled metropolises.

In developing countries, jitney-type modes leading toward minibus systems serve in many ways to economize on automobiles, highways, bridges and structures, thus saving both scarce foreign exchange and capital. Although verbal reports based upon short term observation are common, no study seems to have been made of the role of the jeepney of Manila in the organization of new suburban communities. The interdependencies seem to be at their most advanced stage there. The earliest stages of development are to be found in the growing function of the lambrettas in Saigon and the equivalent 3-wheeler DPV in Kanpur. The back soi of Bangkok is connected into the metropolitan activity by a light, speedy, open van fitted out with removable, padded benches, most likely a 12-passenger Nissan of about 3000 lbs. net weight, called a baht-bus because it carries people for 2-10 km. for a single baht (U.S. $0.05). This kind of vehicle is being introduced into New Delhi, but maintenance suffers from the stringency of foreign exchange controls imposed by the bureaucracy on the importation of spare parts. The minibus has great difficulties in penetrating an over-regulated society.

In most developing countries minibuses are likely to get started within the ambiguities of the vehicular code and the enforcement of traffic law, whereupon the evolution of the system may be expected to unfold new routes and services in a manner similar to that exhibited in the New Territories of Hong Kong. Often, however, a number of unique situations will intervene and put their own imprint on the route structure. It is to the advantage of owners, drivers and especially passengers if the government legalizes earlier than in Hong Kong, because any system operating on the edge of the law will collect poor data, and will therefore be less able to learn from past experience.

Ultimately both developed and developing countries must point toward a thoroughly integrated transport system, which includes a variety of telecommunications and data processing systems, and a fare differentiated according to the service provided. Thus a door-to-door service with less than five minutes' waiting time and 99% reliability co-ordinated by telephone and radio-dispatcher might well charge up to U.S. $0.05 per mile and be heavily used for moving the middle classes to work from the suburbs, relieving them of the need for a second car. Much lower cost services could link up the rest of the city with the mass transit system; they would obviate the need - not the desire - for a first car. From the transport planner's point of view a minibus system with intercommunication would be able to reduce passenger delays attributable to the lumpiness of some modes, such as inter-city trains, ships or the new aircraft.

9-4 SUMMARY

The "public light bus" system launched in Hong Kong in 1969 appears to be the forerunner of systems expected to evolve in other metropolitan areas. A fleet of 3,750 minibuses, each with a capacity of fourteen passengers and operated by owner-drivers, supplies about 22% of all trips while charging 1.5-2.5 times the standard bus fare. Principal advantages to the user have been (a) a guaranteed seat, (b) convenient pick-up, and (c) less noise with greater cleanliness. The system is more adaptable to crises of public order and to the continuous modification of the arterial road network. With reinforcement from telecommunications such systems can operate as automobile substitutes in developing countries, and as substitutes for the second car in the household in high income metropolitan areas.

Chapter 10

THE NORTHEAST CORRIDOR TRANSPORTATION PROJECT: FEDERAL FOLLY OR REGIONAL SALVATION

R. A. Nelson

The Northeast Corridor Transportation Project was begun in 1964 at a time of great optimism about the capabilities of systems analysis to deal with problems which up to then had been thought quite intractable. Triumphs in the space and defense fields fostered the belief that no problem could remain unsolved before the systematic onslaught of the systems analysts. It was in such a mood of high and unrestrained confidence that the Corridor project was launched. Contributing to the rosy view was the circumstance that the Corridor project originators were split about equally between systems people who were quite sure that transportation wasn't all that difficult, and transportation people who for lack of knowledge were totally awed into faith in the magical powers of the systems approach. Together they squared, if not cubed, each others' misconceptions so that at the outset the Corridor project embraced objectives which

---

The author is, Federal Executive Fellow, Brookings Institution, Washington, D. C.

were extraordinarily challenging to say the least. Indeed if we had gone only halfway in achieving these objectives, there would be little left to do in the future in the two fields of transportation and systems analysis. Suffice to say that we were somewhat less than successful. Nevertheless some progress was made in both fields--to some extent in showing what could not be done as well as what could be--which may be worth relating.

Perhaps it should be recalled that the genesis of the Corridor project lay in the proposals for renovation of rail passenger service in the Northeast Corridor made by Senator Claiborne Pell and others. All of the proposals contemplated the Federal government's spending very substantial sums of money--on the order of $1.5 to 2.5 billion. Understandably the guardians of the Federal till were skittish about laying out such amounts for an undertaking whose economic worth was dubious to say the most. Too, the proposals came at a time when many questions were being raised about the Federal highway aid program and its singleminded preoccupation with laying down more miles and more lanes of concrete. It wasn't that building railroads was preferred to building more highways but rather the uncertainty about how much should be invested in highways, how much in airports and airways, and how much in railroads, if any. The thought began to emerge at that time too that one mode properly planned should make other modes more effective. Moreover the more economically complex a region, the more difficult it was to vizualize the best mix of transportation modes. The very real space limits in the Corridor seemed to suggest that realistically neither highway nor air transportation, nor both together, could meet all the Corridor's transportation needs. But also, it was not at all clear that rail passenger service could fill out the remaining requirements. In this setting, systems analysis was advanced as the hope of analyzing and predicting the behavior of the large numbers of simultaneously interacting variables, influencing and being influenced by transportation. It is not surprising that those in the Federal government responsible for investment decisions seized upon systems analysis as a means of dealing with the enormous complexity of transportation in the Corridor.

The alternative which systems analysis posed for deciding on Northeast Corridor transportation investment was not simply a matter of system vs. non-system, i.e. cognition vs. intuition. The traditional way, both at the state and Federal level, of coordinating investment in transportation had been to rely largely on the marketplace for guidance.

The Federal government has always had the primary role in interstate transportation but it has met this responsibility with a minimum of coordinated planning and a very heavy emphasis and reliance on leadership and initiative from the private sector. The number and characteristics of automobiles built by automobile manufacturers has had far more influence on the highway system than has public policy; much the same can be said about air transportation. For example, the decisions to build the "big jets"--the Boeing 747 and the Lockheed Skybus--were essentially private, yet these decisions will have great impact on the amount and staging of public investment in airports in the next several decades.

The rationale for past Federal government policy, more or less explicitly propounded by academics and similar types, and more or less officially adopted in the executive branch, has been that if the public sector would simulate the private sector's valuation of resources and respond to market need only where user revenues promised to cover resource costs, adequate investment guidance would thereby be provided. Highways, air ports and other transportation facilities could then be planned and built by separate agencies in the Federal government or elsewhere provided their sources of funds were based on user revenues. Private sector investments in transportation such as in railroads would be on a basis of equality with public investments.

This system hasn't worked despite its elegance and appeal and for the following reasons a decision was made in 1963 to turn from the marketplace approach and make a more deliberate effort at planning transportation investment in the Northeast Corridor.

(1) Because it has envisioned indirect or "non-market" benefits of some investments as exceeding those measurable in the marketplace, Congress has refused to rely on market values alone to determine benefits flowing from transportation investments, particularly in air and highway. In neither case has it required users to pay full cost.

(2) Even in 1963 non-market costs of transportation had begun to appear significant. Accidents, noise, pollution, and community disruption were increasingly being regarded as undesirable side-effects of transport development.

(3) By providing large amounts of aid to highway and air transportation the Federal government was the major source of investment in transportation facilities. In this respect it began to have the planning problems of a large multi-product seller. Demand and supply had to be estimated, just as by any seller, as a basis for investment policy.

Taken together these three circumstances pointed toward needed change in the Federal government's traditional approach to transportation investment. The Northeast Corridor Transportation Project represented a start at this change; the establishment of the Department of Transportation constituted a much greater step although its rationale was understandably much less clearcut than the Corridor project.

I should like to discuss the five year experience represented in the Corridor project and the report on Corridor transportation released by the Department of Transportation early this year.

10-1 THE STRUCTURE OF THE NORTHEAST CORRIDOR PROJECT

The charge given to the Corridor project was to determine the transportation facility requirements of the Northeast Corridor through 1980. It was widely agreed at the outset, however, by those associated with the Corridor project that no single determination could be made. Changes in transportation would have major and varying effects on different segments of the Northeast Corridor community which could not be measured, and even more important, could not be labelled either as costs or benefits. They included such effects as settlement patterns, patterns of economic growth, resource requirements, etc. In order to assure that decisions on these matters would be made by political representatives of the Community, the Corridor project decided that a number of possible transportation systems would be depicted responding in different ways to policy options. This approach adopted right at the start of the Corridor project, expunged the notion so dear to systems analysts and operations researchers that some <u>optimum</u> transportation system for the Corridor would be devised.

Another major early decision was that the Northeast Corridor would be treated as a closed system so far as transportation and its effects were concerned. The major significance of this decision was that a concentration would be put on Corridor origin-destination traffic flows and the Corridor transportation system would be designed primarily to serve Corridor transportation, as distinguished from Corridor to/from the rest-of-the-world. Additional significance of the "closed system" treatment of the Corridor was that the possible effects of economic developments in the rest of the world were neither considered nor predicted. For example, given Corridor populations were redistributed depending on transportation changes in the Corridor but population totals were not changed. Transportation changes in the Corridor which might improve the Corridor's absolute or comparative advantages vis-a-vis the rest-of-the-world were ignored.

In my judgment, with the present state-of-the-art of economic analysis and the quality of data available, there was no alternative to treating the Corridor as a closed system. There is a question as to whether such an approach can be valid for freight transportation in the Corridor. And it makes doubtful the usefulness of the models for corridors such as Cleveland-Chicago where through traffic beyond the corridor is very heavy. For passenger traffic in the Northeast Corridor however, the assumption was not overly heroic.

A third much less Procrustean but nonetheless hazardous decision was made to treat transportation technology as dynamic. This meant forecasting technological developments among the modes for the next couple of decades. This had seldom been done before in transportation model-building.

In 1964 and 1965 the Corridor project set about structuring the tasks which would have to be carried out to bring the project to culmination.

These tasks are set forth in schematic form in Figure I. Their nature, I'm sure, is not surprising.

Demand Forecasting: Demand for the services of any one transportation mode between two points depends partly on general circumstances of distance and population, income, employment, etc., in the economy, and partly on performance of the mode in competition with other modes. Predicting how travelers will be affected by changes in income level is not easy but it is less difficult than predicting how they will react to a new transportation service. In both cases, lots of good data are necessary for the formulation of satisfactory analytical models. Anticipating such need and recognizing the paucity of transportation flow data, the Northeast Corridor rail passenger demonstrations

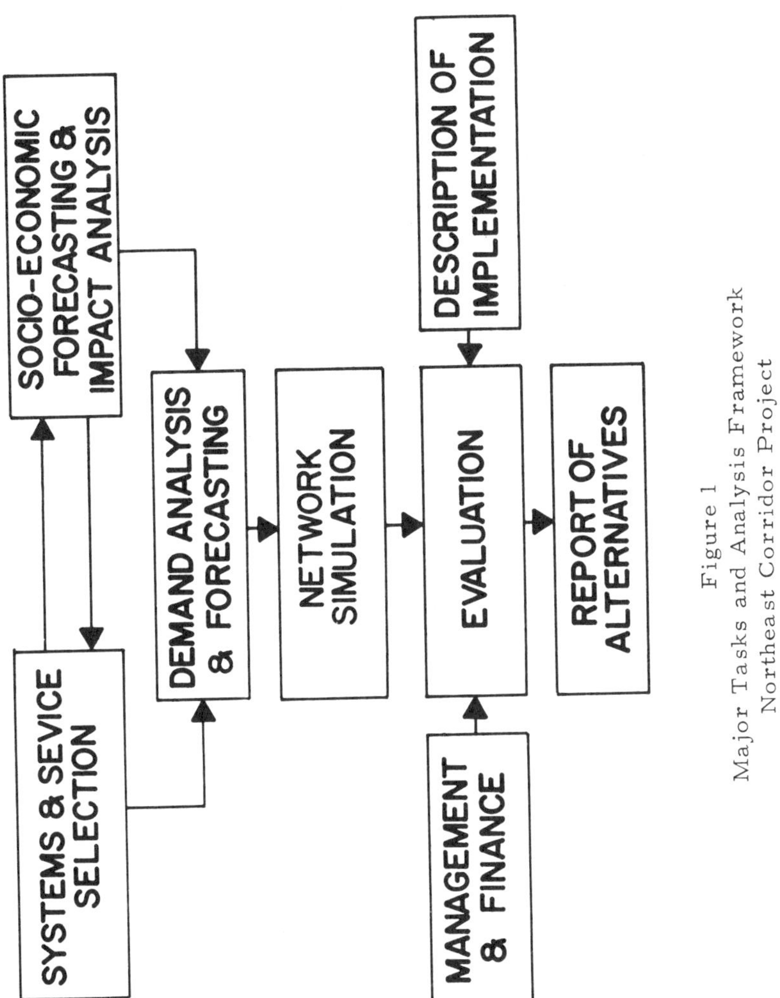

Figure 1
Major Tasks and Analysis Framework
Northeast Corridor Project

were conceived of as providing knowledge and information about travel behavior. In the light of Monday morning, it must be conceded that the demonstrations took on an importance to the public which ran beyond their role as producers of transportation information. On the other hand it is also true that the data collection procedures set up and applied to Corridor rail travel prior to the start of the demonstrations produced far and away the best data available to the Corridor project.

In 1965 Congress endorsed the proposition that data were essential to analysis of transportation needs and provided both authority and funding for the collection of data for the Corridor project. Through a bureaucratic foulup in the executive branch, however, the funds were not available and the project consequently suffered profoundly from a lack of good data. This leads me at this point to stress the importance of good data for analysis of social and economic phenomena. There is simply no substitute except extraordinary luck.

Despite the problems, we developed demand models with a high capability for reproducing the past. By emphasizing transportation service characteristics such as trip time, user cost, and service frequency and formulating the models on that basis, we hoped to be able to predict demand for new modes as well. Only time can give the answer to whether or not we were successful. The demand analysis showed quite conclusively that the primary basis of differences in transport mode performance are in time, cost and frequency. For other aspects of service the modes tend to reach competitive parity. With respect to the three significant variables, trip time was far and away the most important with elasticity coefficients ranging between -2.0 and -4.0 depending on whether a stratification was made for business and non-business travel. Fare elasticity ranged about -1.0. Elasticity of demand with respect to frequency was, as expected, positive, and varied depending on the service already being provided. Service beyond 20 frequencies per day seemed to add little to patronage.

These results seem to emphasize the importance to travelers of trip time and frequency and the somewhat lesser importance of fares. The data resulting from Metroliner operation seem to suggest that these elasticities will obtain over a range of changes in fares and trip times up to 25 per cent.

The project spent a good deal of time on developing a model to predict freight demand. While considerable progress was made on the theory of demand for freight transportation, lack of data prevented any model formulation. There is some doubt in my mind that the desired data on such variables as shipment value can ever be made available.

System Selection: The Corridor project was intended to consider various transportation alternatives for the Northeast Corridor including advanced highway and air modes as well as rail. It was clear that reasonably satisfactory prognoses of air and highway development could be made in 1964 based on quite well established trends, but that such projections could not be made for rail transportation in which there had been only small technological change for many years.

The administration in 1964 decided to support a request for legislative authority for research and development in what came to be called high speed ground transportation. This request was coupled with the proposed rail passenger demonstrations and became in 1965 the High Speed Ground Transportation Act which provided the funding and the authority for the systems engineering work in the Corridor project.

A great deal of systems engineering and research and development in high speed ground transportation provided extremely important information for the Corridor project, and also constituted the basis for an expanded research and development program in high speed ground transportation which is currently being implemented. A significant achievement for the Corridor project and the high speed ground transportation program was to be able to subject research proposals to the discipline of the travel market. This was possible particularly after the Corridor model system became operational. It must be said, however, that because of budget cuts and other slowdowns in the high speed ground program only one advanced system could be defined with any degree of accuracy for the Northeast Corridor project. This was the major reason why, as we shall see, the Corridor alternatives depicted in the 1970 report are limited to technology which could be operational in a system by 1980.

A good deal of advanced work was done too to try to develop models for systems selection. The objective was to analyze the nature of transportation markets and their relationships to the characteristics of different transport modes and then by means of a quantitative model select the mode providing the "best fit". This work was esoteric, some progress was made, but not enough to permit use of a model for selection of the Northeast Corridor alternatives. It remains a challenge for systems analysts.

The new modes which were actually run through the Corridor model system were defined in the same terms as the primary variables in the demand model, i.e., trip time, costs and frequency. Cost estimating relationships were developed for a range of volumes for different parameters of trip time and frequency. In the case of the ground modes because of the high initial capital requirements the cost functions had a pronounced negative slope; the air modes less so.

System Simulation: The most seminal achievements in the Corridor project were the development of Corridor modal networks, a ground mode supply model, and an air mode supply model. Linked with the demand model, this model system permitted the simulation of different modal combinations in operation in the Corridor.

The ground mode supply model can optimize service or profit over cost plus a 10 per cent capital charge based on fares and terminals being exogenous inputs. The air modes supply model is not a linear programming model but it converges rapidly with iteration at a point where all profit is exhausted (See Figure 2). This permits fares and terminals to be variable in the model and has produced very interesting results for terminal location.

The ground mode supply model and the air mode supply model when linked, permit through iteration a supply-demand equilibrium, i.e., the working out of competitive interactions among the modes.

The simulation in operation needs a great deal of tender care, requiring manual transfer between one part of the simulation and another. Recently the Corridor project has attempted with some success to package the simulation and make it more widely usable. Properly used, it is a decision-making tool of considerable power.

Impact Models: The Corridor project's ambitions for impact models were rather considerable at the start. Studies were carried out which penetrated some of the most difficult aspects of the subject and a model was developed which predicts the effects of changes in accessibility. The model has worked well with existing models; how well it will predict the effects of new modes is uncertain.

Evaluation: Here also the Corridor project's ambitions were considerable. They were, however, largely unrealized. We were able to quantify the costs of pollution, accidents, noise, etc., but they were not introduced into the modals as system costs.

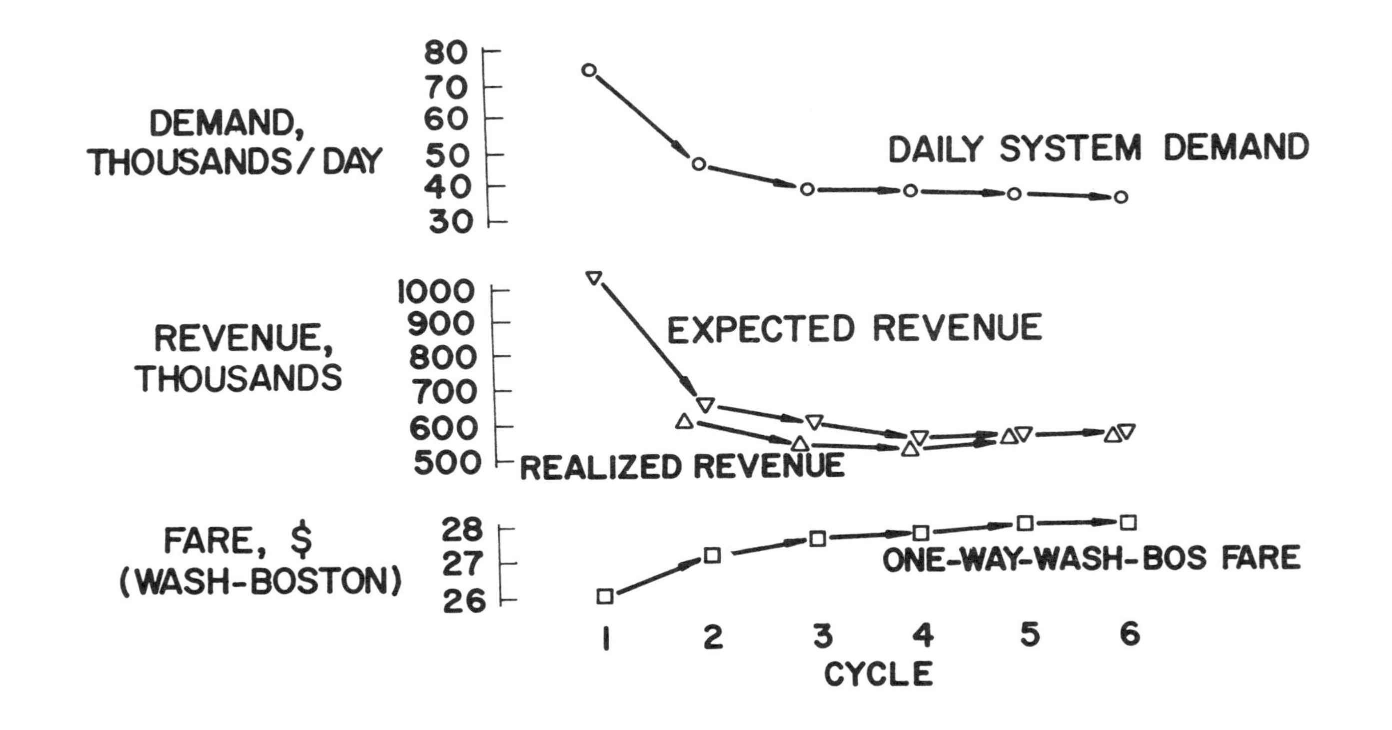

Figure 2
STOL Convergence Iterations

The simulation was and is able, however, to pump out great detail on user revenue and operator cost relationships making possible evaluation of the market performance of both existing modes and new modes.

10-2 THE NINE CORRIDOR ALTERNATIVES

After what seemed to be an inordinately long time the Corridor model system began to fall into place and to offer prospect of being operational. To make it operational, however, required abandonment, at least temporarily, of many of the ambitions held at the outset of the project. The alternative was not to produce a report for several more years. In cutting off further development in the fall of 1968, we decided to strive for as much accuracy as possible even at the expense of being less comprehensive. This resulted in the following compromises with the real world.

(1) Freight would be eliminated from consideration and the models would be run only on passenger loadings. The existing rail system was assumed to have sufficient capacity for substantially increased passenger patronage without major effects on cost.

(2) Quantitative evaluation would be based primarily on user revenues and operator costs. Indirect benefits and costs would be calculated but not made inputs to the model system.

(3) We decided not to impose capacity constraints and thereby disregarded congestion cost. This was a necessary decision if the simulations were to be run. Although this decision has been criticized, it does not, in my judgment, seriously prejudice the results, indeed, they may be better for purposes of comparing system performance.

(4) It was apparent that the supply models were not yet able to handle the life cycle of a system. To have attempted a 15 to 25 year cycle would have complicated the simulation intolerably. We decided to focus on one year far enough in the future to permit the building of new facilities but near enough so that demand projections could be reasonably accurate. The year chosen was 1975. The simulation would compare on an annualized basis the operation of the alternatives in that year. No attempt would be made to calculate life cycle costs or streams of revenues.

(5) The system selection models were not operational, therefore for the ground modes we arbitrarily decided that they would serve the center cities of nine metropolitan areas strung out in the Northeast Corridor between Washington and Boston.

(6) Fare demands were slightly inelastic. This meant that in a simulation, without constraints on demand, fares would be driven up to unrealistically high levels. We decided to input fares to the ground supply models at their current levels and then do sensitivity tests at different fare levels. We decided to run the air modes supply model until it exhausted profit above 10 per cent before taxes. We also decided in the case of the air modes to treat each mode, i.e. CTOL and VTOL as one operating firm.

The systems selected for simulation were nine including in various combinations four new modes and two more or less improved ones. They are shown in Table I.

All of the system alternatives were run through the simulation a number of times with varying assumptions about fares and interest charges. Changes in fares had relatively little effect; changes in interest charges did not very much affect the air modes, but they did have major effect on the ground modes reflecting the high capital requirements of the ground modes. For the main set of runs, fares for the ground modes were set at $1.50 as a fixed element in each fare and 7.5 cents per mile variable. Interest charges were set at 10 per cent in accordance with Bureau of the Budget directives. Tables 2 and 3 summarize the results of the runs.

Several significant phenomena should be noted in the results:

(1) The difference in passenger patronage between the existing system--Alternative I--and the highest performance system--Alternative IX--is only about ten percent. This can be attributed to the relatively small impact that high line-haul performance has on trip time and emphasizes the dependence of the intercity system on the urban systems. In that respect clearly the urban systems are subsystems of the Corridor system. No major improvement in intercity systems can be achieved without improvements in urban transportation. It is also possible that the Corridor models put too much emphasis on inter-modal competitive relationships and were not sufficiently sensitive to induced demand. If this were true, it would mean that the demand for the high performance modes is understated.

Table 1

NINE PASSENGER TRANSPORTATION SYSTEM ALTERNATIVES FOR THE NORTHEAST CORRIDOR

ALTERNATIVE	MODAL COMPOSITION
I	Auto, Bus, Conventional Air (CTOL) Demonstration Rail (DEMO)[1] --125 mph
II	Auto, Bus, CTOL, DEMO Short Take-Off and Landing Air (STOL)--370 mph
III	Auto, Bus, CTOL, STOL High Speed Rail "A" (HSRA) --150 mph
IV	Auto, Bus, CTOL, STOL High Speed Rail "C" (HSRC) --200 mph
V	Auto, Bus, CTOL, STOL Tracked Air Cushion Vehicle (TACV)--300 mph
VI	Auto, Bus, CTOL, STOL, DEMO Vertical Take-Off and Landing Air (VTOL)--265 mph
VII	Auto, Bus, CTOL, STOL VTOL & HSRA
VIII	Auto, Bus, CTOL, STOL VTOL & HSRC
IX	Auto, Bus, CTOL, STOL VTOL & TACV

[1] Demonstration rail assumes that the present Metroliner and Turbo train services will be expanded and extended through 1975.

Table 2

SUMMARY OF MAJOR CHARACTERISTICS
OF NECTP TRANSPORTATION SYSTEM ALTERNATIVES

Alternatives	New Modes	Sustainable Top Speed, mph	Average Speed*		Total Corridor Intercity Travel,** billion pass. miles
			Terminal to Terminal, mph	Door to Door, mph	
I	DEMO	125	72	46	19.4
II	DEMO STOL***	125 370	72 141	46 63	20.3
III	HSRA	150	109	58	21.1
IV	HSRC	200	152	71	21.7
V	TACV	300	198	79	22.3
VI	VTOL	265	147	74	20.3
VII	VTOL HSRA	265 150	151 109	70 57	20.8
VIII	VTOL HSRC	265 200	152 157	70 70	21.5
IX	VTOL TACV	265 300	144 205	70 78	22.1

* Statistical averages computed for each mode by dividing total passenger hours into total passenger miles. Note the controlling influence of access-egress time on door-to-door speeds.

** Includes auto

*** STOL is included in alternatives II through IX.

Table 3

SUMMARY OF FINANCIAL CHARACTERISTICS OF
NECTP TRANSPORTATION SYSTEM ALTERNATIVES

			New Modes		
Alternatives	New Modes	Total Capital Cost, $ x 10^6	Incremental Annualized Costs $ x 10^6	Annual Revenues $ x 10^6	Annualized Surplus or (Deficit) in 1975* $ x 10^6
I	DEMO	70	$ 61	$144	83
II	DEMO	69	60	141	81
	STOL**	195	244	244	0
III	HSRA	1590	240	213	(27)
IV	HSRC	2600	355	288	(67)
V	TACV	3340	452	349	(103)
VI	VTOL	1060	318	318	0
VII	VTOL	966	310	310	0
	HSRA	1580	230	175	(55)
VIII	VTOL	971	292	292	0
	HSRC	2590	340	240	(100)
IX	VTOL	966	291	291	0
	TACV	3330	440	292	(148)

* STOL and VTOL service and fare levels were set to achieve break-even operation at a ten percent return on investment; HSRA, HSRC and TACV service levels were set to maximize profits (revenues less costs); DEMO figure represents the difference between incremental revenues and incremental costs to provide DEMO service. It does not reflect any allocation to DEMO service of costs currently borne by the railroad.

** STOL is included in alternatives II through IX.

Table 4

RELATIONSHIP BETWEEN TRANSPORTATION SYSTEM ALTERNATIVES AND PUBLIC POLICY OPTIONS

Alternatives	New Modes*	Policy Options					
		Degree of Technological Innovation	Orientation to Metropolitan Area	Capital Cost	Service Character-istics	Public Support Required	Institutional Change Required
I	DEMO	None	Center City	Low	Fixed Linear	No	Little
II	DEMO & STOL	None	Center City & Suburbs	Low	Mixed	No	Little
III	HSRA	Some	Center City	Medium	Fixed Linear	Yes	Large
IV	HSRC	Some	Center City	High	Fixed Linear	Yes	Large
V	TACV	Much	Center City	High	Fixed Linear	Yes	Large
VI	VTOL	Some	Suburbs	Low	Flexible Dispersed	No	Little
VII	VTOL & HSRA	Some	Center City & Suburbs	Medium	Mixed	Yes	Large
VIII	VTOL & HSRC	Some	Center City & Suburbs	High	Mixed	Yes	Large
IX	VTOL & TACV	Much	Center City & Suburbs	High	Mixed	Yes	Large

* Auto, bus and conventional air are included in all alternatives; STOL is included in alternatives II through IX.

(2) Both the ground modes and the air modes can be made to pay for themselves at capital charges of 10 percent. Self-sufficiency performance for the ground modes, however, is at much lower level than for the air modes. This is because the capital costs of saving a minute on the ground become very high at upper levels of performance.

(3) VTOL performs quite well in the Corridor largely because VTOL terminals were put at suburban locations where a large amount of Corridor traffic originates and terminates. It should be noted that the 1970 decennial census shows a continued outward shift of population from the center cities of metropolitan areas.

Table 4 shows the relationship between the systems simulated in the Northeast Corridor and the physical and social environment of the Corridor. Whether one or another of these policy options is to be exercised should be a matter of public decision. None of the transportation alternatives for the Northeast Corridor is neutral in its impact.

10-3 CONCLUSIONS

The work of the Northeast Corridor Project probably stands on middle ground between those for whom impeccably rigorous methodology is essential regardless of outcome and those for whom experience and intuition are the safest guides to action. Like all middle ground positions, the Corridor project results will undoubtedly be attacked by both sides; by one because too many arbitrary decisions were made, and by the other because too many of the results depend on fancy models and simulations. Personally, I think the Corridor project has made significant contributions to the science of transportation analysis and to the art of transportation planning. Whether it will be used by decision-makers or not, I can not be assured. Time, however, will most certainly provide a verdict on that question.

Chapter 11

MODELS OF A TOTAL CRIMINAL JUSTICE SYSTEM

A. Blumstein & R. Larson

The need to examine the total criminal justice system - police, prosecution, courts, and correction agencies -- in an integrated way constitutes a central problem in improving law enforcement. Also, any such analysis must reflect the feedback into society of offenders released at various stages in the system. This paper formulates a model for the criminal justice system in one state; it depicts the flow of arrested persons through the system as a function of type of crime, and provides a basis for apportioning costs to system components and to types of crime. The model's feedback feature includes the probability of rearrest as a decreasing function of age and a crime-switch matrix reflecting the successive-crime distribution. The results from the model include a cost distribution by crime type, criminal-career costs, an examination of the courses of criminal careers, and estimates of the sensitivities of costs and offender flows within the system to changes in its controllable variables.

The authors are respectively, Professor, School of Urban Administration, Carnegie Mellon University, Pittsburgh, Pennsylvania, and Department of Urban Studies and Planning, Massachusetts Institute of Technology, Cambridge, Massachusetts.

The Criminal Justice System (CJS), comprising the agencies of police, prosecution, courts, and corrections, has remained remarkably unchanged through the significant social, technological, and managerial changes of recent decades. This stability results partly from the insularity of these institutions, and their relative freedom from external examination and influence; but it also results from the independence of the individual components of the system, each of which operates within a set of prescribed rules to approach its own suboptimized objective. Nowhere is there a single manager of a CJS with control over all the constituent parts[1].

In the past few years, there has been an increasing trend toward examining the interactions among the parts of the CJS. The report of the President's Commission on Law Enforcement and Administration of Justice [Ref. 1] urged much closer relations among the parts of the system. The Omnibus Safe Streets and Crime Control Act of 1968 [Ref. 2] provides Federal funds to State planning agencies to develop "a comprehensive statewide plan for the improvement of law enforcement throughout the State." [Ref. 3] Federal subsidy grants are to be provided on the basis of these plans. Thus, there is developing an especially strong need for models permitting one to study a total CJS. Such models are needed only partly for reasons of resource allocation; perhaps even more importantly, they can provide tools for examining the effects on crime of actions taken by the CJS, for most crimes are committed by people who have previously been arrested. Thus, an examination of the feedback process is central to an improvement in the system's performance. In the present state of extensive ignorance on the cause-and-effect relations, the model of this paper will at least identify the data needs and the research questions that will permit analyses of the crime consequences of the actions taken.

---

1 The closest to which the existence of a single manager is approached is in the Federal CJS, in which the police (Federal Bureau of Investigation), prosecutors (US Attorneys), and corrections (Federal Bureau of Prisons) all report to the Attorney General. The courts, however, are completely independent. We do not suggest that a single manager would be desirable: there are strong checks-and-balances reasons for retaining the institutional independence.

11-1 DESCRIPTION OF CRIMINAL JUSTICE SYSTEM

The CJS comprises the public agencies concerned with apprehending and dealing with the persons, both adult and juvenile, who violate the criminal law. The basic structure of the CJS is depicted in Fig. 1; this outline is, of course, a highly simplified version of a very complicated procedure (for a more detailed description [Ref. 4] or for a more condensed version [Ref. 5]).

Society, comprising former offenders (recidivists) and those not previously so identified, gives rise to criminal acts. Of all crimes which are detected (and many like shoplifting go largely undetected) and reported to the police (and many go unreported[2]), only a fraction lead to the arrest of a suspect.

An arrested person may simply be admonished at the police station and returned home, or he may be referred to some social-service agency outside the CJS. An arrested adult is usually brought before a magistrate, who may dismiss the case or formally accuse the suspect of the original or a lesser charge and set his bail.

The district attorney, who is responsible for prosecution of an accused adult, may dismiss the complaint against the defendant at any time prior to the trial. Those defendants who are not dismissed may plead guilty or stand trial either by a jury or a judge. Those who are not acquitted can receive a sentence by a judge that can be of various forms, but usually one of the following:

1. A monetary fine.
2. Probation, usually with a suspended sentence.
3. Probation, following a fairly short jail term.
4. Assignment to a state youth authority.
5. A jail term (usually of less than one year).
6. A prison term (usually of no less than one year at a state institution).
7. Civil commitment for some specified treatment.

2

A Crime Commission survey in three Washington, D. C., precincts found a victimization rate 3 to 10 times (depending on type of crime) that reported to the police. [Ref. 6]

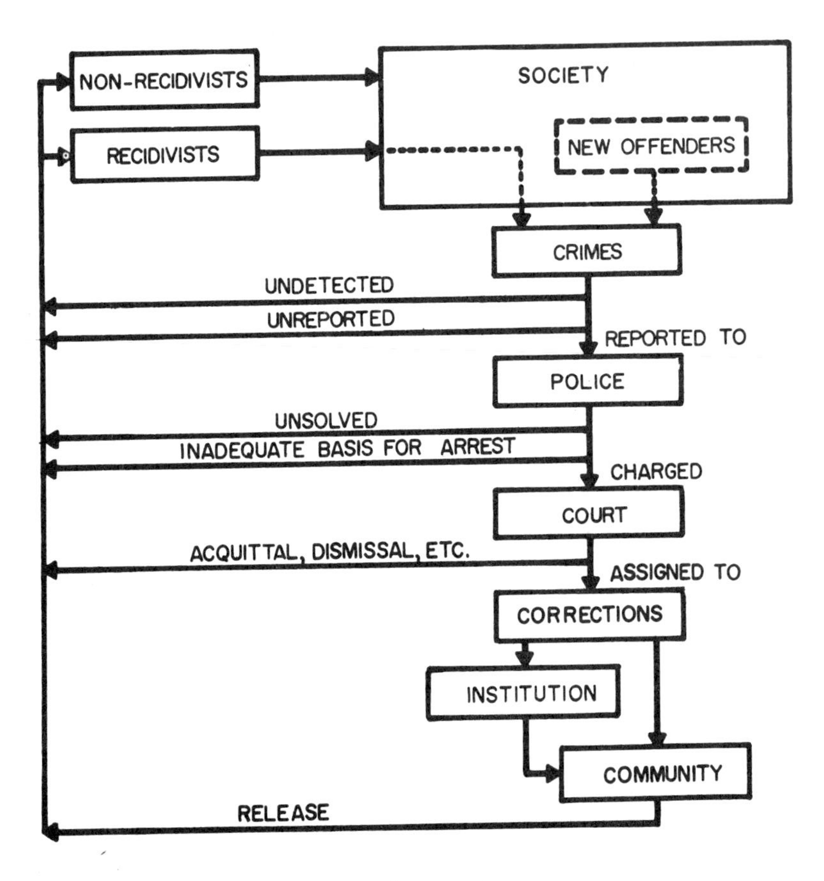

Figure 1
The Criminal Justice System

In addition to newly sentenced offenders from court, prisons can also receive probation and parole violators. Release from prison is usually under parole supervision. Parole violators, if returned to prison, may subsequently be released either on another period of parole, or unconditionally if their sentence has been served.

The processing of juveniles is similar to that of adults, but it is far less formal, with far more freedom of choice exercised by the juvenile authorities.

This processing by the CJS typically involves a series of stages, with the alternatives at most of these stages being either return to the community or further penetration into the CJS. Since virtually all offenders return to society eventually, they are afforded repeated opportunities to return to the CJS by a subsequent arrest, followed by a recycling through the system.

This cursory description suggests two approaches to modeling the CJS. First, there is the simply production process, in which the principal concern is the flow through the system, and the accumulation of costs flowing from a single arrest. Such a linear model provides an opportunity (1) to examine at each stage the workload, the personnel requirements that result, and the associated costs, (2) to attribute these to types of crimes, and (3) to project all of these planning variables as functions of future arrest rates (Roy [Ref. 9] has discussed such models in a preliminary way.

The second is a feedback model, which considers the recidivism probability associated with each released defendant, and his subsequent processing for future arrests after he has once been released by the CJS. Such a feedback model building on the work of the Space-General Corporation [Ref. 7] permits estimating the costs of a total criminal career (considering the succession of rearrests of an individual) and the consequences of alternative actions within the CJS to lower recidivism probabilities.

Some preliminary results with these two models on aggregated US data have been reported previously [Ref. 8]. This paper provides some of the details on the form of these models, and presents some results for California, the single state that comes closest to having an adequate data base[18]. Hopefully, as

---

3

A complete description of the models, the input data, and the results is available in the appendixes to the version of this paper published by the Institute for Defense Analyses.

the use of such models increases, more complete data will begin to become available and the results will increase in reliability and usefulness.

11-2 THE LINEAR MODEL

A steady-state, linear model is used to compute the cost and workloads at the various processing stages and to establish manpower requirements to meet the anticipated workloads. The workload is the annual demand for service at the various processing stages (e.g., courtroom hours, detective man-hours); the manpower requirement is derived from the workload by dividing by the annual working time per man (or other resource); total operating costs are allocated to offenders by standard cost-accounting procedures (these allocated costs are then assumed to be variable costs).

The flow of persons through each processing stage is described by a vector whose ith component represents the yearly flow associated with characteristic type i ($i=1, \ldots, I$). These characteristics can be any attribute associated with individual offenders, their crimes, or their previous processing by the CJS. In most of our studies, there have been seven characteristics (i.e., $I=7$), corresponding to the seven index crimes (the seven types of crimes which the FBI annually tabulates [Ref. 10] to get an "index" of crime in the United States are willful homicide, forcible rape, aggravated assault, robbery, burglary, larceny of $50 or over, and auto theft).

The independent flow vector to the model, which must be specified as input, is the number of crimes reported to police during one year (hereafter, unless stated otherwise, all computed variables and data are considered as seven-component vectors; the flow variables represent annual flow rates). The outputs are the computed flows, costs, and manpower requirements that would result if the input and the system were in steady state.

Each processing stage is characterized by vector cost rates (per unit flow) and branching probabilities (or branching ratios). The input flow at each processing stage is partitioned into the appropriate output flows by element-by-element vector multiplication of the input flow and the branching probability (e.g., $F_{i,n} = F_{i,m}P_{i,mn}$), where

$F_{i,m}$=number of offenders associated with crime-type i entering processing stage m during one year,

$F_{i,n}$=number of offenders associated with crime-type i following route n out of processing stage m, and

$P_{i,mn}$=probability that an offender associated with crime-type i input at stage m will exit through route n ($\Sigma_n P_{i,mn}=1$).

A simple processing stage, representing the verdict of a jury trial, is depicted in Fig. 2. The input N_{t_1} is the number of defendants who receive a jury trial. The outputs N_{tg_1} and $N_{t\bar{g}_1}$ are the numbers found guilty and not found guilty, respectively. The branching probability P_{tg_1} is the probability that a jury trial defendant will be found guilty. With seven crime types, the seven components of P_{tg_1} are required as input data for this stage[4].

Describing the entire model in detail is not warranted here. To illustrate the details, however, we briefly discuss the prosecution and courts submodel. The flow diagram is given in Fig. 3. The input to this part of the model is the vector, N_{ad_1}, the number of adult arrestees who are formally charged with index crimes. This submodel produces seven output vectors corresponding to the seven sentence types. These provide the inputs to the subsequent processing stages. In addition, there are four intermediate output vectors characterizing defendants who never reach the sentencing stage, namely:

1. N_f=number of adults formally charged who do not reach trial stage.
2. N_{td}=number of defendants whose cases are dismissed or placed off calendar at the trial stage.
3. $N_{t\bar{g}_1}$=number of jury-trial defendants not found guilty.
4. $N_{t\bar{g}_2}$=number of bench-and transcript-trial defendants not found guilty.

Clearly, any other intermediate flows can also be calculated, if desired.

4

A more general model would define each branching probability as a function of an offender's prior path through the system and other information which had become known since arrest. The branching probabilities describing the sentencing decision, for instance, would depend on whether the defendant had pleaded guilty, had a jury trial, or a bench or transcript trial. In effect, the possible number of characteristics that could be associated with a flow variable could grow exponentially with the depth of system penetration, the demands for data, of course, grow comparably.

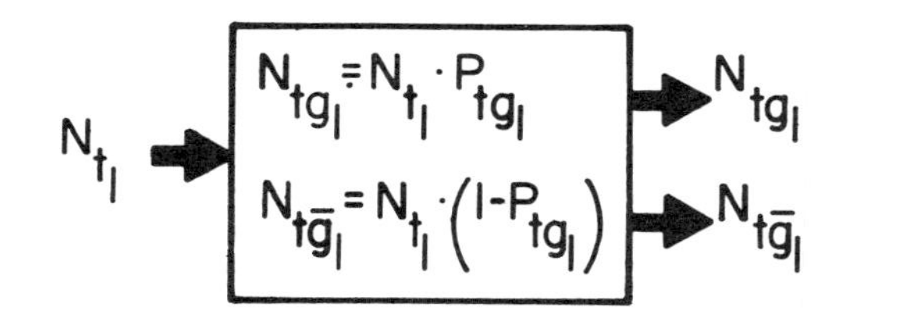

DEFINITIONS:

N_{t_1} = NUMBER OF DEFENDANTS WHO RECEIVE JURY TRIALS

N_{tg_1} = NUMBER OF JURY TRIAL DEFENDANTS FOUND GUILTY

$N_{t\bar{g}_1}$ = NUMBER OF JURY TRIAL DEFENDANTS NOT FOUND GUILTY

P_{tg_1} = PROBABILITY THAT A JURY TRIAL DEFENDANT IS FOUND GUILTY

Figure 2
The Jury-trial Stage

This submodel calls for four classes of branching probabilities. These refer to:

1. Whether or not the defendant reaches the trial stage.
2. The type of trial (or whether dismissed at trial stage).
3. The trial verdict.
4. The sentencing decision.

The definitions of all the flow-and-branching probability variables of Fig. 3 are given in Table I.

Having determined the flow through each processing stage, we can determine the total costs simply as the product of the unit costs and the flow rates. Costs are separated into pre-trial costs, and for each, court and prosecutor's costs[5]. In addition, there is a cost of pre-trial detention.

The flows through the appropriate processing stages permit calculating annual workloads in terms of total trial-days for jury-and-bench (i.e., judge) trials and man-days for pre-trial detention in jail. The annual manpower requirements (e.g., the required number of prosecutors, judges, and jurors) are then calculated on the basis of unit productivity (e.g., annual trial days available per prosecutor).

Some illustrative results were developed based on data principally from California [Ref. 11]. In some cases, where California data were unavailable, data from other jurisdictions were invoked. The input data are presented in Table II.

It is interesting to note, for instance, that P_{t_1}, the probability that a defendant will receive a jury trial, increases with the severity of the offense, but never exceeds 0.25 [the numerical estimate of P_{t_1} is formed by computing the ratio (number of jury-trial defendants/total number of defendants) for a given year]. Regardless of crime type, a majority of those who reach trial pleaded guilty. Probabilities of being found guilty in a trial are roughly three-quarters.

Table II also shows time and cost data. The average jury-trial length, T_1, ranges between 4.6 and 1.6 days (a trial day is typically five hours long), depending on the type of crime. The average cost per day of jury trial was computed by first allocating the total court costs to 'judgeships,' and then dividing the judgeship annual cost by the annual number of judge working days spent in trial. (There are additional court costs to the prose-

---

5 Much of the court-cost data was estimated from other jurisdictions, particularly Washington, D.C., and the Federal Court System.

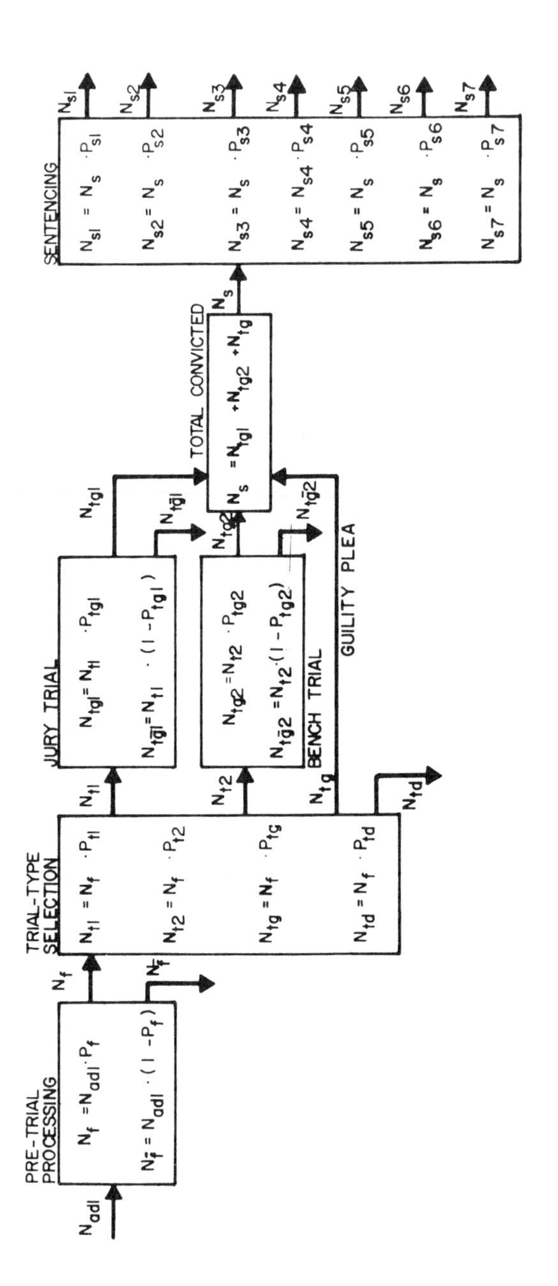

Figure 3
The Prosecution and Courts Flow Model

TABLE I

DEFINITIONS OF FLOWS AND BRANCHING PROBABILITIES IN THE PROSECUTION AND COURT SUBMODEL

(Output flows and corresponding branching probabilities are given as matched pairs. Only the definition of the flow is stated.)

N_{ad_1}	The number of adult arrests who are formally charged by the magistrate.
(N_f, P_f)	The number of adults formally charged who receive a Superior Court felony disposition.
$(N_{\bar{f}}, I-P_f)$	The number of adults formally charged who do not receive a Superior Court felony disposition.
(N_{t_1}, P_{t_1})	Number of defendants who reach trial stage and who receive jury trials.
(N_{t_2}, P_{t_2})	Number of defendants who reach trial stage and who receive bench or transcript trials.
(N_{tg}, P_{tg})	Number of defendants who reach trial stage and who plead guilty.
(N_{td}, P_{td})	Number of defendants who reach trial stage and who are dismissed or placed off calendar.
(N_{tg_1}, P_{tg_1})	Number of defendants who receive jury trials who are found guilty.
$(N_{t\bar{g}_1}, I-P_{tg_1})$	Number of defendants who receive jury trials who are not found guilty.
(N_{tg_2}, P_{tg_2})	Number of defendants who receive bench or transcript trials who are found guilty.
$(N_{t\bar{g}_2}, I-P_{tg_2})$	Number of defendants who receive bench or transcript trials who are not found guilty.
N_s	The number of defendants who are sentenced.
(N_{sj}, P_{sj})	The number of sentenced defendants who receive sentence type j(j=1,2,...,7).

TABLE II

CALIFORNIA INPUT DATA TO THE PROSECUTION AND COURTS MODEL

	Homicide	Robbery	Assault	Burglary	Larceny	Auto Theft	Rape
P_{t1}	0.25	0.18	0.12	0.07	0.06	0.03	0.11
P_{t2}	0.18	0.13	0.25	0.17	0.20	0.16	0.21
P_{tg}	0.50	0.61	0.52	0.67	0.66	0.75	0.58
P_{tg1}	0.81	0.81	0.75	0.78	0.68	0.83	0.54
P_{tg2}	0.68	0.71	0.77	0.71	0.89	0.75	0.61
T_1	4.6	3.1	2.3	2.2	3.0	1.6	2.9
C_j	3680	2480	1840	1760	2400	1280	2320
T_2	1.1	1.1	1.0	1.0	1.1	1.0	0.8
C_b	620	620	560	560	620	560	450

T_1, T_2 = Average number of jury (T_1) and bench trial (T_2) days per case.

C_j, C_b = Average jury (C_j) and bench trial (C_b) cost per trial, including prosecution and court costs.

cutor and to police investigators, attributed before and during trial.) This obviously simplified cost-allocation procedure clearly needs much more refinement when the necessary cost data become available.

11-2.1 Limitations of the Model

However complex this model may appear, it is still a gross simplification of reality. Each processing stage represents a number of detailed processing stages in the real system; the dessciption could have been made more detailed, but the finer data were not available, and little but complexity would have been gained.

The unit costs at each processing stage have been calculated simply by dividing current total yearly cost by current yearly work-load. This implied linear relation between flow and cost (i.e., all costs are variable) ignores the fact that many costs are fixed and independent of flow (e.g., the cost of courthouses). However, this simplification also avoids the problem of having to identify which costs are fixed and which are variable, since many costs that are fixed over a slight variation in flow become variable if there is a large variation in flow. By this costing procedure, certain facilities that may currently be operating well below capacity (e.g., rural courts) would show an excessively high unit cost.

The variables in the model are assumed to be constant over time (a steady-state assumption) and independent of each other or of exogenous variables. There undoubtedly are interactions that limit the validity of this simplification. Certain service times (e.g., detention times) and branching ratios (e.g., probability of prison sentence) are probably functions of the magnitudes of demands. Such interactions need further examination.

Despite these limitations, the model does permit a reasonable first estimate of costs, workloads, and flows, and allocation of these to crime type and processing stage. Furthermore, these planning variables can be projected into the future if the crime or arrest rate can be projected, and if the branching probabilities are either constant or can be projected.

11-3 SENSITIVITY ANALYSIS

An important phase of the analysis is to determine the effects of changes in one subsystem on the workload, costs, and manpower requirements of another subsystem; for instance, if there were indications that an improved fingerprint-detection system would increase the burglary arrest rate (i.e., arrests per burglary), it would be necessary to plan for the increased cost-and-workload effect on the subsequent court and corrections subsystems. In addition, the allocation of costs to various subfunctions is of interest in considering possible reallocation of resources. A sensitivity analysis permits an examination of this distribution.

Given any two system flows, C_i and $N_i (i=1, 2, \ldots, I)$ we find it useful to define the following two quantities:

$\partial C_i / \partial N_i$=incremental change in C_i per unit change in N_i (first partial derivative of C_i with respect to N_i);

$(\partial C_i / \partial N_i)/(C_i/N_i)$=incremental fractional change in C_i per unit fractional change in N_i ('elasticity' of C_i with respect to N_i); a 'unit fractional change' could be, for instance, a 1 per cent change.

To indicate the interpretation of these two quantities, suppose C_i represents the cost at stage 12 associated with processing individuals charged with crime i. Consider that N_i represents the flow of persons into stage 6. In terms of N_i, suppose C_i is linearly related to N_i, i.e., it can be written as follows: $C_i=A_i + B_iN_i$. Then,

$\partial C_i / \partial N_i = B_i$=average additional cost incurred for processing at stage 12 per additional individual charged with crime i inserted at stage 6.[6]

$(\partial C_i / \partial N_i)/(C_i/N_i) = B_iN_i/(A_i + B_iN_i)$=average fractional increase in cost incurred at stage 12 for processing individuals charged with crime i per unit fractional increase in individuals charged with crime i inserted at stage 6.

---

6 This cost could be calculated directly as the product of the unit cost of processing at stage 12 and the probability that an individual inserted at stage 6 will reach stage 12; this latter probability is not explicitly calculated.

More succinctly, the first partial derivative in this case is an incremental cost per person and the elasticity is the fractional increase in cost per unit fractional increase in the number of persons.

As an example, we may be interested in the incremental change in total system direct operating cost (C_t) caused by the addition of one robbery defendant in the flow N_{ad_1}, the number of adults who are charged with a felony in magistrate's court. For this case, the incremental cost per additional robbery defendant (i.e., $\partial C_t / \partial N_{ad_1}$ for robbery) is calculated to be \$4800. This means that an average robbery defendant charged by a magistrate's court costs the system \$4800 (for the current offense) in addition to costs already incurred in previous stages. The value of \$4800 is the expected value of the total subsequent costs (i.e., the sum of each of the unit costs at the magistrate's court and later stages weighted by the probability that the defendant passes through each particular processing stage).

If C_i is a flow, then $\partial C_i / \partial N_i$ is an incremental flow per additional person inserted. For instance, if we let C_i be the number of jury trials for robbery defendants (the robbery component of N_{t_1}) and N_i be the robbery component of N_{ad_1} (the number of adults charged with a felony in magistrate's court), then the incremental number of robbery jury trials per additional robbery defendant from magistrate's court is calculated to be 0.10. This figure can also be interpreted as the probability that a randomly selected robbery defendant from magistrate's court will proceed to and have a jury trial.

Now let us consider an example involving elasticity. Suppose that C_i is the number of burglary defendants placed on straight probation, the burglary component of N_{s_1}, and that N_i is the number of defendants found guilty of burglary in jury trials, the burglary component of $N_{t_{g_1}}$. We calculate that $(\partial C_i / \partial N_i) / (C_i / N_i) = 0.07$. This means that a one per cent increase in the number of burglary defendants found guilty in jury trials would cause a 0.07 percent increase in the number of burglary defendants placed on straight probation.

Other illustrative calculations made for the 1965 California CJS system are shown in Tables III and IV. Table III shows various incremental costs per additional reported crime. Of the

TABLE III

INCREMENTAL COSTS PER REPORTED CRIME

	Robbery	Assault	Burglary	Auto theft	Rape
C_t	$1083	$437	$169	$170	$904
C_{co}	760	197	87	58	534
C_{ct}	59	34	9	8	108
C_p	82	52	37	25	124
C_{pd}	71	44	25	16	108

C_t = total system cost. C_{co} = cost of the correction system. C_{ct} = cost of the prosecution and courts system. C_p = cost of police. C_{pd} = cost of police detectives.

TABLE IV

INCREMENTAL FLOWS PER ARREST

(including juvenile arrests)

	Robbery	Assault	Burglary	Auto theft	Rape
N_1	0.41	0.09	0.09	0.04	0.16
N_p	0.10	0.02	0.03	0.02	0.03
N_{t2}	0.02	0.04	0.02	0.02	0.06
N_{tg}	0.12	0.07	0.09	0.08	0.16
N_f	0.19	0.14	0.14	0.11	0.28

N_1 = number of adult-years served in prison. N_p = number of adults sentenced to prison directly from Superior Court. N_{t2} = number of adults having bench trials. N_{tg} = number of adults who plead guilty. N_f = number of adults who receive a Superior Court felony disposition.

crimes presented[7], robbery costs are highest ($1083), primarily because of the high increment in correction costs; the incremental costs for auto theft are lowest. These calculated costs combine many factors, including the probability of apprehending a suspect, the dismissal probabilities along the way, and the costing procedure[8].

Table IV presents incremental flows resulting from one additional arrest. The first-row entry (additional number of adult-years in prison) is the average man-years served in prison per additional arrest. This can also be interpreted to be the incremental prison population per additional arrest. All other entries have a probabilistic interpretation; for instance, entries in the second row indicate that 10 per cent of those arrested for robbery are sentenced to prison from Superior Court and only 2 per cent of those arrested for assault.

7

No entries are given in Table III for homicide or larceny because of the lack of uniformity of definition of these two crimes in the various processing stages. For instance, police report the incidence of "grand theft, except auto" whereas most (but not all) other processing stages report the number of defendants associated with "theft except auto," a larger category which includes petty theft with prior and receiving-stolen-property offenses (see Ref. 11, 1965, pp. 207-209). Even for the five crime types considered here there are minor deviations of definitions in various parts of the system.

8

The procedure for calculating police costs was a product of time components and time pay rates. For detectives, the time components were preliminary investigation, arrest, and case development. Cost assignment for the police patrol force is somewhat more troublesome. The force spends a large fraction of its time on "preventive patrol," and it is difficult to apportion this time to individual crimes. In the current model, a lower bound on patrol costs was used. The time allocated to crimes was taken as twice the average time to service a call.

11-4 ESTIMATION OF FUTURE REQUIREMENTS

Administrators of the CJS at all levels, from state attorneys general, crime commissions, and budget directors to planners in the various local agencies, require projections of future workloads, costs, and manpower requirements. These projections are needed for earlier decisions that must be made in anticipation of future changes in workload. For instance, new buildings (e.g., courts or correctional institutions) can be designed and constructed or additional personnel can be hired and trained.

In this section we report two applications of the model, using data from the State of California. First, we investigate the degree to which the branching probabilities are constant. Following that, we project for California workloads, costs, and manpower requirements into the year 1970, on the basis of data collected through 1965. Since the number of reported crimes is a basic input to the model, we must independently project the number of crimes that will be reported; a linear extrapolation is used for that projection. Then we develop estimates of the number of arrests per year, and use the model to obtain projections of CJS workloads, costs, and manpower requirements.

11-4.1 Trend in the number of arrests per reported crime

A comparison of system branching ratios over a five-year period indicated that system workload is most sensitive to changes in the average number of arrests per reported crime.

The branching probabilities P_{AC} (the number of arrests per reported crime[9] for California in the years 1961-65 are shown

---

[9] Numerical values for P_{AC} are computed simply by dividing the total number of arrests (adults and juveniles) by the total number of crimes reported. Strictly speaking, it is an estimate of the average number of arrests per reported crime. We often refer to it as the 'arrest probability,' knowing that some crimes generate more than one arrest and that the suspect arrested may not be the perpetrator of that particular reported crime.

in Fig. 4 for aggravated assault, robbery, auto theft, grand theft (in California, larceny of $200 or more), and burglary. (The crimes of homicide and rape are not included because the definition of these crimes changes from the crime report to the arrest stages.) Each rate exhibits a negative slope, with robbery showing the greatest rate of decrease. Indeed, arrests for robbery have shown a marked decline of about 32 per cent from 0.83 per reported crime in 1961 to 0.57 per reported crime in 1965. The burglary arrest probability has decreased by approximately 20 per cent[10]. The general downward trends could be caused by a combination of several factors.

1. More frequent reporting of crimes to or by police.
2. More accurate police classification of reported crimes.
3. Fewer arrests of individuals not associated with the crimes.
4. Saturation of limited police manpower resources.
5. Greater difficulty in solving crimes, caused by such problems as mobility of criminals, lowered citizen cooperation, etc.

Many other possible reasons could be advanced. Without having to attribute cause, however, it is possible to project P_{AC} somewhat into the future. This parameter describes the system's first processing stage of arrest, and its value linearly affects workloads and costs in all other system stages.

11-4.2 Trend in final disposition percentages

To test the constancy of the branching ratios further, a linear extrapolation was performed to estimate trends in the other branching ratios for California. Specifically, for each of the years 1960-65, the ratios of final disposition of adult felony arrests to total arrests were investigated. The final dispositions were:

---

10 More recent data that have since become available indicate a continuation in these trends. For the year 1966, the number of arrests per reported robbery dropped to 0.52, per burglary to 0.21, and per assault to 0.59. Auto-theft and grand-theft probabilities remained about constant.

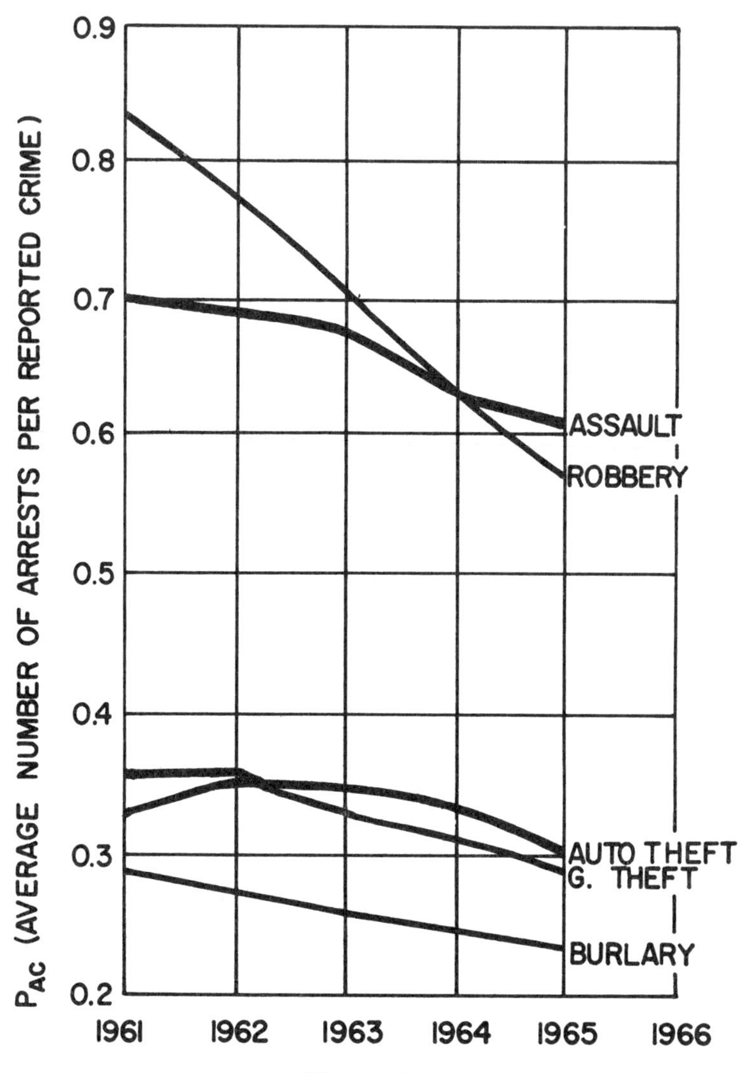

Figure 4
Values of P_{AC} as a Function of Time

1. Released.
2. Assigned to other jurisdiction.
3. Dismissed.
4. Acquitted.
5. Misdemeanor prosecution.
6. Superior court conviction.
    a. Civil commitment.
    b. Prison.
    c. Youth Authority.
    d. Probation.
    e. Jail and fine.

The most significant trend ($t=5.3$) was found in the fraction receiving probation. (Significance was tested with a Student's t-test of the difference from zero of the linear time term; the value $t=5.3$ causes us to reject, even at the $=0.001$ level of significance, the hypothesis that there is no linear time trend in the fraction receiving probation.) During 1960-65, a fraction of approximately 0.13 of felony arrests received probation at the sentencing stage, and this value is increasing 0.00631 per year. No other trends were significant (at the 0.05 level), and none was as important as the trend in P_{AC}.

Although not all of the individual branching ratios were examined in detail, the steady-state assumption appeared justified for all important branching ratios except P_{AC} and those relating to the probation decision.

In making projections with the model, it was especially important to consider the downward trend in P_{AC}, since changes in this fraction propagate throughout the entire system. It was felt that for short-range projections, it would not be necessary to adjust the probation or other branching ratios.

For short-range projections, it was decided to compute output in two ways:

1. Linearly extrapolate the trends in P_{AC} and use the resulting projection of P_{AC}.
2. Use the 1965 value of P_{AC}.

These two projections can be expected to bound the actual future values. In our calculations we use the average of the two projections.

11-4.3 Crime projection

The future numbers of crimes reported to police were projected using a linear time extrapolation of the reported crimes for the years 1958-1966[11]; the results of this analysis are shown in Table V. All the correlation coefficients (except for the crime of forcible rape[12]) exceed 0.95, indicating that the linear fit is a good one. Particularly important to criminal justice system administrators in this table are the yearly growth coefficients in the next-to-last column. Note that the number of reported burglaries is increasing by the largest magnitude at 16,534 per year (with 95 per cent confidence, the yearly growth coefficient is between 13,000 and 20,000 burglaries per year).

11-4.4 Arrest projection

Using the projections of reported crimes from the linear extrapolation, we obtained the approximate upper and lower estimates (keeping P_{AC} constant and projecting its trend, respectively) for the number of arrests in 1970 given in Table VI. The results are expressed as percentages of the numbers of arrests in 1965. The upper estimate indicates about a 30 per cent increase in system workload during this five-year interval while the lower estimate indicates that the increasing trend in reported crimes is almost compensated by the decreasing trend in arrest

---

11 UCR figures for California were used. The definitions of some of the seven crimes are different from the "seven major offenses" of California. Most notably, larceny of $50 and over is counted by the FBI as an index offense whereas "grand larceny" in California requires theft of property valued at $200 and over.

12 In contrast to a simple linear relation, the number of reported rapes was found to remain approximately constant (about 3000 per year) until 1964, when it jumped to 3621, and then to 4432 in 1966.

TABLE V

LINEAR PROJECTION OF INDEX CRIMES
REPORTED TO POLICE IN CALIFORNIA

Offense	Mean no. of reported crimes 1958-1966 (N=9)	Standard deviation of reported crimes 1958-1966	Standard error of linear estimate	Correlation coefficient of estimating equation	Constant term in equation	Linear Coefficient in equation (yearly increment)	T-value of linear term
Criminal homicide	677	128	42	0.956	440	47	9
Forcible rape	3309	528	302	0.863	2427	177	5
Robbery	16501	3402	1144	0.955	10209	1259	9
Aggravated assault	21724	4402	1046	0.978	13388	1667	12
Burglary	168022	43408	8910	0.983	85351	16534	14
Grand larceny	97146	27795	7948	0.968	45060	10417	10
Auto theft	62059	14957	4565	0.963	34163	5579	9

Data Source: Uniform Crime Reports for 1958 through 1966.

probability, and so system workloads will remain about constant (with some fluctuations by crime type, of course). If the declining trend in robbery arrest probability were to continue, the robbery-arrest workload in 1970 would be about half that of 1965. On the other hand, it appears that the arrest probability for auto theft has almost kept pace with the increasing number of reported auto thefts; auto theft exhibits the largest lower estimate in Table VI.

To project a numerical value for arrests in 1970, we arbitrarily average the upper and lower bounds in Table VI; these results are shown in Table VIa.

11-4.5 Projections of system variables

Using these arrest projections we can compute, using the steady-state model, projected values of system variables in 1970; several of these calculations are shown in Table VII. We see that a projected total of 119 additional detectives and 73.9 additional patrolmen will be required to handle increases in the seven major crimes. A projected total of 1393 additional defendants will be placed on probation in 1970. The additional yearly cost to California's criminal justice agencies for increases in the seven major crimes is computed to be $17.3 million. About 41.6 per cent of this additional cost is due to additional burglary workloads, about 22.5 per cent to additional auto-theft workloads. In the 1965 calculations, burglary costs accounted for 31 per cent of the total and auto theft costs 10 per cent. Grouping auto theft, burglary, and larceny as the 'property crimes,' we see that they accounted for 54 per cent of the cost in 1965, but are projected to account for 57 per cent in 1970.

11-4.6 Extensions and further analyses with the Linear model

These projections can be expected to deviate from the future observations. The differences will result from inadequacies of the current model, errors and incompleteness in the reported data, and basic changes in the operation of the California CJS. As actual results are compared with past projections, calibration of the model and the data sources will result, leading to an improved projection methodology.

TABLE VI

PROJECTED NUMBER OF ARRESTS BY CRIME TYPE IN 1970

(Expressed as a percentage of the number of arrests in 1965)

	Homicide	Forcible rape	Robbery	Aggravated assault	Burglary	Grand larceny	Auto Theft
Upper estimate	129.6	124.2	129.8	132.0	138.0	140.4	134.2
Lower estimate	--	--	55.7	109.0	100.0	93.5	121.0

TABLE VIa

PROJECTED NUMBER OF ARRESTS IN 1970

(Obtained by averaging the upper and lower estimates)

Crime type	Projected number of arrests in 1970, expressed as a percentage of the number of 1965 arrests
Homicide	130
Forcible rape	124
Robbery	93
Aggravated assault	120
Burglary	119
Grand larceny	117
Auto theft	128

TABLE VII

PROJECTED INCREASES IN VALUES OF CJS VARIABLES
IN CALIFORNIA FROM 1965 TO 1970

	Homi-cide	Rape	Robbery	Assault	Burglary	Theft	Auto theft	Total
N_{ad1}	900*	1700	4200	4600	13700	4400	4400	33900
	+270	+400	-300	+920	+2600	+750	+1200	+5840
M_d	24	22	85	65	310	115	75	696
	+7	+5	6	+13	+60	+19	+21	+119
U_p	3600	9100	35000	31000	415,000	45000	115,000	654,000
	+1000	+2200	-2600	+6200	+78000	+7400	+32000	+124,000
N_{t1}	180	100	420	260	500	195	77	1730
	+54	+25	-30	+53	+94	+33	+22	+251
M_p	2.1	5.3	21	18	240	26	67	379
	+0.6	+1.3	-1.5	+4	+46	+4.5	+19	+73.9
N_{t2}	130	190	290	570	1200	700	410	3490
	+40	+50	-20	+110	+220	+120	+115	+635
N_s	590	700	1950	1800	5900	3000	2300	16200
	+180	+170	-140	+360	+1200	+520	+640	+2930
N_{s1}	90	280	95	600	1200	1000	500	3770
	+30	+70	-7	+120	+230	+170	+140	+753
N_{s2}	140	150	290	420	1400	660	440	3500
	+40	+40	-20	+75	+270	+110	+125	+640
N_4	320	110	1200	340	1450	420	400	4240
	+100	+30	-85	+70	+280	+70	+110	+575
C_t	8.1	3.3	23	11	38	15	14	112.4
($million)	+2.4	+0.8	-1.7	+2.2	+7.2	+2.5	+3.9	+17.3

TABLE VII (CONTINUED)

*For each pair of entries, the projected increase is given below the 1965 value.

N_{ad1} = number of adult felony arrests which result in a felony charge.

M_d = total number of detectives required.

U_p = total number of patrolman manhours allocated (to these crimes).

M_p = total number of patrolmen required (for these crimes).

N_{t1} = number of jury trial defendants.

N_{t2} = number of bench or transcript trial defendants.

N_s = number of convicted defendants.

N_{s1} = number of convicted defendants granted straight probation.

N_{s2} = number of convicted defendants granted probation with jail as a condition.

N_{s4} = number of convicted defendants sentenced to state prison.

C_t = total system direct operating costs.

As the model is improved, other useful analyses can be performed. The effects on CJS operations of significant changes in system branching ratios can be explored. For instance, introduction of new police hardware (e.g., an electronic automobile-license-plate scanner or automated fingerprint files) might dramatically change one or more branching ratios (e.g., the probability of arrest for auto theft or burglary) and thus affect the workloads at subsequent stages. More widespread provision of free defense counsel, especially for juveniles as a result of recent court decisions, might provide additional strain on prosecution and court workloads. Greater use of nonadjudicative treatment (e.g., use of social service agencies as an alternative to prosecution) will require the introduction of additional flow routes in the model and can be expected to reduce court workloads. A change in sentencing policies (e.g., more use of community treatment or longer sentences) might affect decisions on construction of new correctional facilities or hiring and training of additional parole and probation personnel.

Crime projections can be improved by taking into account changes in such demographic characteristics as age, income, education, and urbanization. Similarly, since many of the branching ratios also depend on these characteristics, they can be used for more accurate estimation throughout the system.

In our model, the branching ratios were assumed to be mutually independent. In a number of cases, interaction can be expected. For instance, if the number of convictions increases, and if prisons operate near capacity, one might expect a reduction in probability of prison sentence or the time served. Such interaction must be explored to improve the model.

11-5 FEEDBACK MODEL

This section summarizes a feedback model that describes the recycling through the CJS during the course of an individual's criminal career. The model has several important applications. First, given the age of an offender at first arrest and the crime for which he is arrested, the model computes his expected criminal career profile (i.e., the expected crimes for which he will be arrested at each age). Second, using the cost results of the linear model, the model computes the average costs incurred by the CJS over a criminal career. Third, recidivism parameters (e.g., rearrest probabilities) can be varied to assess how each parameter affects criminal careers and cost. For instance, we can study the effect of an intensive rehabilitative program that reduces arrest probability by a specified amount. Fourth, and most fundamental, the model provides a unified framework in which to study the process of recidivism and in which to test the effects of proposed alternative CJS policies on recidivism.

11-5.1 Overall structure of the model

As in the linear model, flows are distinguished by crime type. In addition, each flow variable is broken down by the offender's age. The input to the model, rather than crimes reported to police, is the numbers of arrests during a year, by crime type and by age, of individuals who have never previously been arrested for one of the crimes being considered. In the model, these 'virgin' arrests are added to recidivist arrests (i. e., arrests of individuals who have previously been arrested) to obtain the total arrests during the year[13]. The total arrests

---

13 Although reported crimes are a more adequate variable upon which to compute police workloads and the over-all magnitude of the crime problem, arrest is the first event linking crime to a specific individual. Statistics describing recidivism often use arrest as the index of recidivism, even though the arrest may not necessarily indicate that one or more crimes have been committed by the individual arrested. In this model, recidivism is consistently measured by rearrest. Using arrest as the basis for measuring recidivism introduces two types of error: crimes for which no offender is arrested are not counted, and offenders who are erroneously arrested are counted. Using a later stage for counting (e. g., conviction) would introduce the additional, more serious error of omitting the many crimes for which evidence is insufficient to warrant conviction. In much of the criminological literature, where the concern is principally on the correction process (e. g., Glaser [Ref. 12]), recidivism is often defined in terms of the imprisonment-to-imprisonment cycle. It should be clear that, for the same amount of crime repetition, the measured probability of recidivism decreases as one measures it at stages of successively deeper penetration into the CJS. Thus, FBI estimates [Ref. 13] of rearrest recidivism of about three-quarters are consistent with Glaser's [Ref. 12] estimate of reimprisonment recidivism of about one-third due to the arrests that do not result in imprisonment. A simple Markov model, using a reasonable value of 0. 75 for arrest-to-imprisonment attrition probability, shows this compatibility.

then proceed through the CJS just as they do in the linear model.

Since the offender flows comprise individuals who cycle back into the system after dismissal or release from the CJS, it is necessary to compute the number that do recycle, when they are rearrested, and for what crime. At each possible dismissal point, the offender is characterized by a probability of rearrest that is, in general, a function of his age and his prior criminal record. The expected number who will be rearrested at some later time is computed by multiplying the number in the flow by the appropriate rearrest probability. Then, the age at rearrest is computed by using the distribution of delay between release and the next arrest. Finally, the crime type of the next arrest is computed from a rearrest crime-switch matrix, where the matrix element p_{ij} is the conditional probability that the next arrest is for crime type j, given that rearrest occurs and the previous arrest was for crime type i ($1 \leq i, j \leq I$). A flow diagram of the model is given in Fig. 5.

There are two different interpretations of the computed flows: as a cohort-tracing model or as a population-simulation model. In the first, a cohort of virgin arrests can be inserted at some age and the aggregate criminal career of that cohort can be traced. For a 15-year-old cohort, for instance, the model will compute the expected number of arrests by crime type incurred at ages 16, 17, etc. Alternatively, in the second case, we can input as virgin arrests the total present distribution of such arrests, by age and by crime type; in this case, invoking a steady-state assumption, the computed flows represent the current distribution of all individuals (including recidivists) processed by the CJS. With this interpretation, the computed number of arrested 20-year olds, for instance, represents arrests of both virgins and recidivists. If the virgin-arrest distribution were known for the US, this use of the model would be a good check on the validity of the model.

11-5.2 Feedback branching ratios

Many details treated explicitly in the linear model are aggregated in the feedback model. Only four branching probabilities are required to determine flows through the trial stage:

1. P_C=probability that an arrested adult is formally charged with a felony.

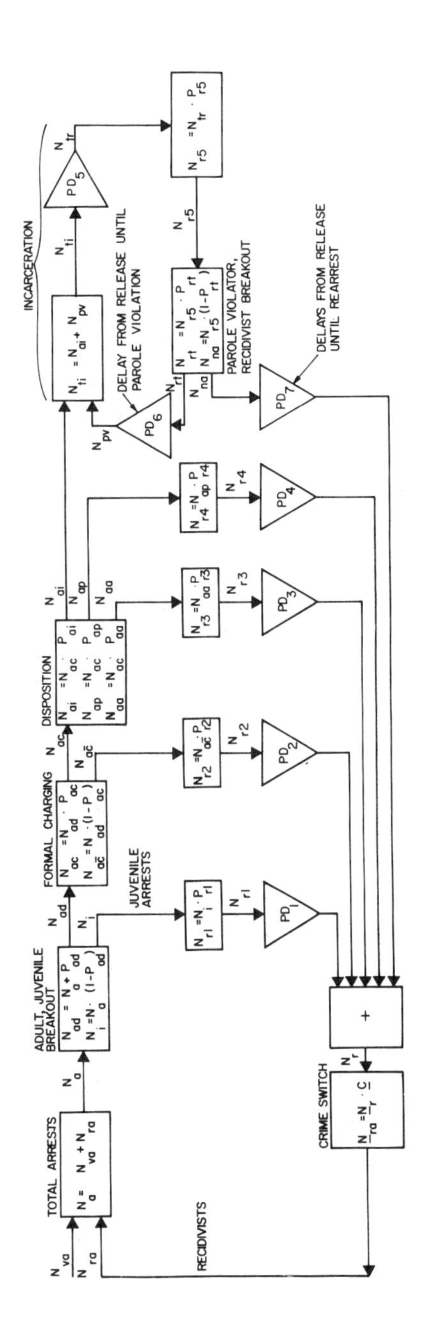

Figure 5
Flow Diagram of Feedback Model

Figure 5
Flow diagram of the feedback model.

VARIABLE OR PARAMETER NAME	DEFINITION
N_{va}	Number of virgin arrests
N_{ra}	Number of recidivist arrests
N_a	Total number of arrests
N_{ad}, P_{ad}	Number (proportion) of arrests which are adult arrests
N_j	Number of arrests which are juvenile arrests
N_{ac}, P_{ac}	Number (proportion) of adult arrests formally charged
$N_{a\bar{c}}$	Number of adult arrests not formally charged
N_{ai}, P_{ai}	Number (proportion) of charged adults incarcerated
N_{ap}, P_{ap}	Number (proportion) of charged adults granted probation
N_{aa}, P_{aa}	Number (proportion) of charged adults released or acquitted
N_{ti}	Total number of adults who are incarcerated
N_{tr}	Number of adults released from incarceration
N_{r1}, P_{r1}	Number (proportion) of arrested juveniles who are rearrested
N_{r2}, P_{r2}	Number (proportion) of adults arrested but not formally charged who are rearrested
N_{r3}, P_{r3}	Number (proportion) of adults released or acquitted who are rearrested
N_{r4}, P_{r4}	Number (proportion) of adults granted probation who are rearrested

Figure 5 (Cont'd)

VARIABLE OR PARAMETER NAME	DEFINITION
N_{r5}, R_{r5}	Number (proportion) of adults released from incarceration who recidivate*
N_{rt}, P_{rt}	Number (proportion) of adults released who violate parole and are reincarcerated
N_{pv}	Number of adult parole violators who reenter prison
N_{na}	Number of adult releases who are rearrested
N_r	Total number of those who will be rearrested
$\underline{C}$	Rearrest crime-switch matrix
PD_1	Distribution of time until rearrest of juvenile recidivists
PD_2	Distribution of time until rearrest of adults not formally charged and who are rearrested
PD_3	Distribution of time until rearrest of adults acquitted or released and who are rearrested
PD_4	Distribution of time until rearrest of adults granted probation and who are rearrested
PD_5	Distribution of time from entrance until release from prison
PD_6	Distribution of time from prison release until parole violation, for those adults who violate parole
PD_7	Distribution of time until rearrest of adults released from prison and who are rearrested

* Adults released from incarceration who recidivate either violate parole or are rearrested.

2. P_i=probability that an adult who is charged will be incarcerated in a state correctional institution.
3. P_p=probability that an adult who is charged will be placed on probation or in a local jail.
4. P_a=probability that an adult who is charged is dismissed before or during trial or is acquitted.

The values of these probabilities that were used in the current model are given in Table VIII, based on California statistics. [Ref. 11]

One of the facts noted from these data is the assault charges, most of which result from attacks on relatives or acquaintances, frequently result in dismissal and only rarely in incarceration. A similar situation exists for rape charges. Larceny charges, probably many of which are against first offenders, most often lead to probation.

11-5.3 Rearrest probabilities

Rearrest probabilities are specified at each point of dismissal and are functions of age and crime of last arrest[14]. The variation with age of the offender is typically a gradual decrease after about 30 years of age. To approximate this decrease, we allowed the rearrest probability to be the following function of age:

$P_R(a)$=probability that an offender dismissed at age a would be rearrested for an index crime

$=P\min \{1, [1/(T-C)]\max(T-a, 0)\}$.

This function is plotted in Fig. 6. The three parameters of this function have intuitive definitions:

P=probability of rearrest of individuals released who are less that C years of age at time or release.

C=age at which the rearrest probability starts declining linearly to zero.

T=age beyond which rearrest does not occur.

Table IX shows values of these parameters for two types of dispositions:

---

14 Rearrest probability data (e.g., the data on criminal careers in UCR, 1966) exhibit a marked variation by type of crime of the last arrest and the type of disposition.

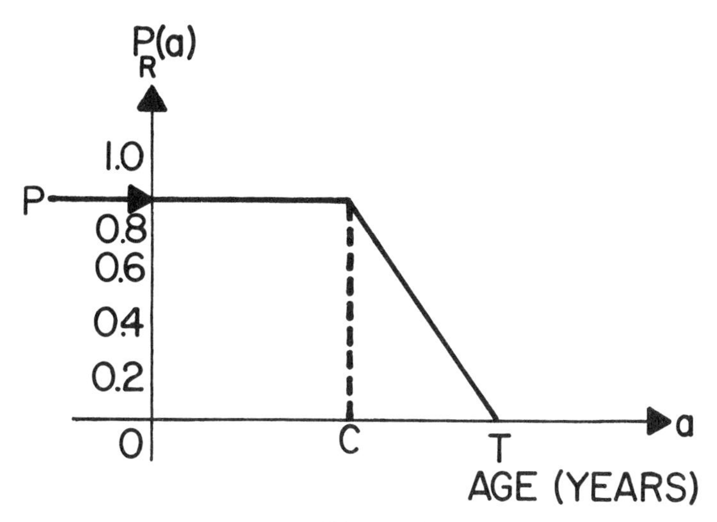

Figure 6
Rearrest Probability as a Function of Age

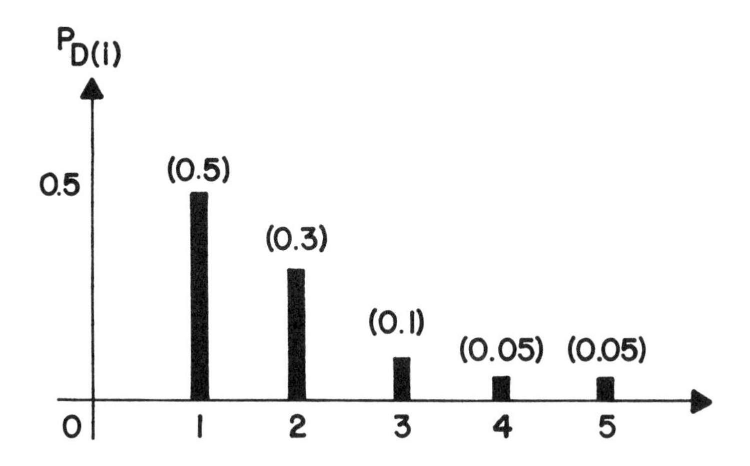

$P_{D(i)}$ = PROBABILITY THAT THE DELAY FROM DISMISSAL UNTIL REARREST IS i YEARS, GIVEN THAT REARREST OCCURS(i IS A POSITIVE INTEGER).

Figure 7
Illustrative Distribution of Delay
From Dismissal Until Rearrest

TABLE VIII

BRANCHING RATIOS FOR THE FEEDBACK MODEL

	Homicide	Robbery	Assault	Burglary	Larceny	Auto Theft	Rape
P_c	0.68	0.41	0.34	0.50	0.53	0.42	0.59
P_i	0.43	0.35	0.09	0.15	0.12	0.17	0.10
P_p	0.29	0.22	0.31	0.27	0.55	0.35	0.30
P_a	0.28	0.43	0.60	0.58	0.33	0.48	0.60

TABLE IX

PARAMETER VALUES FOR THE REARREST PROBABILITY FUNCTION

		Homicide	Robbery	Assault	Burglary	Larceny	Auto Theft	Rape
Disposition I	P	0.65	0.80	0.785	0.833	0.770	0.833	0.65
	C	40	35	40	35	40	60	25
	T	100	80	65	80	75	100	55
Disposition 2	P	0.25	0.573	0.375	0.572	0.539	0.675	0.33
	C	35	30	30	30	35	40	25
	T	100	80	64	75	75	100	55

Disposition I: Adults who are formally charged but not found guilty

Disposition 2: Adults who are found guilty and who are placed on probation or in a local jail.

1. Adults who are formally charged but not found guilty.
2. Adults who are found guilty and who are placed on probation or in a local jail.

These values were estimated from data presented in UCR/1966, pp. 33-42. There is a marked decrease in likelihood of recidivism for those placed on probation, even though they were found guilty[15].

11-5.4 Time between release and rearrest

Data describing times between release and rearrest are sketchy, at best, and the distributions which were used were chosen to have a mean of about two years[16]. An illustrative delay distribution function of this time interval is given in Fig. 7.

11-5.5 Rearrest crime-switch matrix

In the present model, the same crime-switch matrix is used for all recidivists, regardless of age and number of prior arrests. Even with this simplification, 42 independent probability estimates are required to specify the matrix for seven types of crime. Thus, a relatively large sample of recidivists is required for accurate estimation. Those few studies that have reported

---

15

It may be that supervision during the probationary period provided a relatively successful rehabilitative environment. Part of the effect noted, however, must be attributed to the selection of probationers, since those granted probation were judged good risks during the pre-sentence investigation.

16

A mean of two years was chosen to match the UCR/1966 statistics which showed that about 0.5 index arrests per year occurred from the start of an individual's criminal career. Delay-distribution data for time from release on parole until parole suspension for parole violation are published for California [Ref. 15]. These data, because of many unique characteristics about the parole process, are inadequate for the model.

data from which a crime-switch matrix can be developed have either had an inadequate sample size or their sample was biased in some important sense. Table X presents the rearrest crime-switch matrix that was used in most of our studies. This matrix was based primarily on a sample of about 500 recidivists who were studied by the Minnesota Board of Corrections[17]. In this matrix, none of the on-diagonal terms is greater than 0.50, indicating a strong tendency to commit (or at least to be arrested for) different types of crimes.

Table XI presents a rearrest crime-switch matrix based on a sample of several thousand recidivists; it was computed primarily from the Federal Bureau of Prisons statistical tables for the years 1961-65 [Ref. 17][18]. The sample was biased in the sense that a disproportionate number of offenders had been arrested for federal offenses, the definitions of which often differ from those of local jurisdictions (an example is inter-state auto theft, the perpetrator of which is prosecuted under the Federal Dyer Act). In this matrix, the on-diagonal terms for both burglary and auto theft are greater than 0.50, the burglary probability being higher at 0.63. We will compare results computed from the model using each of these matrices to see how the matrix affects the criminal careers depicted.

---

17 The data were obtained from Crime Revisited, [Ref. 16] Minnesota Board of Corrections. The estimates for murder and nonnegligent manslaughter, forcible rape, and aggravated assault were best estimates based on inadequate data. The Federal Bureau of Prisons statistical tables [Ref. 17] for fiscal year 1965 were also used in estimating the matrix where the Minnesota sample was too small.

18 The entries for robbery, burglary, grand larceny, and auto theft were calculated from the Federal Bureau of Prisons statistical tables for the years 1961-65. The entries for forcible rape and aggravated assault were estimated from reference 14. The row for murder and nonnegligent manslaughter was set equal to the row for aggravated assault.

TABLE X

REARREST CRIME-SWITCH MATRIX I[a]

If arrested again for an index crime, the the probability it will be for--

Last index arrest for:	Murder and Non-negligent man-slaughter	Forcible rape	Robbery	Aggravated assault	Burglary	Larceny ($50 and over)	Auto theft
Murder and nonnegligent manslaughter[b]	0.025	0.025	0.150	0.400	0.200	0.100	0.100
Forcible rape[b]	0.020	0.150	0.110	0.260	0.200	0.140	0.120
Robbery	0.015	0.010	0.350	0.060	0.350	0.115	0.100
Aggravated assault[b]	0.025	0.040	0.150	0.300	0.085	0.200	0.200
Burglary	0.010	0.020	0.135	0.063	0.459	0.282	0.031
Larceny ($50 and over)	0.010	0.020	0.140	0.025	0.400	0.275	0.130
Auto theft	0.010	0.027	0.045	0.028	0.390	0.222	0.278

[a] Based on data from Crime Revisited: Minnesota Board of Corrections; 1965 "Uniform Crime Reports," pp. 29-31; and Federal Bureau of Prisons, statistical tables, fiscal year, 1965.

[b] Best estimates based on inadequate data.

TABLE XI

REARREST CRIME-SWITCH MATRIX 2

If arrested again for an index crime, the probability it will be for--

Last index arrest for:	Murder and Non-negligent man-slaughter	Forcible rape	Robbery	Aggravated assault	Burglary	Larceny ($50 and over)	Auto theft
Murder and nonnegligent manslaughter[a]	0.03	0.03	0.12	0.31	0.26	0.14	0.11
Forcible rape[b]	0.03	0.10	0.08	0.30	0.21	0.20	0.08
Robbery[c]	0.03	0.00	0.42	0.06	0.34	0.04	0.11
Aggravated assault[b]	0.03	0.03	0.12	0.31	0.26	0.14	0.11
Burglary[c]	0.02	0.00	0.15	0.04	0.63	0.04	0.12
Larceny ($50 and over)[c]	0.01	0.01	0.12	0.06	0.40	0.15	0.25
Auto theft[c]	0.01	0.00	0.10	0.03	0.29	0.06	0.51

[a] Set equal to the row for aggravated assault.

[b] Forcible rape and aggravated assault based on District of Columbia data, reference 4, appendix, p. 605.

[c] Robbery, burglary, grand larceny and auto theft based on Bureau of Prisons statistical tables for the years 1961, 1962, 1963, 1964, 1965.

11-5.6 Simplifying the assumptions of the current model

Before this feedback model can be used confidently to make decisions regarding rehabilitative programs and over-all allocation of resources, appropriate data must be collected and analyzed. Limitations of existing data have required that we make a number of simplifying assumptions in our model such as the following:

1. Future criminal behavior is determined solely by the age of the offender, the crime for which he was last arrested, and the disposition of his last arrest.
2. The crime-switch matrix depends only on the crime type of the last arrest, not upon age, disposition, or otherwise upon prior criminal career.
3. CJS branching ratios are not a function of age or prior criminal career.
4. Delay until rearrest is a function only of disposition.

Because of these assumptions, the numerical results must still be treated with caution. The model, however, has identified the required data and provides the framework in which to use them once they become available.

11-6 SOME RESULTS FROM THE FEEDBACK MODEL

Recognizing these limitations, we computed some illustrative results by using the feedback model. In the first set of runs, 1000 20-year-olds are first arrested for crime $i(i=1, 2, \ldots, 7)$ and their criminal careers are traced. Table XII presents the mean number of subsequent career arrests for crime type j (the columns) among the population of 1000 people first arrested at age 20 for crime type i (the rows). This matrix was computed using the rearrest crime-switch matrix of Table X.

Those who are initially arrested for auto theft have the greatest average number of career arrests (3.76) and represent the only type of initial arrests that has an off-diagonal term greater than one (i.e., those initially arrested for auto theft will commit an average of 1.084 burglaries). Table XII also presents the total average number of career arrests for the seven crimes, the career costs using results from the linear model.

TABLE XII

CAREER MATRIX FOR 1000 20-YEAR-OLD NEW ARRESTEES
(Using rearrest crime-switch matrix I)

Crime of original arrest	Total number of career arrests							Total career arrests per person	CJS direct operating costs, $
	Homi-cide	Rob-bery	As-sault	Bur-glary	Theft	Auto theft	Rape		
Homicide	1038	330	426	645	412	262	57	3.17	8100
Robbery	28	1486	154	816	427	230	41	3.18	4500
Assault	43	379	1402	687	561	395	78	3.55	3600
Burglary	28	371	176	2021	634	200	56	3.49	3500
Theft	26	336	128	900	1574	261	51	3.28	4000
Auto theft	31	309	157	1084	657	1455	70	3.76	3500
Rape	34	296	326	656	437	269	1144	3.16	3400

TABLE XIII

CAREER MATRIX OF 1000 20-YEAR-OLD NEW ARRESTEES
(Using rearrest crime-switch matrix 2)

Crime of original arrest	Total number of career arrests							Total arrests per person	CJS direct operating costs, $
	Homi-cide	Rob-bery	As-sault	Bur-glary	Theft	Auto theft	Rape		
Homicide	1052	338	353	860	209	404	32	3.25	8100
Robbery	52	1569	145	909	117	384	5	3.18	4400
Assault	61	395	1413	1005	245	472	37	3.63	3500
Burglary	52	435	148	2385	138	475	5	3.64	3400
Theft	39	365	151	1060	1211	576	13	3.42	3900
Auto theft	46	416	146	1162	177	1993	5	3.95	3400
Rape	52	302	355	810	256	377	1081	3.23	3400

For comparative purposes, we show in Table XIII the career arrest matrix for the same cohort, but using the rearrest crime-switch matrix of Table XI. Over-all, the total number of career arrests appears to be only slightly greater; the number of career grand theft and rape arrests appears to be significantly less. As we would expect, the total numbers of arrests (which depend principally on the rearrest probability) are much less sensitive to the crime-switch matrix than are the crime-type distributions.

In another run, 1000 15-year-old virgin arrestees were taken as the cohort. The distribution of initial arrests, by crime type, was made to approximate the actual distribution of total 15-year-old arrests reported in UCR/1965. Because of low age, this distribution is probably based largely on virgin arrests. The output distributions are shown in Table XIV for ages 16 and 20. Also shown in Table XIV is the arrest distribution of all arrests of 20-year-olds as reported in the UCR/1965 (this distribution is made up of virgin arrestees as well as recidivists with various lengths of prior criminal careers). Even though the model-derived distribution is only for those with five-year-old criminal careers, and the UCR distribution includes all arrestees, we would expect a similarity in the two distributions to be a modest validation check. We see that the distributions are roughly similar, with only the fraction that are assaults deviating significantly from the UCR value.

The recidivism model also permits examination of a crucial question confronting CJS administrators: How does reduction of recidivism probability affect a criminal career? Many experimental programs have been run to try to discover how various rehabilitative programs affect recidivism probability. For instance, one study of youthful offenders, which was part of California Community Treatment Project, included randomly separated treatment and control groups. During a 24-month period, the institutionalized control group had a failure probability of 0.61 and the Community Treatment Group had a rate of 0.38, or about a 1/3 reduction in recidivism probability [Ref. 18]. To investigate what a factor of a 1/3 reduction of recidivism probability implies in terms of criminal careers, the model was run with 20-year-olds first arrested for crime type i, with the rearrest crime-switch matrix of Table X and with all of the rearrest probabilities reduced by 1/3. The results are given in Table XV. The total career arrests are reduced by about a factor of 2 by reducing recidivism probability by 1/3.

TABLE XIV

ARREST DISTRIBUTIONS OVER CRIME TYPE
FOR A 15-YEAR-OLD COHORT

	Age 15 input distribution from UCR/1965	Model-derived distributions for Ages		Arrest Distributions for all 20-year-old arrests
		16	20	
Homicide	0.002	0.011	0.01	0.01
Robbery	0.047	0.115	0.15	0.11
Assault	0.045	0.054	0.07	0.14
Burglary	0.335	0.398	0.39	0.35
Grand theft	0.246	0.248	0.24	0.19
Auto theft	0.317	0.149	0.11	0.17
Rape	0.008	0.024	0.02	0.03

TABLE XV

CAREER MATRIX FOR 1000 20-YEAR-OLD NEW ARRESTEES

(Assuming a one-third reduction in rearrest probabilities)

Crime of original arrest	Total number of career arrests							Total career arrests per person	(a)
	Homicide	Robbery	Assault	Burglary	Larceny	Auto theft	Rape		
Homicide	1017	124	223	205	128	96	22	1.81	$6600
Robbery	11	1223	55	301	136	85	13	1.82	2900
Assault	19	142	1202	188	200	168	33	1.95	1800
Burglary	10	136	63	1400	248	57	20	1.93	1800
Larceny	9	123	38	340	1222	102	18	1.85	2400
Auto theft	11	88	46	402	239	1209	27	2.02	1600
Rape	14	103	159	209	145	103	1079	1.81	1900

[a] CJS direct operating costs per arrestee career.

11-7 SUMMARY

This paper has described means of modeling the CJS-both in a detailed way with the linear model and in a more aggregated way using feedback to account for recidivism. Clearly, the focus here was on the CJS itself, and so we did not address the many public and private means outside the CJS by which criminal behavior is controlled, nor did we address the deterrent effects of the CJS. Our goal has been to describe in a quantitative way the operation of the system that tries to apprehend, adjudicate, and rehabilitate offenders, and to assess some of the effects of this system on their future criminal behavior. Within the constraints of the available data, these models allow us to study questions regarding the CJS, its costs, workloads, and resource requirements, and the effects of alternative rehabilitative procedures on criminal careers.

Future studies can include more realistic assumptions within the framework of these models, and more complete and accurate data for performing the calculations. The end goal of such studies should be to improve the management of the system, including appropriate allocation of public resources to minimize the total social and dollar costs of crime and its control. The models also provide a research tool for examining the behavior of the CJS in order to understand its impact on the problem of crime.

REFERENCES

1. The Challenge of Crime in a Free Society, President's Commission on Law Enforcement and Administration of Justice, U.S. Government Printing Office, 1967.
2. Omnibus Safe Streets and Crime Control Act of 1968, Public Law 90-351, enacted on June 19, 1968.
3. Ibid, Section 203(b) (1).
4. Donald M. McIntyre, Law Enforcement in the Metropolis, American Bar Foundation, Chicago, Illinois, 1967.
5. Geoffrey C. Hazard, "The Sequence of Criminal Prosecution," Proc. National Symposium on Science and Criminal Justice, June 22-23, 1966, Government Printing Office, Washington, D.C.

6. Task Force Report: Crime and Its Assessment, President's Commission on Law Enforcement and Administration of Justice, U.S. Government Printing Office, June 1967.
7. "Prevention and Control of Crime and Delinquency in California," Space-General Corporation, El Monte, California, July 29, 1965.
8. A. Blumstein, R. Christensen, S. Johnson, and R. Larson, "Analysis of Crime and the Overall Criminal Justice System," Chap. 5 of Task Force Report: Science and Technology, President's Commission on Crime and Administration of Justice, U.S. Government Printing Office, Washington, D.C., June 1967.
9. Robert H. Roy, "An Outline for Research in Penology," Opns. Res. 12, 1-12 (1964).
10. Crime in the United States: Uniform Crime Reports, published annually by the Federal Bureau of Investigation, U.S. Department of Justice, Washington, D.C. (U.S. Government Printing Office). Hereafter referred to as UCR or UCR/j, where j is the year of publication.
11. State of California, Department of Justice, Division of Criminal Law and Enforcement, Crime and Delinquency in California, Bureau of Criminal Statistics, published annually.
12. Daniel Glaser, The Effectiveness of a Prison and Parole System, Bobbs-Merrill Co., New York, 1964.
13. UCR/1966.
14. Report of the President's Commission on Crime in the District of Columbia, Government Printing Office, Washington, D.C., Appendix, p. 605, 1966.
15. Department of Correction, Research Division, Administrative Statistics Section, California Prisoners 1961, 1962, 1963, (summary of statistics of felon prisoners and parolees), Sacramento, California, p. 128, 1963.
16. M.G. Mundel, et al., Crime Revisited, Minnesota Department of Corrections, 1963.
17. Federal Bureau of Prisons, statistical tables, published each fiscal year.
18. M. Warren, et al., "Community Treatment Project, An Evaluation of Community Treatment for Delinquents, Fifth Progress Report," State of California, Youth and Adult Corrections Agency, Department of the Youth Authority, Division of Parole and Division of Research, p. 59, August 1968.

Chapter 12

SYSTEMS ANALYSIS OF CRIME CONTROL AND THE CRIMINAL JUSTICE SYSTEM

A. Blumstein

12-1 INTRODUCTION

Systems analysis, which has made important contributions in dealing with technological systems, also has an important role to play in dealing with a variety of social system problems that occur within the city. The problem of crime, and especially of the operation of the criminal justice system is one of the most fundamental of these problems, and one which in many respects typifies the operation of a wide class of social systems. This paper first presents a brief description of the criminal justice system, a discussion of some attempts to model that system, an enumeration of some of the difficulties inherent in developing and using such models, and then an indication of some approaches to dealing with those difficulties. This leads to a description of JUSSIM, an interactive computer model for criminal justice planning and system design. Finally, we indicate how that model can be applied to other social systems.

The author is Professor, School of Urban Administration, Carnegie Mellon University, Pittsburgh, Pennsylvania.

12-2 THE CRIMINAL JUSTICE SYSTEM

The criminal justice system, (CJS), comprising the agencies of police, courts, and corrections, is charged by society with trying to deal with the problems of crime through a process of arrest, adjudication, and "correction". A general summary of this system is shown in Figure 1. Here, a society gives rise to crimes, some of which get reported to the police, the normal entry point to the CJS. When the police solve the crime (and that happens most frequently for murders, least frequently for property crimes like larceny and burglary) and arrest a suspect he may then be brought to court. If the court convicts him (and only a small portion of court cases involve a trial) the court may sentence him to corrections. In the corrections subsystem, a small portion of those convicted are sent to an institution for imprisonment. Most find their sentences suspended, pay a fine, or are placed on probation in the custody of a typically overworked (caseloads of 100-150 are common) probation officer.

Thus, we see a system characterized by a downstream flow from reported crimes to corrections, with the flow through any stage being a subset of the flow at a previous stage. And we also note that in virtually all cases (except those in which an individual dies while in the process), the individuals leave the control of the criminal justice system and return to society, perhaps to commit further crimes (as "recidivists").

The traditional operation of the CJS has called for independent and autonomous operation of the individual subsystems, subject to independent judicial review. This process has been developed principally to provide checks and balances within the system, and to prevent any single "system manager" from being able alone to decide the fate of an individual who flows through the process.

An important contribution of the President's Crime Commission [Ref. 1] in 1967 was the recognition that even though autonomy in handling individual _cases_ was necessary, the operating _policies_ of the separate parts of the system interacted intimately, and so must be examined and dealt with in an integrated way. The arrests by police represent the principal input into the courts, the convictions by the courts provide the input to corrections, and the failures of corrections provide the bulk of the subsequent input to the police.

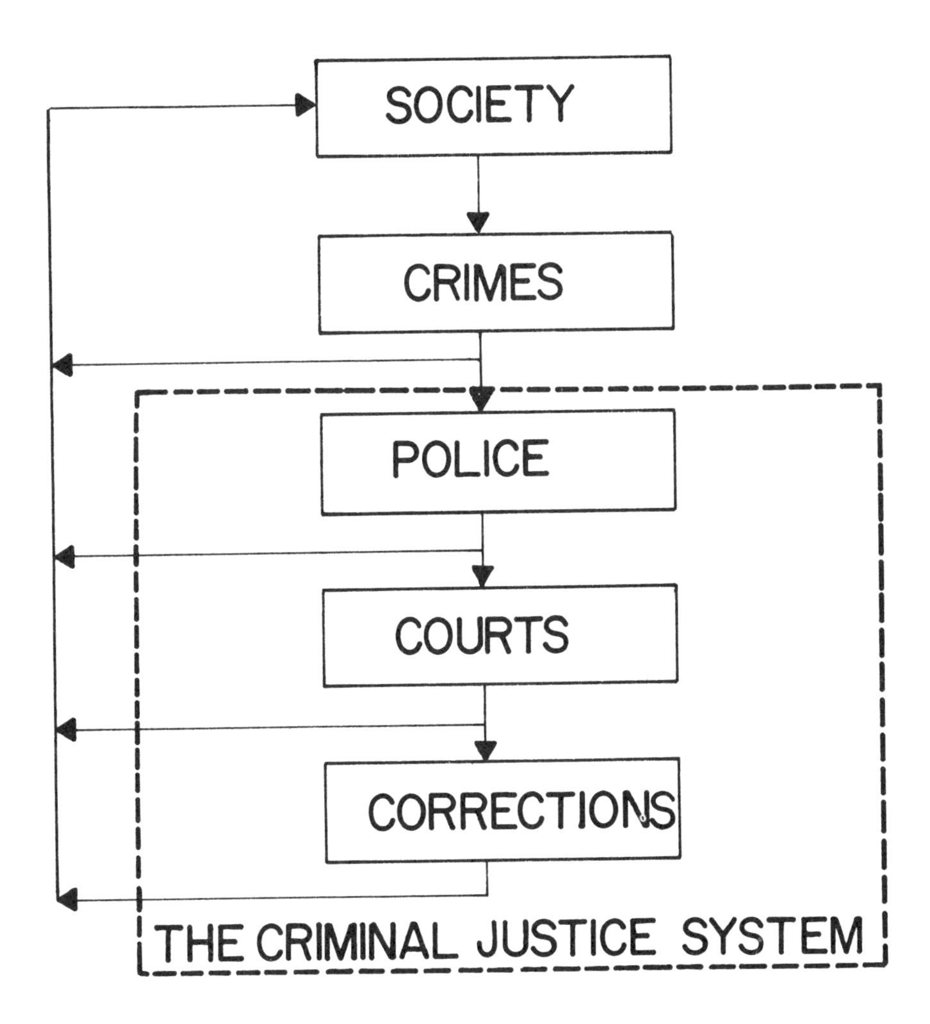

Figure 1
The Criminal Justice System

There are important interactions and tradeoffs among the parts of the system that should be considered in any examination of the total system. For example, in many jurisdictions, arrested offenders without bail money spend extended periods in detention cells awaiting trial. In many of these cases, it may be very wise to put additional resources into the courts to speed up the court processing of these cases, thereby reducing the time spent in pre-trial detention, and thereby saving more money that it would cost to speed the processing through the courts. This, of course, would be in addition to the other social values of speedier trial - avoiding undue imprisonment of the innocent, excessive detention of the guilty, and increasing deterrence of potential criminals.

In view of the recognition of the need to deal with this total system, the Crime Commission's recommendations were effectuated in the Omnibus Crime Control and Safe Streets Act of 1968, which led to the creation in each state of a "State Criminal Justice Planning Agency" (SPA) charged with planning for the total criminal justice system. Furthermore, these planning agencies have the financial power to do more than merely plan, for they are the disbursers of the Federal funds appropriated under the Act. Thus, here was an extremely valuable opportunity to deal with the total CJS effectively and in an analytical way. The effort of these state planning agencies can be characterized much better as "grant administration" than "planning". This lack of planning was due in part at least to the inadequacy of the methodology available to the SPA's for undertaking such planning.

This was a surprise and somewhat of a disappointment, since there is in the literature a paper [Ref. 2] that appeared to have laid out the methodology for undertaking such planning.

The models in that paper can be used to allocate costs by subsystem or by type of crime, to project future resource requirements (i.e., for judges, cells, probation officers, etc.) based on a projection of future crimes or arrests, or to change the system parameters (branching ratios, unit costs, etc.) to see the impact on flows, costs, and resource requirements elsewhere in the system. All these considerations are important parts of the planning process. Nevertheless, they have not yet been extensively applied by the SPA's.

12-3 CHARACTERISTICS OF SOCIAL SYSTEMS

To explore the reasons for the limited use of such models and approaches, it is necessary to examine some of the basic characteristics of decisions, decision making, and decision makers in social systems like the criminal justice system.

12-3.1 Measure of Effectiveness

A fundamental aspect of all such systems is that their measures of effectiveness are complex vectors rather than the simple scalars much more characteristic of industrial systems. This problem of dealing with a multi-dimensional criterion is characteristic of most problems in the public sector, where government tries to perform some social good at some cost. In most cases, the social good (e.g., reduce crimes) is not easily measured in dollar terms (what is the "value" of a rape avoided), even though economists often try to force some such fit. Thus, there is at least one effectiveness component reflecting this social objective of the system. In addition, there are typically several cost components reflecting governmental costs and costs to various sectors of the society, and these various costs are not easily compared, nor are they easily related to the effectiveness measures.

In these circumstances, with at least two components in the objective function, all the mathematics of optimization becomes particularly difficult to apply. If one could specify all but one component as constraints, then he could optimize the remaining one. But that implies a zero-one loss function at the constraint, which is rarely the case. Typically, a decision maker is much more concerned with exploring the rich collection of trade-offs available to him. There is rarely a natural specification of the constraint set that will permit the formulation of a straightforward optimization problem.

Not only is the quantifiable portion of the measure of effectiveness large and complex, but for most social systems the <u>non</u>-quantifiable considerations tend to be very significant in a decision. It is perhaps because of this that many public decisions have been declared to be "irrational". The apparent irrationality of course derives from inability of the observer to comprehend the rich utility function of the decision maker. An official who weights some of the components that are not readily quantified (e.g., protection of privacy, prospects for reelection, defusing a tense situation), may appear to be "irrational" on simple cost or even cost-effectiveness grounds, but making such a judgment requires a much deeper knowledge of his goals than can possibly be subsumed in any model.

12-3.2 Uncertainty of Causal Relationships

Fundamental difficulties in dealing with social systems arise from the widespread uncertainty about cause-and-effect relationships. This derives, in large part, because the systems are based on a large (but not very large) number of human actions and decisions, and such decisions are typically characterized by capriciousness, adaptation to changed environment, and uniqueness to the specifics of any particular situation. Available and reliable social science theory typically covers only a limited set of circumstances in the highly multi-variate environment, but the theoretical formulation rarely provides adequate guidance to the boundaries of the space where it is applicable. Furthermore, it can reasonably be expected that any specific case of concern is not specifically accounted for in any available theories. Thus, a decision maker in a social system who looks to social science theory (as opposed to the often valuable insights of social scientists) rarely finds the theory adequate to answer his question, and so he must make his own intuitive guess of the consequences of actions he is considering.

The problem of estimating effects is further complicated by the fact that the behavior of the participants in social systems changes over time in response to changes in public policy, changes in public values, changes in what is regarded as good professional practice, adaptation to differing capacity constraints, and simply new fads. Thus, even if there were a good theoretical construct that would describe, say how the plea bargaining process changes with the introduction of additional public defenders, it is not clear that this response would be the same in separate jurisdictions or even that the response in one jurisdiction would be consistent over a period of several years. Here again, the decision maker is forced usually to rely on his own judgment and intuition.

12-3.3 Technical Backgrounds

It is rare for managers and planners of public systems to have a strong technical or analytical background. Much more typically, they reach their positions from backgrounds of politics, law, social work, or possibly one of the social sciences pursued no further than a master's degree. Thus, in view of these backgrounds, elaborate mathematical models would be treated at best with a kind of distant and uncertain respect and at worst with a xenophobic hostility. These public officials can reasonably be expected to relate only to a model which they can comprehend and for which they understand and help develop the assumptions. They know very well all the ignorance about the operation of the system, and so are properly skeptical of any model that would presume to function as an oracle, dispensing conclusions and recommendations from a complex and mysterious "black box."

In this setting, then, it would appear that many of the classical models involving optimization or even batch-processed simulation might have only limited applicability. The fact, however, that decisions are necessary in these large social systems provides a need for such models if they can be properly coupled to the decision makers. The models must be comprehended by individuals who know very well the systems they run and operate but who have no extensive technical background. The assumptions involved in the models must be understood by the decision maker, and he must participate in making them. Then, if the model provides him outputs, he can use it easily to gain insight into some of the quantifiable aspects of his decision, and then it could make a valuable contribution to his operations.

12-4 THE JUSSIM MODEL

It is with these considerations in mind that we developed the JUSSIM model of the CJS.[1] This model is intended to permit criminal justice system managers and planners to test the downstream resource and cost implications associated with contemplated changes within their CJS.

JUSSIM is an interactive computer program, which represents the flow of "Units" (crimes, arrestees, cases, offenders, etc.) through the CJS and the application of resources

---

1
I am pleased to acknowledge the many contributions of Jack Belkin and the valuable programming of William Glass in this development.

to process these units as they flow through the system. The CJS is characterized by a flow diagram and the associated branching ratios p_{ij} representing the proportion of the flow from stage i going to stage j. The resources are characterized by the stage or flow paths at which they are applied, an associated annual availability, and a cost per unit of workload. Each processing operation is characterized by a unit processing time. Furthermore, multiple parallel flow paths are provided, each path having its own parameter values, so that type can be treated as a separate parallel channel of flow.

12-4.1 Single Stage Analysis

We can describe the basic notions of the JUSSIM model by considering a single stage, say the Circuit Court. If $N_i (i=1, 2, \ldots, m_c)$ is the number of defendants charged with crime type i appearing in Circuit Court for disposition in one year, and if p_{ik} is the proportion receiving disposition k $(k=1, 3, \ldots, q)$, and if t_{ik} is the judge-days used to dispose of a case of type i by disposition k, then $N_i p_{ik}$ defendants receive disposition k, and this requires $(N_i p_{ik} t_{ik})$ days of judge time. If judges work T_1 days per year, and if the cost of a judgeship is c_1 dollars per day, then

$$W_{1i} = \sum_{k=1}^{q} N_i p_{ik} t_{ik}$$

days of judge time and

$$N_{1i} = \left(\sum_{k=1}^{q} N_i p_{ik} t_{ik} \right) \Big| T_1$$

judges are required for dealing with crime type i, and the cost of these judgeships in Circuit Court is

$$C_{1i} = c_1 \left(\sum_{k=1}^{q} N_i p_{ik} t_{ik} \right) \quad \text{dollars per year.}$$

Then, the total number of judges required for all crime types, N_1, and the judgeship costs, C_1, are found from:

$$N_1 = \sum_{i=1}^{m_c} N_{1i} = \frac{1}{T_1} \sum_{i=1}^{m_c} \sum_{k=1}^{q} N_i p_{ik} t_{ik} = \sum_{i=1}^{m_c} W_{1i}/T_1$$

$$C_1 = \sum_{i=1}^{m_c} C_{1i} = C_1 \sum_{i=1}^{m_c} \sum_{k=1}^{q} N_i p_{ik} t_{ik} = \sum_{i=1}^{m_c} c_1 W_{1i}$$

Similarly, we can calculate the number of prosecutors required, N_2, and the prosecutorial cost, C_2, and the total Circuit cost is $C_1 + C_2$.

12-4.2 Basic Operation

The operation of the JUSSIM model begins with a "base case" reflecting the current operation of the system. All the data on the base case parameters must be collected and stored. The user of JUSSIM - the criminal justice system planner himself - then creates a "test case" by making changes in any of the base case parameters. The program then reports to him the changes in flows, costs, workloads, and resource requirements resulting from the changes he introduced. He thus uses the model as a very flexible design tool by making contemplated changes, rapidly getting an assessment of the effects of those changes, and then trying another change suggested by the feedback from the previous try. If the designer does not like the implications of a proposed change, then he can reject it immediately and try another.

Operating the model to assess the consequences of a system change, a user necessarily has to make assumptions about the detailed micro-consequences of the change as they are reflected in changes in the system's parameters. For those changes that initially appear attractive, he may then want to explore the assumed consequences more carefully. Several system planners can each explore the same system changes, each using his own assumptions. If a threshhold of acceptance or rejection lies outside the range of calculated macro-consequences by this group, then acceptance or rejection is clear and further exploration is not necessary. If it lies within the range, the closer examination is required to assess the validity of the various assumptions made. In other terms, if E (possibly a vector) is a measure of effectiveness, and if E_{aj} and E_{rj} ($E_{aj} \geq E_{rj}$) are acceptance and rejection values of E for a proposed system j (i.e., accept the change if $E_j \geq E_{aj}$ and reject it if $E_j \leq E_{rj}$), and if planner i in evaluating change j makes assumptions leading to a value E_{ij}, then accept if $E_{aj} \leq \text{Min } E_{ij}$, reject if $E_{rj} \geq \text{Max } E_{ij}$, and explore further if either these inequalities is reversed.

One of the virtues of this process is that it forces the planners into a debate on their assumptions and micro-consequences rather than on the generalized goodness of a possible change. Thus, the model serves the same function as any other model, it lowers the level of argument to more fundamental and empirically testable issues.

The operation of the JUSSIM model can best be described by enumerating its inputs, the outputs, and the relationships between them.

12-4.3 Inputs

The basic inputs for JUSSIM are enumerated below:

1) A vector $\underline{V}$ of crime types considered ($\underline{V}=v_1, \ldots, v_m$)
2) A vector $\underline{S}$ of system stages ($\underline{S} = s_1, s_2, \ldots, s_n$)
3) An n x n matrix $\underline{\underline{P}}$ of branching ratios characterizing the proportion of flow from stage $s_i (i=1, 2, \ldots n-1)$ to stage $s_j (j=2, 3, \ldots, n)$ which succeeds it in the CJS flow diagram. Here, $p_{ij} = 0$ if there is no flow from stage s_i to s_j and $p_{ij} = 0$ if $i \leq j$.
$\left[\sum_{j=1}^{n} p_{ij} = 1 \right]$
4) A vector $\underline{R}$ of resources ($\underline{R} = r_1, r_2, \ldots, r_k$), and associated vectors $\underline{A}$ of the annual unit availability of the resources ($\underline{A} = a_1, a_2, \ldots, a_k$), and C of the cost per unit time ($\underline{C} = c_1, c_2, \ldots, c_k$)
5) A k x n matrix $\underline{\underline{T}}$ of the unit workloads, or times for processing a unit of flow at stage s_j by resource r_i represented by t_{ij}
6) A reference flow, (N_0), typically the number of reported crimes or the number of arrests that sets an absolute level of flow throughout the rest of the system when the branching ratios are specified.

The branching ratios, the unit processing times, and the reference flow are all functions of crime type, and so each element in these is a vector with a component for each crime type.

12-4.4 Outputs

The outputs of the JUSSIM model are presented to the user in whatever order and organization he specifies. The potential outputs include the following variables:

1) Flow through each processing state, a vector $\underline{N}$ ($\underline{N} = N_1, N_2, \ldots, N_n$)
2) Costs at each stage, a vector $\underline{D_s}$ ($\underline{D_s} = D_{s1}, D_{s2}, \ldots, D_{sn}$). Costs can be given for any aggregation of stages into specified subsystems, including a complete aggregation into a single total system.
3) Resource costs, a vector $\underline{D_r}$ ($\underline{D_r} = D_{r1}, D_{r2}, \ldots, D_{rk}$) indicating the costs associated with each of the k resources.
4) Resource workloads, a vector $\underline{W}$ ($\underline{W} = W_1, W_2, \ldots, W_k$) of the manhours per year of workload imposed on each of the resources.
5) Resources required, a vector $\underline{Q}$ ($\underline{Q} = Q_1, Q_2, \ldots, Q_k$) indicating the numbers of each of the specified resources that would be required to handle the workload.

All of these output variables are functions of crime type, and so each of the components of the output vector can be presented as a vector with a component for each crime type, or as a scalar summed over the crime types.

12-4.5 Basic Relationships

Assuming knowledge of all the input parameters, one can then calculate the output variables by the following relationships.

Knowing the basic input flow (N_1) as a reference level and the branching ratio matrix ($\underline{\underline{P}}$), one can then iteratively calculate the flow at each stage by:

$$N_j = \sum_{i<j} p_{ij} N_i$$

and create the flow vector:

$$\underline{N} = (N_1, N_2, \ldots, N_n)$$

The ith stage processing costs are:

$$D_{si} = N_i \sum_k c_k t_{ik} \quad (i=1, 2, \ldots, n)$$

or, in vector notation,

$$\underline{D}_s = (\underline{C}\ \underline{\underline{T}}) * N$$

where the vector operation $*$ is the component-by-component product, i.e., if $\underline{C} = \underline{A} * \underline{B}$, then $c_i = a_i b_i$. The subsystem costs are simply the sum over the constituent-stage costs. The resource costs for the use of resource r_k are:

$$D_{rk} = c_k \sum_i N_i t_{ik}$$

or:

$$\underline{D}_r = (\underline{N}\ \underline{\underline{T}}') * C.$$

The workload on resource r_k is:

$$W_k = \sum_i N_i t_{ik} ,$$

or, as a vector:

$$\underline{W} = \underline{N}\ \underline{\underline{T}}'$$

The number of resource r_{kl} required per year is then:

$$Q_k = a_k \sum_i N_i\ t_{ik}$$

or

$$\underline{Q} = (\underline{N}\ \underline{\underline{T}}') \div A$$

where the vector $\div$ operation is the component-by-component ratio, i.e., if $\underline{C} = \underline{A} \div \underline{B}$, then $c_i = a_i / b_i$ for $b_i \neq 0$.

12-4.6 Operation of a Run

In the operation of a JUSSIM run, the user's basic role is to create a "test case" to compare with a "base case" already stored. Initially, the test case is made identical to the base case, so the user enters only changes to the base case. The user, sitting at a terminal, is asked a sequence of questions about what changes he wants to make. Each of these questions is an entry gate to a "phase".

A separate phase is provided for changing each of the following parameters:

1) Branching ratios ($\underline{P}$)
2) Unit Workloads ($\underline{\underline{T}}$)
3) Annual unit resource availability ($\underline{A}$)
4) Resource unit costs ($\underline{C}$)

Once a phase is entered, further detailed questions permit the user to specify precisely which parameters he wants to change and the crime type(s) for which he wants to make the change. All the questions are in clear language, and the answers regarding stage numbers, resource numbers, crime-type codes, and other codes are based on a code used in creating a base-case data file. The program then displays the base-case value for the parameters the user identified and asks him to type in the new, or test-case values.

In dealing with the multiple parallel channels of flow for the m crime groups,[2] most users, at least initially, do not want to sit through the complete detailed output for each crime type. In a separate phase, JUSSIM permits them to choose one of a number of standard complete partitions of the crime types (e.g., into felonies and misdemeanors, Part I and Part II crimes, etc.), or to specify his own crime groupings. Thus, he may want the complete details on one or a few specific crime types, and may wish to aggregate the remainder. The complete crime-type vector and the standard groupings are specified in the data file.

In another phase, the user specifies the output tables he displayed. The output tables present calculated results on flows, costs, workloads, and resource requirements for the base case, the test case, and the absolute and percentage change in going to the test case. These results can be presented by crime group or summed over all crime groups. The choices to be made available are specified in the data file, reflecting various aggregations of stages into subsystems as well as resource partitions or aggregations. Presumably, in the early stages of an exploration, he will want to conserve time and will examine results only for the total system or for a few critical subsystems. At the end of an exploration, the user is more likely to want more detail.

The user is also asked if he wants to specify a reference flow. This is usually the number of reported crimes or arrests, although specification of any flow variable anywhere in the system completely specifies all the other flow variables.

At the end of a run, the user is asked if he wishes to rerun the model. In addition to using this to explore a new issue, he will do this iteratively if he is dissatisfied with the implications of some of the assumptions he has made and would like to

---

[2] The FBI's Uniform Crime Reports organizes its data into 29 separate crime types.

reconsider some of those. In a re-run, he is given considerable flexibility in respecifying his new base case (e.g., calling a data file, using his latest test case), and then begins again to create a new test case.

12-4.7 An Illustrative Run

To illustrate the operation of JUSSIM, we indicate in Table 1 two potential improvements in police operations, more intensive detective investigation of Part I crimes and the use of summons for street arrests of Part II crimes. The user must identify the system parameters affected, and then specify the specific parametric changes. These are shown in Table 1.

Figure 2 presents a reproduction of the computer terminal printout developed in running this case. In the figure, the user inputs to the program are underlined and indicated by a mark in the left-hand margin.

The total computer charges for performing this run at prime-time commercial rates was under $4.00.

12-4.8 Limitation and Extention

The most fundamental limitation of the JUSSIM model is that it can now provide only downstream consequences of an upstream change. As a result, the output measures are all in terms of flows, resources and costs. However important these considerations may be they are unquestionably secondary to an ultimate concern over crimes as an output measure. That model would have to deal with feedback and recidivism. Unfortunately, however, so little is known quantitatively of the impact of almost any CJS actions on crimes. In operating the feedback model, it is necessary to make important distinctions between the recidivist arrests and the virgin arrests, a distinction not easily nor normally made by most users. We are now extending the model in these directions to provide this added capability.

12-4.9 Implementation

The response to the interactive JUSSIM model was, as expected, far more positive than to a batch-process model. In teaching criminal justice planners how to use the model, it was impressive to note how quickly and easily they learned to sit at the terminal and respond to the program's interrogations, to translate their project ideas into judgments about model parameter changes, and to operate the model iteratively as a design tool.

As a result of the introduction to the model in September 1970, two states and two cities have already organized data collection efforts to describe their system as a base case for the JUSSIM model, and plan to use a version of the model for their own planning.

12-5 SUMMARY

This paper has tried to emphasize the virtues of an interactive computer model as a planning and design tool that could be very valuable to decision makers involved with social systems. Their direct involvement in making the necessary assumptions assures both that they become familiar with the model and its limitations, and that the analysis draws on their judgment and experience in the many social-system areas where no scientific knowledge exists. We have described here one such model for the criminal justice system, and have indicated how it has been used.

It should be clear to readers concerned with other social systems that there is nothing in the basic operation of JUSSIM that confines it to the CJS. All of the inputs and outputs are described generally, and so the model can be used for any system describable by a non-feedback flow diagram where concern is over issues of flows, costs, and resources. Hopefully, the feedback limitation will also be removed shortly.

REFERENCES

1. THE CHALLENGE OF CRIME IN A FREE SOCIETY, President's Commission on Law Enforcement and the Administration of Justice, Government Printing Office, Washington, D.C., 1967.
2. Blumstein, A. and Richard C. Larson, "Models of a Total Criminal Justice System", Operations Research, Vol. 17, No. 2, March-April 1969.

TABLE 1

CHANGES IN VALUES OF SYSTEM PARAMETERS FOR TEST CASE

PROGRAM IMPROVEMENT	SYSTEM PARAMETERS CHANGED*	BASE CASE VALUES	TEST CASE VALUES
Additional detectives for more intensive investigation	↑ Detective investigation time per report (Part I)	1.54 hrs.	2.0 hrs.
	↑ Proportion of arrests per report (Part I)	.228	.300
	Lower Court disposition type		
	↓ -nolle	.569	.500
	↑ -guilty plea	.418	.485
	↑ -bench trial	.013	.015
	↑ -bindovers	.212	.250
	Superior Court disposition type		
	↓ -nolle	.178	.150
	↑ -guilty plea	.721	.747
	↑ -bench trial	.039	.041
	-jury trial	.062	.062
Use of summons for street arrests	↓ Patrolman time per arrest (Part II)	1.00 hrs.	0.70 hrs.

* An increase in the parameter is indicated by ↑ and a decrease is indicated by ↓.

Figure 2

AN ILLUSTRATIVE RUN WITH THE JUSSIM MODEL

```
RUN CJS

WELCOME TO VERSION 1 OF THE CJS MODEL

ENTER PHASE 1 - SPECIFICATION OF CRIME GROUPING
ENTER CODE NUMBER OF DESIRED GROUPING
2
                                        TEST CASE 1

DO YOU WISH TO SPECIFY NEW BRANCHING RATIOS...
Y

ENTER PHASE 3 - SPECIFICATION OF BRANCHING RATIOS

ENTER STAGE NUMBER, CRIME GROUP
1,1

STAGE 1 - REPORTED
CRIME GROUP 1
CURRENT VALUES ARE...
ARRESTED       22.8
NO ARREST      77.2
ENTER NEW VALUES
30,70

ENTER STAGE NUMBER, CRIME GROUP
5,1

STAGE 5 - L ARRAIGNED
CRIME GROUP 1
CURRENT VALUES ARE...
JUVENILE CT.     10.7
BIND OVER        21.2
L FOR DISP       68.2
ENTER NEW VALUES
10.7,25,64.4

ENTER STAGE NUMBER, CRIME GROUP
6,1

STAGE 6 - L FOR DISP
CRIME GROUP 1
CURRENT VALUES ARE...
NOLLED           56.9
G PLEA           41.8
BENCH             1.3
JURY              0.0
VIOL. BUREAU      0.0
ENTER NEW VALUES
50,48.5,1.5
```

```
ENTER STAGE NUMBER, CRIME GROUP
17,1

STAGE 17 - S FOR DISP
CRIME GROUP 1
CURRENT VALUES ARE...
NOLLED          17.8
G PLEA          72.1
BENCH            3.9
JURY             6.2
ENTER NEW VALUES
15,74.7,4.1

ENTER STAGE NUMBER, CRIME GROUP
*

DO YOU WISH TO SPECIFY NEW WORKLOADS...
Y

ENTER PHASE 4 - SPECIFICATION OF WORKLOADS

ENTER INDEX NUMBER, CRIME GROUP
2,2

CRIME GROUP 2
CURRENT VALUE FOR PATROL AREST IS 1.00 HRS.
ENTER NEW VALUE
0.7

ENTER INDEX NUMBER, CRIME GROUP
3,1

CRIME GROUP 1
CURRENT VALUE FOR DETECT REPRT IS 1.54 HRS.
ENTER NEW VALUE
2.00

ENTER INDEX NUMBER, CRIME GROUP
*

DO YOU WISH TO SPECIFY NEW TIME RESOURCES...
N

DO YOU WISH TO SPECIFY NEW COSTS...
N

DO YOU WISH TO SPECIFY DESIRED OUTPUT...
Y
```

```
ENTER PHASE 7 - SPECIFICATION OF DESIRED OUTPUT

ENTER SUBSYSTEM NUMBERS OF DESIRED TABLES
1,3,4,15

ENTER WORKLOAD INDEXES OF DESIRED TABLES

DO YOU WANT A BREAKDOWN BY CRIME GROUP...
N

DO YOU WISH TO SPECIFY NEW LEVELS OF REPORTED CRIMES...
N

DO YOU WISH TO REDO ANY PHASES...
N
```

TEST CASE I RESULTS

SUMMARY OF RESULTS FOR POLICE

	BASE	TEST	CHANGE	0/0 CHANGE
COSTS IN THOUSANDS				
PATROLMAN	689.4	575.4	-114.0	-16.5
DETECTIVE	3343.3	4076.4	733.1	21.9
TOTAL	4032.7	4651.8	619.1	15.4
WORKLOADS				
PATROLMAN HRS.	106064.4	88522.3	-17542.1	-16.5
DETECTIVE HRS.	324591.9	395768.3	71176.4	21.9
RESOURCE REQUIREMENTS				
PATROLMAN	62.4	52.1	-10.3	-16.5
DETECTIVE	190.9	232.8	41.9	21.9
TOTAL	253.3	284.9	31.5	12.5
FLOWS				
PATROLMAN	73239.9	77691.8	4451.9	6.1
DETECTIVE	73239.9	77691.8	4451.9	6.1

SUMMARY OF RESULTS FOR LOWER COURT

	BASE	TEST	CHANGE	0/0 CHANGE
COSTS IN THOUSANDS				
PROSECUTOR	753.5	789.8	36.3	4.8
JUDGE	643.5	740.7	97.2	15.1
TOTAL	1397.0	1530.5	133.5	9.6
WORKLOADS				
PROSECUTOR CASE	59805.0	62683.3	2878.3	4.8
JUDGE DAYS	1650.6	1900.0	249.4	15.1
RESOURCE REQUIREMENTS				
JUDGE	7.3	8.4	1.1	15.1
FLOWS				
PROSECUTOR	59805.0	62683.3	2878.3	4.8
JUDGE	59805.0	62683.3	2878.3	4.8

SUMMARY OF RESULTS FOR SUPERIOR CT.

		BASE	TEST	CHANGE	0/0 CHANGE
COSTS IN THOUSANDS					
PROSECUTOR		945.4	1235.3	289.9	30.7
JUDGE		538.2	759.3	221.2	41.1
TOTAL		1483.5	1994.6	511.1	34.4
WORKLOADS					
PROSECUTOR	CASE	3468.0	4531.4	1063.4	30.7
JUDGE	DAYS	748.6	1056.2	307.7	41.1
RESOURCE REQUIREMENTS					
JUDGE		5.2	7.3	2.1	41.1
FLOWS					
PROSECUTOR		3468.0	4531.4	1063.4	30.7
JUDGE		3468.0	4531.4	1063.4	30.7

SUMMARY OF RESULTS FOR TOTAL SYSTEM

	BASE	TEST	CHANGE	0/0 CHANGE
COSTS IN THOUSANDS				
POLICE	4032.7	4651.8	619.1	15.4
COURT	2880.5	3525.1	644.6	22.4
CORRECTIONS	6936.6	8945.3	2008.7	29.0
TOTAL	13849.9	17122.2	3272.4	23.6
RESOURCE REQUIREMENTS				
POLICE	253.3	284.9	31.5	12.5
COURT	12.5	15.7	3.2	25.9
TOTAL	265.8	300.6	34.8	13.1
FLOWS				
POLICE	73239.9	77691.8	4451.9	6.1
COURT	63273.0	67214.7	3941.7	6.2
CORRECTIONS	43714.6	46715.2	3000.6	6.9

DO YOU WISH TO RERUN THE PROGRAM...

Y TEST CASE II

DO YOU WANT YOUR TEST CASE TO BECOME THE NEW BASE CASE...

Y

Chapter 13

URBAN GHETTO REVOLTS AND LOCAL CRIMINAL COURT SYSTEMS

I. Balbus

13-1 INTRODUCTION

This paper summarizes some of the principal findings of a comprehensive study of the response of the American criminal courts to the urban ghetto revolts of the past six years. By "urban ghetto revolts" we mean to include a multitude of disorders since 1964, ranging widely in magnitude from the massive revolt of thousands of blacks -- such as the Detroit rebellion of July 1967 -- to confrontations between ghetto residents and police which in previous years might well have never gained substantial public attention[1]. Similarly, the "American criminal

---

[1] Despite this manifest diversity, all the post-1964 disorders included in this study share certain important elements in common. First, they have all involved a confrontation between ghetto residents and local authorities within the ghetto and almost always have been precipitated by an incident between police and residents. Second, the post-1964 disorders -- in contrast to the "race riots" of previous generations -- involved not indiscriminate inter-racial assaults on persons but rather the destruction of largely white-owned and white-controlled property. Thus we can point to certain behavioral regularities which merit their classification under the same heading.

The author is Professor, Department of Politics, Princeton University, Princeton, New Jersey.

courts" refers to many different court systems through whose interactions justice is allocated for many separate local political systems. This diverse reality of wide variations in the magnitude of revolts and in local political systems inspires two questions which serve to locate the explanatory variables of this study. First, can we establish any relationship between variations in the magnitude of ghetto revolts -- the damage to life and property they have entailed, and the participation they have evoked -- and the nature and severity of court sanctions imposed on the participants? Second, is there any relationship between inter-city political structure variations -- typically expressed, for example, in the distinction between "Machine" and "Reform" -- which have formed the basis for the traditional study of local politics and such variations in sanctioning behavior?

This paper, then, will attempt to explain court responses to ghetto revolts in terms of 1) variations in the magnitude of revolts and 2) variations in local political structures. Los Angeles, Detroit and Chicago have been selected as sample cities in order to permit us to examine systematically the inter-relationship among these variables. To begin with, and as we shall see in detail in the following section, the three cities differ significantly along certain structural dimensions commonly cited as critically important in the study of local politics. Since our conceptualization of the local criminal court system is that of an open system, influenced by, and sharing many of the characteristics of, the local political system of which it may be said to be a subsystem, these inter-city political structure variations might significantly influence the justice allocated by the respective court systems. This focus on inter-city political structure variations as an explanatory variable, then, is intended to forge a much-needed link between the study of the administration of justice and the broader study of urban politics,[2] and places this study squarely within the framework of a growing number of works relating variations in the outputs of local political systems to variations in the structure of local political systems.[3]

2
The few empirical studies of the criminal justice system which do exist tend to treat this system as a closed system and fail either to conceptualize or investigate empirically the linkages between it and its broader political environment.

3
A good sample of this burgeoning literature appears in [Ref. 3].

The three cities have also been chosen, however, because each has experienced at least one major ghetto revolt and at least one minor ghetto revolt over the past six years. This concern with variations in the magnitude of ghetto revolts as an explanatory variable enables us to specify one parameter of the relationship between political structures and outputs, since it leads us to consider the possibility of a differential impact of inter-city political structure variations (on court responses) at different levels of violence. This study is thus intended to complement the analyses of students of American local politics who have focused on inter-city political structure variations as an explanatory variable but have failed to examine the conditions under which the explanatory power of this variable is itself likely to vary. The following section outlines these inter-city political structure variations as a preliminary to just such an examination.

13-2 LOCAL POLITICAL SYSTEMS AND CRIMINAL COURT SUBSYSTEMS

Although the metropolitan criminal court is part of a state court hierarchy, and is therefore not strictly speaking a local institution, the immediate political environment in which it operates -- its constituency -- is urban and local. As McConnel and others have convincingly demonstrated, federalism and the general decentralization and fragmentation of authority within the American system guarantee that any state or federal institution will be powerfully influenced by its local constituency [Ref. 4]. Consequently the structural relationship between the court system and the local political system of which it may be said to be a subsystem is likely to be an important determinant of the character of the justice allocated by the court system.

13-2.1 The Cohesion of Local Authorities

Much of the concern with the structure of local political authority in the literature of American politics has traditionally been expressed in terms of the distinction between "Machine" and "Reform" types of government. Analysts have disagreed widely as to the relative merits of both types, but they tend to agree that the "Machine" centralizes political influence while

"Reform" acts to decentralize and fragment political influence.[4] Consequently in the literature of more recent vintage the dichotomous, qualitative distinction between "Machine" and "Reform" governments has been refined and reformulated into a continuous variable whereby it is possible to conceive of cities as partaking to varying degrees of "machineness" and "reformness". Thus Greenstone and Peterson speak in terms of the relative dispersion of political power within local political systems as indicated, for example, by the success of the dominant local party in monopolizing control of the process of recruitment to public office.[5] As we suggested in the introduction, our three cities have in part been selected precisely because they differ dramatically along this dimension of dispersion of political power, or what we would prefer to call the relative cohesion of authority organization. As we shall see, this variable affects the likelihood of unified action on the part of court authorities, and is therefore likely to be a significant determinant of the nature and severity of court sanctions.

13-2.1.1 Chicago as a case of extreme cohesion

In Chicago the Democratic Party exercises virtually complete control over recruitment to, and the perquisites of, positions of authority within the criminal court system. Access to those positions which are elective (e.g. judicial and prosecutorial positions) is controlled through the process of slating for nomination (which in Chicago is virtually tantamount to election), a process controlled by the Chairman of the Party (Mayor Daley), and access to those positions which are appointive is controlled by the elected superiors through a system of patronage which dispenses jobs to party faithful. Thus for example Associate and Circuit Court judges are elected, but they (in reality the Chief Judge) in turn appoint magistrates who handle the bulk of criminal business (ordinance violations and misdemeanors) and

---

4 For a good summary of this literature see [Ref. 5].

5 [Ref. 5], pp. 279-280.

who serve an indeterminate tenure at the pleasure of the Chief Judge. Similarly, the County Clerk is elected after party nomination, and he in turn controls nearly 2,000 patronage jobs through which the bulk of the court's administrative business and paperwork is funneled. The Public Defender, responsible for the defense of indigents, serves at the pleasure of a committee of judges controlled by the Chief Judge, and his attorneys, in turn, are largely if not entirely recruited through the process of party Committeeman or Aldermanic sponsorship.

In Chicago political party ties thus serve as a powerful solvent through which the centrifugal tendencies of bureaucratic differentiation are held in check. This informal centralization ensures that administrative authority within the court system, although formally in the hands of the Circuit Court judges acting in committees, will in fact rest with the Chief Judge, who is traditionally handpicked by the Chairman of the Party and who exercises budgetary control over the majority of agencies which comprise the court system. The Chicago court system, in other words, is an integral part of the Chicago political machine: the authorities within the court system, although functionally specialized and formally differentiated, owe a common allegiance to, and thus perceive and articulate common interests with, the leadership of the Democratic Party. The Party thus provides a channel of communication to and control over the court system which might, other things being equal, be a powerful determinant of the behavior of court authorities.

13-2.1.2 Los Angeles as a case of low cohesion

In Los Angeles, reformers succeeded in radically decentralizing political authority and at the same time destroyed any possibility that this formal decentralization might be overcome by informal means through a strong local political party. Thus in Los Angeles in contrast to Chicago the court administrative structure is formally decentralized; each division, Municipal and Superior, has its own presiding judge and administrative apparatus, and unlike in Chicago there is no judge or executive officer who presides over the administration of both courts and to whom the individual judges are responsible. Moreover, and crucially, in Los Angeles and in radical contrast to Chicago, there is little or no local party control, and hence little or no centralized control, over access to authority positions within the court system, and thus very little solvent available to dissolve this formal bureaucratic autonomy. Municipal and Superior Court judgeships are typically filled by gubernatorial appointment; control over access to the judiciary is in state rather than local hands, and there is in California little connection between influence at the State level and influence at the local level.[6] Incumbents of other authority positions -- such as the Los Angeles County District Attorney (who prosecutes felonies) and the Los Angeles City Attorney (who prosecutes misdemeanors) are elected by different constituencies, and non-partisanship guarantees that there will be no local political party control over the nominating process which might dissolve the impact of a difference in constituency on their respective perceptions of interest. Finally, the staff of such elected officials is recruited almost entirely through civil service rather than through a system of patronage which dispenses jobs to party faithful. There are, in other words, few if any Los Angeles political actors to whom court authorities owe allegiance and to whose directives they are likely to respond; in contrast to Chicago, where there is a high degree of centralization of control over the disparate bureaucracies which comprise the court system, in Los Angeles these bureaucracies are free from centralized control and thus to an important extent independent from each other. Under these conditions of extremely low court authority cohesion, we can presume that a perception of shared interest which unites the functionally specialized and formally differentiated court authorities and which predisposes them to unified action both among themselves and with Los Angeles political authorities generally would be far less likely to develop than in Chicago.

13-2.1.3 Detroit as a case of Intermediate cohesion

Detroit, like Los Angeles, is a "reform" city where non-partisanship and civil service effectively limit the cohesion of local political authority organization. Unlike Los Angeles, however, this lack of cohesion is not magnified by extreme geo-

---

6

Thus Edward C. Banfield and James Q. Wilson note that "No mayor of either Los Angeles or San Francisco in recent years has been able to climb the taller political ladder." [Ref. 6]

graphic decentralization. Perhaps more importantly, formal decentralization of authority is not as great in Detroit as in Los Angeles. Whereas reform in Los Angeles created a plethora of commissions legally independent from mayoral control, and lodged much of the authority for the governing of the city in the hands of the county government;[7] in Detroit reform did not go that far, thus, for example, the mayor appoints his City Attorney and does not share his executive authority with county officials nearly to the extent of such sharing in Los Angeles.

This difference in the degree of formal decentralization serves to differentiate the organization of court authorities in Detroit from that in Los Angeles. Whereas, for example, the court system in Los Angeles is divided into a Municipal and Superior Court, the latter of which is a court for all felony matters in the entire county of Los Angeles, the Detroit Recorder's Court is a unified court handling both felonies and misdemeanors only for the city of Detroit. Reformers have consistently failed to merge the operations of the Recorder's Court with those of the Wayne County Circuit Court; as a consequence, and in contrast to Los Angeles, where judges are either appointed by the Governor or (rarely) run for elective office on a county-wide basis, Detroit judges run in (what recently have been) closely contested city-wide elections. Thus the process of recruitment to judicial office is far more closely tied to the issues and cleavages of local politics in Detroit than it is in Los Angeles, and the opportunity and necessity to form alliances with other Detroit political authorities is consequently far greater than in Los Angeles. Thus for example a recent Executive Judge of the Recorder's Court was known to be a personal friend of Mayor Cavenaugh, and other judges were viewed as part of the Mayor's "Irish Mafia". Both because formal decentralization is less extreme in Detroit, and because some informal centralization occurs as a result of the need to forge electoral alliances, shared perceptions of interest among court authorities which unite them with political authorities generally are far more likely to develop in Detroit than in Los Angeles. On the other hand, in no sense is the structural relationship between local political system and court subsystem in Detroit characterized by the centralized channels of communication and control which exist in Chicago in the form of a highly cohesive political party.

---

[7] [Ref. 6], p. 110.

13-2.2 The cohesion of local defense community

Our three cities also differ dramatically along another critical dimension which helps define the structural relationship between court subsystem and local political system and which therefore might well be a significant determinant of the behavior of that subsystem. Since court sanctions are not simply a function of the will and organization of political authorities, but rather are at least in part a product of a conflict between political authorities or the "State" on the one hand and the "Defense" on the other, it follows that variables which affect the likelihood of unified action on the part of defense attorneys will also significantly influence the nature and severity of court sanctions. Since defendants in the metropolitan criminal courts are largely black and overwhelmingly poor, we are interested above all in the extent to which the local political system provides channels for the defense and articulation of the legal interests of blacks in particular and low-income citizens in general.

13-2.2.1 Detroit as a case of extreme cohesion

The presence of a strong industrial union in Detroit, and the alliance which that union forged with elements of a large black community in the 1940's and 1950's has profoundly shaped (as it has influenced politics in Detroit generally) the nature of the Detroit defense community. More specifically, it left the legal community of Detroit more prepared than elsewhere to articulate and defend the interests of blacks and poor within the criminal court system.

The United Auto Workers presence in Detroit has produced an abundance of lawyers in Detroit with labor union backgrounds. This fact alone has significantly influenced the ideological tone of the Detroit legal community, leaving it decidedly more progressive than legal communities dominated by corporation lawyers. Those holding positions within the "Establishment" Bar of Detroit -- the Detroit Bar Association -- are typically Liberals with a strong interest in and commitment to Civil Rights; thus, for example, a recent former President of the Detroit Bar Association was co-Chairman of the Michigan Civil Rights Commission.

The consequence of the union orientation of so many Detroit lawyers has not only been a relatively progressive "Establishment" Bar, but also considerable overlap between positions of authority within the Bar Association and positions of responsibility within "non-Establishment" defense groups such as the ACLU and Neighborhood Legal Services. Thus for example the OEO-funded NLS program is largely staffed with lawyers who belong to the radically-oriented National Lawyer's Guild (some of whom have ACLU connections), but has the support of the Detroit Bar Association, whose representatives comprise a large segment of the Board of Governors. This overlap of positions within the defense community, moreover, cuts across racial lines: there is a great deal of inter-racial cooperation within the Detroit legal community. Thus the Black Bar Association -- the Wolverine Bar -- is presided over by young, increasingly militant black lawyers who have close ties with the ACLU and the Guild. In short, the Union-Black alliance in Detroit, and the near majority size of the black population, which in general have produced a rather stronger link between class and race than elsewhere in the United States [Ref. 7], have helped produce a cohesive defense community which is interracial in its composition and inter-racial in its concerns, a defense community with both the propensity and capacity to defend and articulate the interests of low-income defendants within the Detroit court system.

13-2.2.2 Los Angeles as a case of low cohesion

In Los Angeles, a weak labor movement and the dominance of business interests, a relatively small and unorganized black population, and extreme geographic dispersion all have combined to produce a defense community with a relatively low propensity and capacity to articulate the interests of blacks and poor within the Los Angeles court system. To begin with, the Los Angeles County Bar Association, dominated by corporation lawyers, has traditionally been a voice of conservative business interests in Los Angeles; unlike their colleagues in Detroit, Los Angeles "Establishment" lawyers "have not been known to come forth on such matters" as Civil Rights.[8] Moreover, most of the

---

8 Confidential interview with a prominent Los Angeles attorney.

members of the LACBA also belong to separate organizations such as the Santa Monica or Beverly Hills Bar Associations to which they devote the bulk of their efforts; thus even if there were a propensity to mobilize on behalf of low-income defendants, Los Angeles' extreme geographic decentralization militates against the formation of any organizational locus for such an effort.

One consequence of the conservatism of the "Establishment" Bar is that in Los Angeles "non-Establishment" groups such as the ACLU and the Lawyers Guild remain -- in contrast to their colleagues in Detroit -- virtually isolated from positions of authority and influence within the dominant Bar Association. These "non-Establishment" groups, in turn, are themselves affected by Los Angeles' extreme geographic decentralization. Thus there appears to be very little overlap and cooperation between the ACLU and the Guild, and Guild lawyers bemoan their inability, in contrast to Detroit's Guild, to mobilize non-Guild members for their projects. This general lack of cohesion within the "non-Establishment" sector of the Defense community is also reflected in a relative lack of Black-White legal cooperation and coordination; whereas in Detroit the Guild for example is heavily involved in ghetto projects, in Los Angeles the Guild has increasingly become the legal arm of the Anti-War movement. Finally, although younger, more militant members are increasingly assuming positions of responsibility within the Black Bar Association, their numbers are far fewer than in Detroit, and they have far fewer white allies upon which they can rely.

13-2.2.3 Chicago as a case of intermediate cohesion

The articulation and defense of black and low-income interests in Chicago is hampered neither by extreme geographic decentralization nor by a relatively small black population. Although the extent of overlap is by no means as great as in Detroit -- in part because Chicago lacks Detroit's history of strong and progressive industrial unions -- representatives of "non-Establishment" defense groups do play important roles within the Chicago Bar Association; thus, for example, the General Counsel of the ACLU has for some years served on various Bar Association committees and serves as a link between the two organizations. A good deal of cooperation, moreover -- some of it interracial -- exists between the various "non-Establishment" defense groups themselves, and in contrast to Los Angeles there has existed since the birth of the Civil Rights Movement a fairly sizeable group of dedicated black and white attorneys who are well known to the black community and who concern themselves primarily with the legal defense of black community organizations.

The cohesion of the defense community, however, is severely limited by one pervasive reality which colors all of Chicago politics the Cook County Democratic Party, the "Machine". The Machine functions not only to centralize the organization of Chicago court authorities (as we have already seen), but also serves to limit drastically the articulation and defense of black and low-income interests by co-opting organizations which are potential vehicles for the expression of these interests and purposely hampering those organizations which it is unable or unwilling to co-opt. Thus the Black Bar Association -- the Cook County Bar Association -- is even more dominated by considerations of party loyalty than is its "Establishment" counterpart, the Chicago Bar Association. The political fortunes of those holding positions of responsibility within the Black Bar have typically been intimately associated with the fortunes of the Machine; in past years, for example, the President of the CCBA has typically been rewarded in return for "faithful service" with a judgeship by the Party, and the general membership of the Association includes many who hold lesser patronage jobs within the Party. In this sense the Party, in contrast to the Union in Detroit, serves to isolate black lawyers from white liberal lawyers and thus to reduce the capacity of both to defend and articulate black and low-income interests. Those legal groups specifically concerned with the defense of indigents which cannot be "bought off", moreover, face continual Machine harassment; thus the one Chicago legal service agency funded by OED outside the Mayor's anti-poverty "umbrella" recently faced a threat of congressional investigation and dismissal of its VISTA staff initiated by a Machine Congressman. A strong party organization in Chicago, in other words, acts to reduce both the propensity and capacity of the defense community to mobilize on behalf of blacks and poor; nevertheless that segment of the defense community which is prepared to articulate and defend these interests is substantially larger and more cohesive than in Los Angeles.

13-3 COURT RESPONSES TO GHETTO REVOLTS

13-3.1 Major revolts

The Watts rebellion of August 1965, the Detroit rebellion of July 1967, and the Chicago major revolt of April, 1968 all produced extensive damage to life and property and evoked massive resident participation, and each presented court authorities with comparable and virutally overwhelming problems. On the one hand, each confronted court authorities with the necessity of operating while the respective cities were in flames, while the threat of continued or even spiraling violence made them acutely sensitive to the demands for "law and order" and the preservation of the local political regime as a whole. On the other hand, each produced a volume of arrests many times greater than the norm for comparable periods and far greater than each court system's normal absorption capacity, thus threatening to innundate the court system, to occupy the time and energy of court authorities for months to come and to disrupt totally the processing of ordinary, "non-riot" criminal cases[9]. As we shall see, in each city these overriding and conflicting pressures enforced common perceptions of interests and common actions and led -- notwithstanding the wide variations in political structures which we outlined in the previous section -- to a virtually uniform court response across all three cities.

In Los Angeles and Detroit the response of prosecution officials to the twin threat of the "fires in the street" and the deluge of arrestees entailed the virtual abandonment of normal "gatekeeping" activities crucial to the stability of the court system. In both cities police requests for prosecution were routinely accepted (or not even read), with the result that in each city there was an unusually high ratio of prosecution arrests to total arrests. Thus whereas under normal conditions the prosecution in both Los Angeles and Detroit rejects at least 50% of police requests for felony prosecutions, during the major revolts the ratio of felony arrests was 86% in Los Angeles and 71%

---

9 For an analysis of arrest procedures, which in each city could only be described as haphazard, see [Ref. 8], ch. 3-5.

in Detroit.[10] Given the huge volume of arrests -- in Los Angeles some 3,400 adult arrests during a week's time and in Detroit over 6,000 adult arrests during a five day period -- even the maintenance of the normal ratio of prosecution arrests to total arrests would have produced a number of prosecutions sufficient to swamp court facilities and personnel; with the abandonment of normal gatekeeping and the prosecution of the overwhelming majority of those arrested, however, both court systems were faced with the threat of total collapse! This threat was made all the more imminent by the fact that in both cities the overwhelming majority of charges were felony charges (thus reversing the normal preponderance of misdemeanor over felony cases) which did not admit, of immediate disposition but which would rather require costly and time-consuming prosecution, prosecution which would be extremely difficult given the overwhelming evidence problems entailed in mass arrests.

The prosecution response in Chicago was somewhat different, in part reflecting what Chicago court authorities had learned from the experience in Los Angeles and Detroit.[11] In fact, the prosecution in Chicago abandoned its normal passivity at the initial charging stage[12] and screened charges at the police precincts, charging the majority of defendants with misdemeanors rather than felonies. Although as in Los Angeles and Detroit the overwhelming majority of those arrested were prosecuted, in Chicago the decision to employ mainly misdemeanor charges for which all metropolitan court systems are far better geared

---

10

The 29% non-prosecution rate in Detroit actually vastly overestimates the amount of prosecution screening which took place during the revolt. In fact, a full 98% (or 4,260 out of 4,367) of police requests for warrants were granted by prosecution officials. The bulk of the non-prosecutions, then, resulted from police decisions not to request warrants. There is evidence, moreover, that the overwhelming majority of these non-prosecutions involved persons arrested well after the peak days of the revolt. See [Ref. 8], ch. 4.

11

For a detailed description and analysis of this learning experience and the plans which evolved from it, see [Ref. 8], ch. 5.

to process in large volume,[13] meant that there would be considerably less strain on court facilities and personnel than in Los Angeles and Detroit. Nevertheless, the approximately 900 felony prosecutions -- or an influx of a normal three week load of felony cases in a three day period -- represented a very real threat to the stability of the court system, and in this sense decision making at the initial charging stage in Chicago conformed to the pattern of, and was to result in a similar threat of disruption as, decision making at the charging stage in Los Angeles and Detroit.

In each of the three cities, bail was set so as to ensure that arrestees would be "kept off the streets" at least for the duration of the revolt and in each city both judges and prosecutors publicly articulated this policy of preventive detention. In Los Angeles felony bail was set at $3,000 above normal schedule and misdemeanor bail generally at $1,500 above schedule; in Detroit bail was generally set at or around $10,000 irrespective of the nature of the offense. In Chicago, where court authorities were acutely aware of the enormous detention problems which bonds set at these levels created in Los Angeles and Detroit, cognizant of past criticism from the defense community, and aware from prior experience that most defendants could not quickly meet even normal felony bonds, the prosector recommended, and judges set, felony bonds at or below normal levels and misdemeanor bonds above normal levels. In all three cities, however, bail was not set on an individualized basis. Thus in each city the level of bail set bore no relation to standard determinants such as prior arrest record, age, employment status, educational level, roots in the community, etc.[14] [Ref. 10, 11].

In each city, this concerted effort to "keep them off the streets" was matched by a pronounced hostility on the part of

---

[12] In Chicago, in contrast to Los Angeles and Detroit, the prosecution normally plays no role in the charging process until after the defendant makes his first, formal appearance in court.

[13] Thus for example the Chicago court system normally handles only slightly more than three hundred felony case preliminary hearings per week, as compared to over four thousand misdemeanor and ordinance violations case dispositions per week. [Ref. 8], ch. 5.

[14] [Ref. 8], ch. 3-5. See also [Ref. 10 and 11].

court authorities to those defense efforts which took place at the bail hearings to secure the release of defendants. Access to court rooms was made difficult and in some cases impossible, and volunteer attorneys were not permitted to represent defendants at this stage of the proceedings. Motions for reductions of bond were routinely denied or ignored, and judges in each city refused to cooperate with volunteer efforts to interview defendants for the purpose of obtaining information requisite to eventual bail reduction. Prosecution officials in all three cities insisted publicly that arrestees would be punished "to the limit of the law".

Defense efforts at this early stage in each city were undertaken almost solely by representatives of "non-Establishment" groups such as the ACLU or the local legal aid agencies, and with the exception of the activities of these attorneys the response of the defense community in all three cities can only be described as moribund during the period of the revolt, i.e., the period while bail was being set. Following this period, however, "non-Establishment" groups played a crucial role in evoking the participation and support of the "Establishment" Bar Association, thereby initiating in each city a loose and highly conflictual alliance for the purpose of effectuating a rapid reduction in bonds and the release of the several thousands of incarcerated arrestees. In each city, however, the participation of the "Establishment" Bar in bail reduction efforts served to moderate defense community demands and to discourage class actions (legal actions taken on behalf of a class of defendants, rather than on behalf of a single defendant) desired by the left-wing of the alliance, in favor of informal negotiations with individual judges and prosecutors and/or formal requests for bond reductions on behalf of individual defendants.

These efforts were in each city largely ineffectual until well after the revolt had ended and until the emergency had been declared over. The effort of court authorities to "keep them off the streets" at least for the duration of the revolt was successful; thus, for example, in Los Angeles only 8% and in Detroit only 2% bailed out while the revolt was in progress.[15] Even after this point, however, defense attorneys in each city tended to minimize the impact of their efforts, emphasizing instead the

---

15 [Ref. 8], ch. 3-4.

crucial impact of the pressure on already overcrowded detention facilities created by the unprecedented influx of prisoners. In all three cities, but especially in Detroit and Los Angeles (whose jails quickly filled to double or more their maximum capacity), these pressures threatened the stability of the entire detention system and carried with them the ever-present possibility of jail revolts. Once the "fires in the street" were "extinguished", the need to relieve these detention pressures became a paramount concern of court authorities, and led in Detroit to the release on recognizance on half of all felony defendants, in Los Angeles to the reduction to normal schedule of a similar proportion of felony defendants, and in Chicago to the release on recognizance of the majority of curfew violators and somewhat later to bail reductions for virtually all defendants still in custody -- all marked departures from normal practices in these cities.[16]

As a consequence of these releases and reductions most felony defendants secured their release far sooner than do felony defendants charged with comparable crimes in "non-riot" times. Thus for example whereas normally in Los Angeles only between 40 and 50% of felony defendants bail out prior to coming to trial, following the revolt more than 70% were able to secure their release prior to trial. Similarly, whereas in Detroit in 1965 a full 26% of arraigned felony defendants were unable to bail out before their trials, following the July 1967 rebellion scarcely more than 1% of all felony defendants were in custody by the first day of the trials. Finally, in Chicago whereas normally less than half the defendants charged with state offenses bail out within ten days of their arrest, almost two-thirds of this category of "riot" defendants secured their release within ten days.[17] In each city, then, while bail releases were far <u>less</u> frequent during the first few days following arrest than normally, releases subsequent to this initial period were dramatically <u>more</u> frequent than normally, with the result that the average pre-

---

16 For normal practices with respect to bail and bail reductions, see [Ref. 12].

17 These figures on "riot" and "non-riot" bail-out patterns come from [Ref. 8], ch. 3-5.

sentence detention time of "riot" felony defendants was measurably lower than that of "non-riot" defendants charged with comparable crimes. Figure one graphically portrays the overall similarities in the detention flow pattern following major revolts in all three cities.

In each city the massive detention problems were matched by the pressure of thousands of largely unsupportable cases which required disposition within statutorily prescribed time limitations and which thus placed an almost unimaginable burden on court facilities and personnel. These cases carried with them, moreover, the implicit threat of disruptive jury trial demands which would render their disposition literally impossible: as numerous students of the criminal process have observed, a mere 5% increase in jury trial demands in any metropolitan criminal court would be sufficient to bring the administration of justice to a complete halt. [Ref. 13] Consequently court authorities in each city sought to meet the threat of jury trial demands and other disruptive defense tactics which might cause widespread mandatory dismissals (due to failure to meet statutory deadlines) by ensuring through a variety of means that the defense of the "rioters" would rest firmly in the hands of defense attorneys upon whose cooperation in a rapid and efficient processing of cases they could rely. Thus, with the help of court authorities, in each city the Public Defender or his functional equivalent in Detroit, a group of police court lawyers known affectionately as the "Clinton Street Gang", secured control of the defense operation, and effectively frustrated the effort of those "non-Establishment" attorneys who had hoped to develop a disruptive, jury trial strategy.[18] With control of the representa-

---

18

In Los Angeles, volunteer attorneys did handle the cases of some 300 defendants, but given the overwhelming preponderance of Public Defender representation a jury trial strategy, which requires solidarity among most if not all defendants, became unworkable. Similarly, in Detroit an even larger number of volunteer attorneys provided representation at the preliminary hearing stage, but their effectiveness was limited by the fact that subsequent to that stage representation was largely dominated by the Clinton Street Gang. In Chicago, finally, the overwhelming majority of volunteer attorney efforts occurred at the bail reduction hearings only. For a detailed analysis of defense community efforts in all three cities, and of court authority efforts to control or nullify them, see [Ref. 8], ch. 3-5.

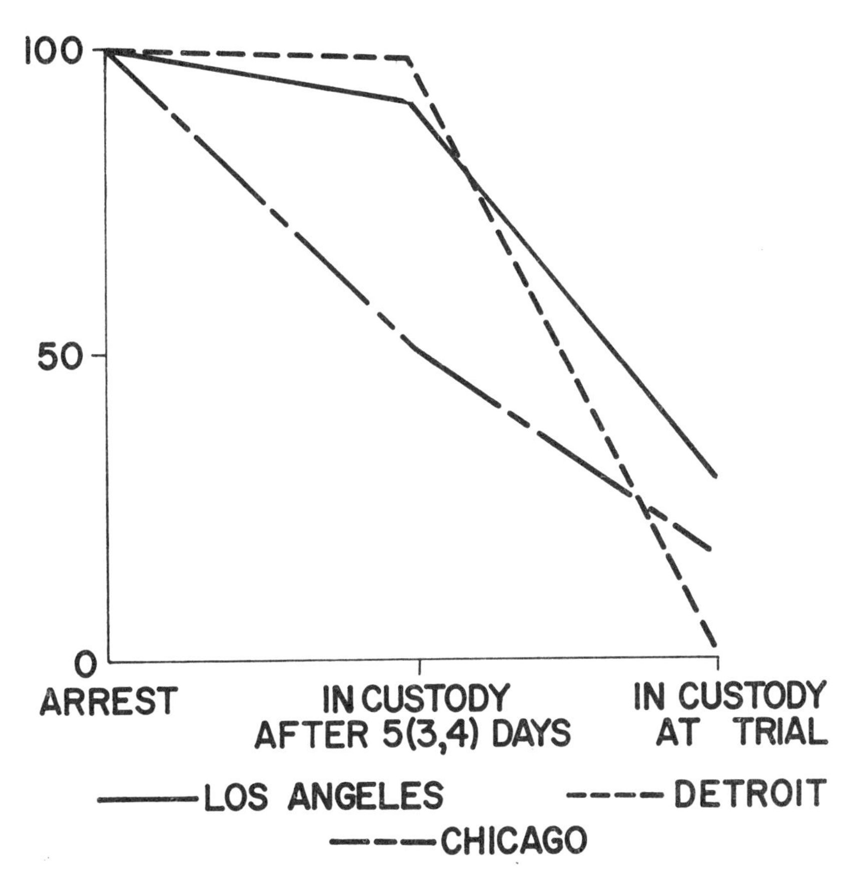

Figure 1
Comparative Major Revolt Detention Flows

tion process in "friendly hands", court authorities were able to secure continuances which rendered statutory time limitations nugatory and thus eliminated the threat of widespread mandatory dismissals.

Even with the burden of time limitations essentially eliminated, however, court authorities were still faced with a dramatic increase in their normal workload which threatened to disrupt seriously the processing of normal, "non-riot" criminal cases for months if not years to come. The first response to this threat on the part of court authorities in each city was to evolve a policy of sentencing convicted misdemeanor defendants to the time they had already served awaiting trial and/or to minor fines or probation. In each city, court authorities admitted that a conscious effort was made to "clear up the docket" and to avoid jury trials by assuring that in return for a plea of guilty or even a bench trial that sentences would be minimal.[19] The result of this policy was in fact a rapid and efficient disposition of misdemeanor cases in all three cities.

The massive influx of felony cases confronting the three court systems, however, entailed a far more serious threat to their stability than did the influx of misdemeanor cases. Whereas the misdemeanor cases represented only a marginal increment to the total yearly misdemeanor quota of the respective court systems, the felony volume with which they were each confronted represented a substantial proportion of their normal yearly quota of felony cases -- in Detroit nearly a six month's quota of felony prosecutions, in Los Angeles almost a three month's quota of felony prosecutions, and in Chicago almost a three month's quota of felony indictments.[20] Many of these cases, moreover, especially in Los Angeles and Detroit, were, as we have seen, relatively unscreened according to normal stan-

---

19

One exception to this pattern occurred in Los Angeles, where a relatively small proportion of misdemeanor defendants pleaded guilty <u>during</u> the revolt when prosecution and judicial tempers were high and received rather severe sentences. In general, however, misdemeanor sentences in all three cities were far less severe than normally.

20

[Ref. 8], ch. 3-5.

dards, carrying with them serious evidence problems which promised to make prosecution extremely difficult.

In Detroit and Los Angeles court authorities responded to this twin problem of a massive volume of cases and an unusual proportion of "bad" cases by increasing significantly the normal rate of dismissals first at the preliminary hearings and then at the felony trials. Thus in Los Angeles only 59% of those prosecuted, and only 77% of those indicted, were convicted, whereas normally some 75% of those prosecuted and almost 90% of those indicted, are convicted. Similarly, in Detroit only 54% of those prosecuted, and 75% of those indicted, were convicted whereas normally 78% of those prosecuted, and 85% of those indicted, are convicted.[21] Thus we found in both cities a striking reversal of the standard model of the criminal process which posits a series of screens whose holes progressively diminish in size and from which the defendants thus find it increasingly difficult to escape [Ref. 14]; following the Los Angeles and Detroit major revolts, in contrast, the "holes" became progressively larger, and it was much easier to "escape" at the preliminary hearing and trial stages than it was at the earlier prosecution stage.

In order to ensure a smooth and efficient disposition of cases and to minimize the risk of jury trials, moreover, court authorities in Los Angeles and Detroit were also forced to reduce drastically the severity of charges confronting those defendants who did come to trial, making reductions far greater than those entailed in the normal process of "plea bargaining" in these cities. Thus, in return for a plea of guilty or even a bench trial felony defendants in each city were routinely convicted of misdemeanor charges; in Los Angeles some 60% were convicted of misdemeanors, compared to a norm of less than 10%, while in Detroit almost 100% were convicted of misdemeanors, compared to a normal reduction rate of 71%.[22] In both cities, then, both the convictions rates (convictions/prosecutions and convictions/indictments) and the felony conviction rate (convictions on fel-

---

21 [Ref. 8], ch. 3-5.

22 See [Ref. 8], 3-5. Statistics in Detroit normally do not distinguish between conviction on a reduced felony charge and conviction on a misdemeanor charge. In all likelihood only a small proportion of the total reduction rate is accounted for by reductions to misdemeanors.

ony charges/total convictions in felony cases) decreased from "non-riot" to "riot" in proportion to the increase in the prosecution rate (prosecutions/arrests) from "non-riot" to "riot", reflecting the twin impact of the volume of cases and the lack of police and prosecution screening accompanying them, and representing an effort on the part of both systems to compensate for it.

In Chicago this compensation did not take place. Since the proportion of "bad" cases was markedly lower than in Los Angeles and Detroit, due to an increased screening role on the part of the Chicago prosecution, both the conviction rate and the felony conviction · rate approximated normal levels. The sheer volume of cases nevertheless compelled Chicago court authorities, as it did their colleagues in Los Angeles and Detroit, to offer lenient sentences in return for the assurance of a predictable and efficient disposition of cases. Thus, as table one illustrates (see below), sentences for convicted felony defendants in all three cities were minimal, entailing far less deprivation than is normally imposed on felony defendants charged with comparable crimes. Despite the character of the initial court response -- bail set so as to ensure preventive detention, hostility to defense efforts, and judicial and prosecutorial assertions that lawbreakers would be punished as severely as the law allows -- in each city only a small minority of felony defendants served any time in jail following the determination of their guilt. Moreover, those few who were committed were rarely committed for more than one year (i.e., received felony sentences), and probationary and fine sentences were far less severe than normally.[23]

23

Although it is true that "riot" defendants generally had less serious records and more stable educational and employment backgrounds than the typical "non-riot" felon, the minimal sentences following major revolts cannot be attributed to the atypicality of the "input" population. Thus those "riot" defendants who did have serious records and unstable employment and educational backgrounds nevertheless fared much better than the typical "non-riot" felon; indeed "riot" defendants with "bad" records and backgrounds scarcely fared worse than "riot" defendants with "clean" records and backgrounds, indicating that the standard determinants of sentence variation tended to break down in the face of a massive volume of cases. See [Ref. 8], ch. 3-5.

Table One: Comparative Sentence Distribution, "Non-Riot" and Major "Riot" Felony Cases

	Los Angeles		Detroit		Chicago	
	Non-Riot	Riot	Non-Riot	Riot	Non-Riot	Riot
% Peniten-tiary (Felony) Sentences	30%	3%	33%	1%	59%	9%
% Jail Sentences	35%	18%	24%	2%	17%	18%
% Committed	65%	21%	57%	3%	76%	27%

Figure two (see next page) presents a visual comparison of the three major revolt case flows from point of arrest through final disposition and sentencing. With the single exception of Chicago's unique ability to maximize convictions through advance planning, the remarkable similarity in the slope of each case flow at each stage of the criminal process testifies dramatically to the decisive commonalities in the court response to major revolts across all three cities. We have tried to show that this commonality was a product of the fact that in each city the revolts created two overriding pressures: 1) an obvious threat to the entire local political regime (of which court authorities are a part); and 2) overwhelming volume pressures on the court system itself -- two pressures which enforced common perceptions of interest and common actions in each city, and which served to differentiate sharply the administration of justice following major revolts from its administration under normal conditions. Initially, in all three cities court authorities responded to the threat to the local political regime and the demands of "riot control" by prosecuting and incarcerating far greater numbers and proportions than are normally prosecuted and incarcerated, thus creating overwhelming pressures on court and detention facilities and manpower which threatened the very survival of the court system. To put out the "fires", police, prosecutors and judges alike "innundated" the court system. Once the obvious threat to the local regime ended with the end of the revolts, these volume pressures determined the subsequent course of events by making it imperative -- in order to

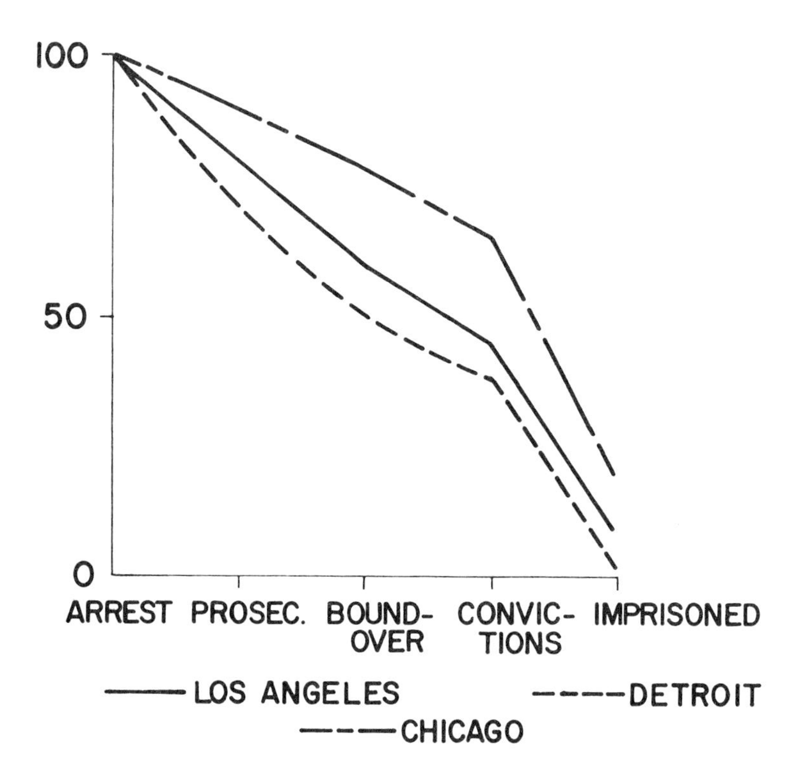

Figure 2
Comparative Major Revolt
Felony Case Flows

eliminate the "flood" -- to reduce drastically the normal severity of court sanctions. Thus in a curious way court authorities in each city were hoist on their own petard; the choice to perform an essentially executive function, i.e., to aid in the management of the revolt, ultimately prevented them from imposing their normal level of formal judicial sanctions.

The foregoing considerations suggest that inter-city variations in the cohesion of authority organization and the cohesion of the defense community were insignificant determinants of the court responses to major revolts. Thus in all three cities the strategy and actions of court authorities were similar in decisive respects despite wide variations in the organizational cohesion of political authorities. In both Chicago, where the Machine provides a high degree of organizational cohesion among court authorities, and in Los Angeles, where extreme formal decentralization exacerbated by fragmented political parties creates extremely low organizational cohesion, for example, court authorities united initially to an equal extent in the name of "riot control" and subsequently to an equal extent in the name of court system efficiency. Similarly, despite wide variations in the organizational cohesion of the defense community across the three cities, the structure of defense community alliances and their policies were remarkably similar. Thus both in Detroit, where the predominance of Labor-oriented white lawyers and the alliance they have formed with black legal groups produces high defense community cohesion, and in Los Angeles, where due among other things to corporation lawyer predominance defense community cohesion ranks lowest, "non-Establishment" defense groups were joined by "Establishment" defense groups to protest the high bail policy primarily by means of informal negotiations with court authorities, and later acquiesced in, and joined, the cooperative process by which "state" and "defense" achieved a smooth and efficient disposition of the huge volume of cases. In short, we observed a remarkably similar court response across all three cities despite major differences across the three cities along political dimensions commonly cited as critically important determinants of the behavior of local political actors.[24]

24

The one important exception to this common pattern, moreover -- Chicago's unique ability to maximize convictions and avoid reductions -- was not a product of political characteristics peculiar to Chicago, but rather of a learning experience to which only Chicago was privy prior to its major revolt. In fact, plans which called for procedures virtually identical to those worked out in Chicago were evolved in both Los Angeles and Detroit following their respective major revolts.

13-3.2 Minor revolts

In the Summer of 1966, just one year prior to the July rebellion, a confrontation between Detroit police and several members of two militant community organizations blossomed into a spate of rock throwing and windowbreaking that lasted only two days, produced no major injury to life or property, and resulted in the arrest of some 90 black adults. There is evidence to the effect that police and prosecutor collaborated in a careful effort to isolate the militant, prosecuting them but releasing many of those not affiliated with either organization. Thus only 43 of 89 were prosecuted -- or 48%, as compared to almost 75% the following July; in contrast to the following July then, police and prosecution played their normal gate-keeping role.

The prosecutor made a concerted effort to prosecute the militants. They were virtually all charged with felonies -- common law riot and conspiracy to disturb the peace, instead of simply disturbing the peace or disorderly conduct. Nevertheless the arraignment judges, who apparently did not perceive a threat as great as that perceived by the prosecutor, set either normal bond or released arrestees on their own recognizance. Fifteen or some 41% were in fact released on their own recognizance (as compared to a "non-riot" norm of about 30%) at the initial bail hearings, whereas only 2% were so released the following July. The median bond for the others was $1,000; as a result most were back on the streets within one or two days.

A group of lawyers -- largely black but including several white Guild lawyers -- formed the Metropolitan Defense Committee, which provided free and complete representation for all those prosecuted. These attorneys refused to play by the normal "rules of the criminal court game", refusing to waive Preliminary Hearings and to plead their clients guilty, and in many cases demanding jury trials. Despite an obvious prosecution effort to deal harshly with the defendants -- indicated by his use of rather contrived and serious charges -- only fifteen were convicted, or only 40% of those prosecuted and 52% of those brought to trial.[25] The MDC lawyers successfully challenged the legal-

---

25

These percentages are based on the total number of cases for which we have complete disposition information. Since cases against four defendants were still pending at the time this research was conducted, the percentages are of course subject to minor changes.

ity of many of the counts, forcing the dismissal of all conspiracy charges. They felt, in retrospect, that they "were able to beat the system", and that "the impact of their defense effort in '66 as compared to '67 was like the difference between night and day.[26] Nevertheless, of the fifteen convicted, eleven were given probation for at least two years (six for three years or more), and two defendants receiving probation were also fined $500 each. The remaining four were incarcerated, three of whom were sentenced to from one to five years in the state penitentiary. In all then, some 27% of those convicted were incarcerated, as compared to only 3% following the major revolt one year later, and 20% of those convicted received penitentiary sentences as compared to less than 1% the following July.

Following each of four minor incidents in Chicago in the Summer of 1967 -- the smallest of which resulted in 21 arrests and the largest in around 60 -- virtually all those arrested were prosecuted; moreover, they were multiply charged with from two to four offenses. One particularly vague charge, which virtually made it illegal for two people to congregate on the street (and which carried with it a maximum penalty of one year in prison), was used extensively. Judges set bonds ranging from $5,000 to $20,000 for offenses for which the normal maximum bond would be $1,000. The Public Defender did not object, and those volunteer attorneys who did -- including some black attorneys and some ACLU attorneys -- were met by extreme hostility from court authorities. Judges refused to hear, or routinely denied, habeas corpus petitions, and bond reductions did not take place, if at all, for several weeks, since cases were automatically continued over the objections of defense attorneys. Bail was thus set and remained sufficiently high to ensure that pre-sentence detention time averaged between two and four weeks, depending on the incident, and if not for volunteer efforts following several of the incidents which finally succeeded in obtaining bail reductions it is likely that presentence detention would have been even higher.

The twin pressures of multiple charging and prolonged presentence detention -- along with a relatively unorganized defense

---

26 Confidential interviews with Detroit Neighborhood Legal Services attorneys.

effort which did not provide representation at the trials -- combined to produce unusually high conviction and guilty plea rates. Thus the conviction rate averaged over 75% for the four incidents, while the guilty plea rate averaged a full 85%, both far higher than the normal rate in misdemeanor court in Chicago. Bulking together all those convicted in the four incidents, 17% were incarcerated following their conviction, or approximately the normal rate of imprisonment for ordinance violations. While few sentences were for more than thirty days in jail, extensive and extraordinary use was made of probationary sentences, so that anyone re-arrested during the Summer would automatically be assured of imprisonment.

Finally, on the eve of the third anniversary of the Watts Revolt of 1965, violence flared briefly again in Watts and resulted in the arrest of some 35 adults. The Los Angeles County Prosecutor rejected 60% of the police requests for felony complaints, as compared to a rejection rate of only 14% during the major Watts revolt, and approximating the normal rate of prosecution screening in Los Angeles. The majority of the rejections were recommended to the City Attorney for prosecution as misdemeanors, but only about half of these were actually prosecuted by the City Attorney, who also did not prosecute all of the ten adults originally booked on misdemeanors. Thus in Los Angeles, as well as in Detroit (but in marked contrast to Chicago) following the minor revolt the prosecution performed its normal gatekeeping function.

Rejecting a prosecution motion for a denial of bail, judges set both felony and misdemeanor bail at or around normal levels and on an individualized rather than uniform basis. Although several defendants bailed out within three days, detention time for all defendants averaged fifteen days, and a quarter of the defendants spent at least four weeks in jail -- a distribution closely approximating normal, "non-riot" pre-sentence detention.

No mobilization of the defense community occurred; the Public Defender handled the bulk of the cases, with one ACLU lawyer handling only one case. Consequently there were no jury trial demands, and the great majority of defendants pleaded guilty. Convictions were forthcoming in 80% of the cases which came to trial,[27] higher than the overall conviction rate in Au-

---

27 Two cases had not come to trial by the time this research was undertaken, so this percentage is subject to slight change.

gust 1965 and approximating the normal rate. Most felony defendants who were convicted, moreover, were convicted on the original charges, as is the norm in Los Angeles. Three out of the twelve convicted -- or 25% -- were incarcerated, as compared to 21% following the major revolt. Fines and probationary sentences, finally, were markedly more severe than following the major revolt, and tended to approximate the normal sentence distribution.

Figures three and four compare respectively the minor revolt detention and case flows across all three cities, illustrating wide variations in the court responses to minor revolts. These variations, it should be clear, are intimately related to the inter-city political structure variations which we outlined in the second section (and which failed to play a determinant role following major revolts). In fact, the two variables which determine the structural relationship between local political system and court subsystem -- the cohesion of local authorities and the cohesion of the local defense community -- respectively determined the nature of the effort undertaken by each party to the court conflict. Thus for example in Chicago, where cohesion of local authorities is highest, all the court authorities -- police, prosecution, judges, detention officials and public defender alike -- united to prosecute virtually all those arrested, to ensure their prolonged incarceration by means of extraordinarily high bonds and automatic continuances, and to achieve the highest possible conviction rate. Our research revealed, in fact, that Chicago court authorities were acting on the basis of a coherent plan, carefully elaborated at the highest levels within the Party, to manipulate the judicial system so as to further Mayor Daley's general violence prevention goals. A complete explanation of the unusually severe and repressive Chicago court response would therefore require an analysis of the origins and evolution of this strategy, e.g., an analysis of why even minimal violence was perceived as a serious threat to the Chicago political regime during the Summer of 1967. What is clear, however, is that whatever the Mayor's perceptions, these perceptions could not have been <u>implemented</u> as policy in the absence of coherent channels of communication to and control over the actors in the court system. In this sense, then, extreme authority cohesion is a <u>necessary</u> if not sufficient determinant of such a punitive court authority respons[illegible] low levels of violence.

In Los Angel[illegible] in contrast, where cohesion of local authorities is lowest, no such coordinated and unified court authority effort took place. In fact, the normal amount of intra-

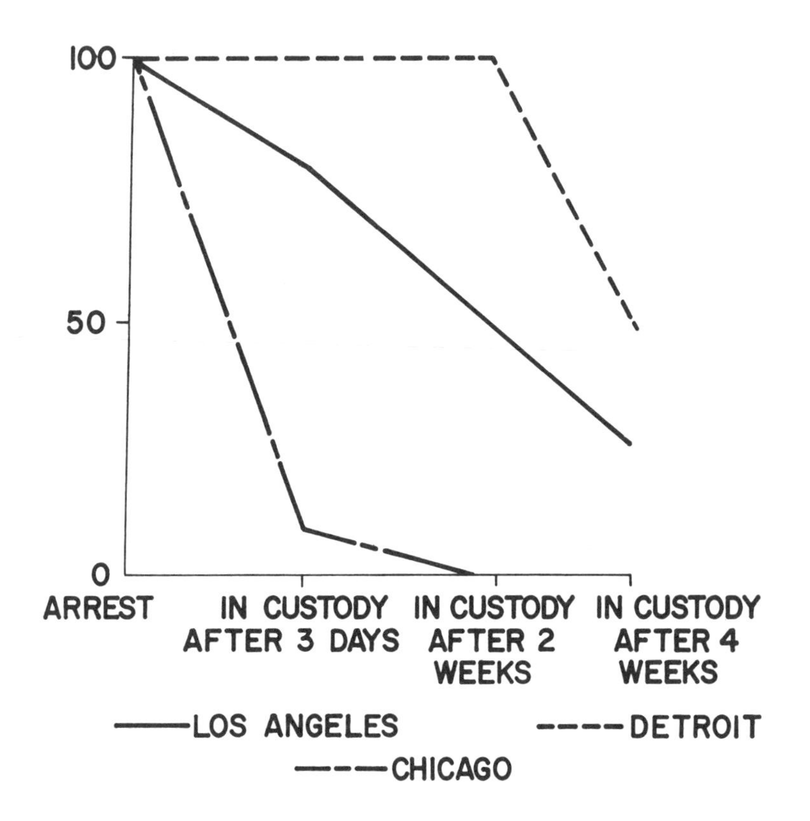

Figure 3
Comparative Minor Revolt Detention Flows

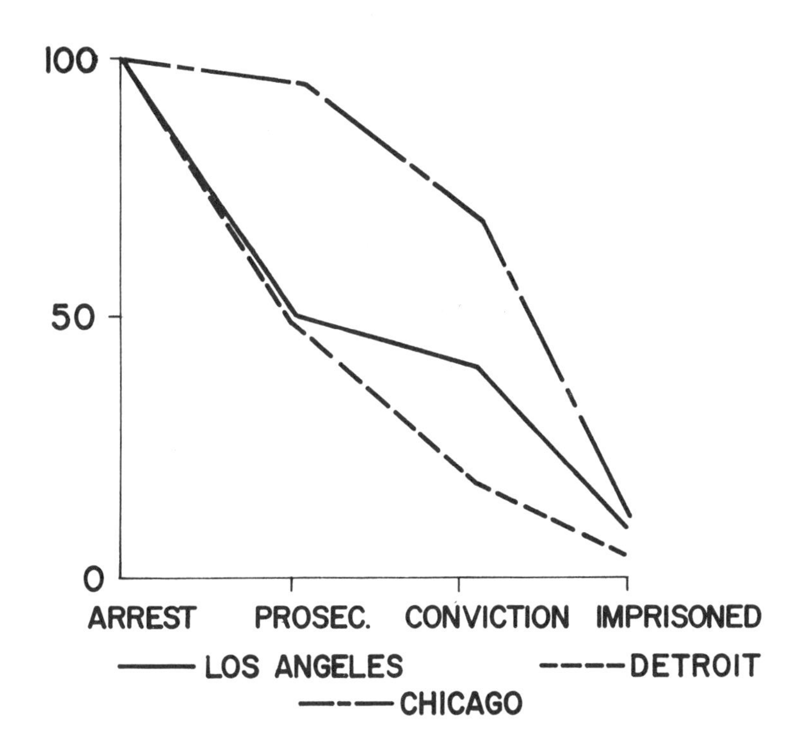

Figure 4
Comparative Minor Revolt Case Flows

authority conflict occurred: the prosecution played its normal gatekeeping role, rejecting at least half of the police requests for felony prosecutions, and judges rejected the prosecution request for a denial of bail. In Detroit, finally, where there is intermediate cohesion of local political authority organization, police and prosecution collaborated in an effort to deal harshly with the militant arrestees (of all the non-prosecutions only one was the result of prosecution denial of a police complaint), but their efforts were frustrated by judges who set bond at or below normal levels.

Similarly, the cohesion of the defense community predicted the strength of the efforts undertaken by defense attorneys on behalf of the "rioters". Thus in Detroit, where cohesion of the defense community is highest, a group of both black and white attorneys evolved a unified defense strategy and refused to play by the normal "rules of the game", with the result that criminal proceedings there approximated much more closely than elsewhere the traditional adversary model of criminal justice. In Los Angeles, in contrast, where defense community cohesion is lowest, no mobilization of the defense community took place, with the Public Defender monopolizing the defense effort. In Chicago, finally, where defense community cohesion is intermediate between that of Detroit and Los Angeles, volunteer efforts were undertaken but were largely confined to the initial bail hearing stages.

If each structural variable separately predicts the respective sanction efforts of the two parties to the court conflict, then the respective efforts determine the outcome of the conflict, and the two variables combined should accurately predict the sanction outcomes. Table two ranks the three cities ordinally in terms of both authority and defense community cohesion as a prelude to this "combining" operation. Since the authority cohesion score

Table Two: Authority and Defense Community Cohesion

	Los Angeles	Detroit	Chicago
Authority Cohesion	1	2	3
Defense Community Cohesion	1	3	2
Conflict Balance Score	1/1=1	2/3	3/2

indexes the structural potential for court authority unity, and the defense community cohesion score indexes the structural potential for coordinated effort among defense attorneys, by dividing the authority cohesion score by the defense community cohesion score we can create a rough composite index which measures the relative structural advantage which one party to the court conflict has over the other. The result of this operation is a "conflict balance" ranking (which increases as the balance moves in favor of court authorities) of 2/3, 1, and 3/2 for Detroit, Los Angeles, and Chicago, respectively, a ranking which, it turns out, corresponds perfectly to the severity of their court sanctions, illustrating the combined explanatory power of our two structural variables.

Thus in Chicago where the conflict balance score is weighted most heavily in favor of the authorities, a concerted authority effort to deal severely with the defendants, combined with a relatively unorganized defense effort, produced sanctions far more severe than normally. In Detroit, on the other hand, a relatively uncoordinated authority effort was confronted with a coherent and rigorous defense effort, with the result that sanctions resulting from full-scale adversary proceedings were less severe than for comparable "non-riot" offenses. In Los Angeles, finally, which ranks midway between Chicago and Detroit on the conflict balance index, an uncoordinated authority effort matched by the almost total absence of a defense community effort produced court sanctions more severe than in Detroit but less severe than in Chicago, sanctions which in fact closely approximated the normal, "non-riot" level in Los Angeles.

13-4 LEVELS OF VIOLENCE, INTERCITY POLITICAL STRUCTURE VARIATIONS AND COURT SANCTIONS

We can now attempt to specify the empirical relationship among the variables which have formed the organizing paradigm of this study: levels of violence, inter-city political structure variations, and court sanctions. To begin with, whereas we described a virtually uniform court response across all three cities at the major revolt level, at the minor revolt level we observed wide variations in the nature and severity of court sanctions. The preceeding discussion has established that these variations are indeed related to the inter-city political structure variations which have been the focus of the study of American local politics and which (we have seen) fail to account for sanctioning behavior at the major revolt level. The data thus suggest a differential impact of inter-city political structure variations at different levels of violence. More precisely, they inspire the following proposition: the importance of inter-city political structure variations varies inversely with the level of violence.

Secondly, we found that with a vast increase in the magnitude of violence there is certainly no corresponding increases in the severity of court sanctions. Indeed, table three, which compares the severity of both pre and post-conviction sanctions across all three cities at both levels of violence, tends to confirm the impression of an inverse relationship between the magnitude of revolts and the severity of court sanctions imposed on participants. Thus following Detroit's minor revolt a signifi-

Table Three: Levels of Violence and Pre-and Post-Conviction Sanctions*

	Detroit		Los Angeles		Chicago	
Level of Violence	Pre	Post	Pre	Post	Pre	Post
High	5-7 days	3%	14 days	21%	3-4 days	15%
Low	2 days	27%	15 days	25%	14-28 days	17%

*

Pre-Conviction sanctions = average detention time for all prosecuted defendants. Post-conviction sanctions = % imprisoned of those convicted. In Detroit and Los Angeles, % imprisoned refers to % felony defendants imprisoned, since in both cities following both major and minor revolts felony charges predominated. In Chicago, % imprisoned refers to % misdemeanor and felony defendants imprisoned, since misdemeanors were used extensively along with felonies. These decisions were based on a desire to achieve the highest possible degree of comparability.

cantly higher proportion of those convicted were incarcerated than following the major revolts in all three cities. In Los Angeles, average pre-sentence detention time following its minor revolt was somewhat greater than following the major Watts revolt, and the proportion incarcerated following conviction was greater than following any of the three major revolts. In Chicago, finally, although post-conviction deprivation following its minor revolt was only slightly greater than following its major revolt, pre-conviction deprivation in the form of pre-sentence detention time was up to ten times greater than following its major revolt and more prolonged than following the Los Angeles and Detroit major revolts.

Neither of the respective measures of pre and post-conviction deprivation utilized in table four, moreover -- average detention time and the proportion convicted who were incarcerated -- is an entirely satisfactory overall index of deprivation, and other indicators would tend to support even more strongly the above hypothesis. Thus for example average detention time obscures the extremes of pre-sentence detention, and fails to reveal that following minor revolts in Los Angeles and Chicago there was a much greater proportion of defendants in custody over thirty days and at the time of trial than following major revolts in all three cities. Similarly, the proportion imprisoned needs to be supplemented by other measures of sentence severity. For example, probationary and fine sentences following the Los Angeles minor revolt were more severe than following the major Watts revolt, and the proportion of those convicted who received penitentiary sentences following Detroit's minor revolt was far greater than following any major revolt. The introduction of additional measures, then, tends to confirm the general hypothesis of an inverse relationship between size of revolt and severity of sanctions.[28]

We are left, then, with the seemingly paradoxical conclusion that a participant in a full-scale ghetto revolt involving widespread participation and destruction of life and property is likely to incur <u>less</u> deprivation from the criminal courts than one arrested for a comparable offense during "normal" conditions (as

---

28

Data from additional minor revolts which occurred in our three cities but which have not been described in this paper also tend to strongly confirm this proposition. See [Ref. 8], ch. 6.

we saw in section iii) or for a comparable offense during a minor revolt which involves but a tiny fraction of the scope and destructiveness of the major revolt. The analysis in the body of this paper suggests that this is perhaps the inevitable result of the use of the ordinary criminal court system to sanction participants in mass revolts. The massive volume of arrests accompanying these revolts represents an overwhelming threat to the permanent interests which the members of the court system share in the stability of the norms and procedures which regulate the sanctioning of ordinary crime. To avert this threat, i.e., to ensure the efficient, predictable, disposition of cases which their long-standing interest in the stability of the court system requires, court authorities are forced to reduce drastically the normal severity of sanctions imposed for comparable offenses in "non-riot" periods. This is another way of saying that American political authorities have been forced to pay a price for their effort to treat political revolts as if they were ordinary crime, the price being minimal deprivation inflicted on the participants in the revolts. If this interpretation is correct -- if minimal sanctions are in fact the inevitable result of the use of the ordinary criminal courts -- then it follows that a complete explanation of the nature and severity of the sanctions imposed on the participants in the urban ghetto revolts of the mid-1960's would require an explanation of why in fact the ordinary criminal courts have been employed.[29] This question, however, is beyond the scope of the present paper.

13-5 CONCLUSION: IMPLICATIONS FOR A SYSTEMS ANALYSIS OF LOCAL POLITICS

This concluding section explores briefly some of the implications of the foregoing analysis for a systems analysis of local politics and, indeed, of politics in general. To begin with, it should be clear that political subsystems such as the local criminal court system can not profitably be studied as closed systems. Those who have applied a systems or functionalist perspective to the court system have, unfortunately, done just that; i.e., they have tended to explain the behavior of this system in

---

29 For the beginnings of such an explanation, see [Ref. 8], ch. 6.

terms of internal self-regulatory mechanisms rather than explore the relationship between the court system and its broader political environment.[30] Our analysis has, hopefully, demonstrated to the contrary that the behavior of the court system cannot be explained without investigating systematically the commitments and interests which actors within the court system share with political actors within the local political system generally. Indeed, we have suggested that court actors play roles which are peculiar to the court system <u>and</u> roles within that system's political environment, and that the behavior of the court system is usefully conceptualized as a product of an effort to strike a balance between these often contradictory roles. Thus for example we saw that major revolts confronted the actors within the court system with a dramatic challenge to both their interests as members of the court system and their interests as political authorities generally -- the volume of arrests represent an "input overload" which threatened to completely innundate the court system, yet at the same time the "fires in the street" threatened the stability of the entire local political regime -- and that the overall court response was a reflection of a balance struck between these conflicting role requirements. Any analysis which focused exclusively on the commitments of court authorities to the stability of their own subsystem, i.e., any analysis which conceptualized the court system as a closed system, would thus have failed to explain fully the behavior of that system.

We have also tried to show, however, that the analysis of the court system as an open system ought not to focus exclusively on the concrete structural relationship between the court system and its broader political environment; i. e., the role expectations which link court actors to other actors within the court's political environment are not simply a reflection of the political structure variables on which students of local politics have typically focused. Thus while local political structure variations such as Machine versus Reform did indeed shape court responses following minor revolts, we saw that these and other inter-city political structure variations played little or no determinative role following major revolts. To conclude from this, however, that following major revolts court actors were

---

30 See Footnote 2. [Ref. 1]

not influenced by their interests and commitments within the broad local political system would of course be totally unwarranted. We have already suggested that the behavior of the court system following major revolts in each of our three cities was radically influenced by these commitments; the point, of course, is that the content of these commitments or role expectations cannot be reduced to the structural relationship between local political system and court sub-system. For example, had we not assumed for heuristic purposes the existence of some common interest or role expectation which unites the members of the disparate bureaucracies of the court system and which they share with local political authorities generally, we would have been totally unable to explain the uniform effort on the part of court authorities in all three cities to apply court sanctions in such a manner as to aid local political authorities generally in managing the revolts. Had we not assumed, moreover, that this common interest or role expectation existed independently of the particular structural relationship between court subsystem and local political system, we would have been at a loss to explain why such a common response occurred in three cities where this structural relationship is so very different. The point is that court authorities, for example, are usefully conceived as possessing certain role interests or expectations in common with political authorities generally whether the structural relationship between court subsystem and local political system is extremely diffuse, as in the case of Los Angeles, or controlled and cohesive, as in the case of Chicago. This structural relationship should properly be understood as only one determinant of the impact of the broader political environment on the behavior of the court system; another, we have suggested, is the degree of threat to, or stress on, the role interest itself. During major revolts (i. e., when this stress was greatest), for example, it hardly mattered whether a judge was hand picked by the Chairman of the local political party or selected by the Governor of the State, or whether a prosecution official was selected by a Party committeeman or recruited through Civil Service: in both cases the judges and prosecution officials acted to defend their interests as political authorities generally. Following minor revolts, where the stress on their political authority role interests was much lower, however, these kinds of structural variations were determinative of the court response. Our analysis thus suggests both that the impact of the local political system as a whole on the court subsystem cannot be understood simply in terms of the structural relationship between the two, and that the importance of this structural relationship is not constant but rather that it is likely to vary depending upon the value of other variables.

Finally, once the court system is conceptualized as an open system the assumption of homeostasis which is generally associated with systems analysis[31] -- the assumption that the members (or at least the authorities) of a system will attempt to cope with stress in such a way as to maintain or restore the equilibrium of that system -- becomes an assumption of questionable heuristic value. Indeed, confronted with a massive influx of cases, court authorities during major revolts acted exactly contrary to the postulate of homestasis; rather than drastically increasing the normal amount of gatekeeping in order to reduce the volume stress on the court system, court authorities decreased and (in Los Angeles and Detroit) virtually completely abandoned normal gatekeeping activities, flooding the courts and detention facilities with a massive, overwhelming volume of cases. In other words, initially court authorities acted in such a manner as to deliberately increase the threat to the equilibrium of their own subsystem when faced with a threat to their interests as political authorities generally. Thus for example the abandonment of normal gatekeeping functions on the part of the authorities of a Detroit court system confronted with over eight times the normal weekly amount of arrests (stress factor = 8+) led to the prosecution of twenty-seven times (stress factor = 27) the normal weekly number of felony cases. Similarly, in Los Angeles court authorities were initially faced with five times the normal, weekly number of police requests for felony complaints (stress factor = 5) but by abandoning normal screening they allowed ten times the normal weekly quota of felony cases to be prosecuted (stress factor = 10). Finally, a Chicago court system confronted with only a three week's normal quota of felony prosecutions (Stress factor = 3) responded by producing a (stress factor = 11) week's normal quota of felony indictments. In each case, then, court authorities following major revolts

---

31 Chalmers Johnson, for example, defines a "system" as "any group of variables which are so arranged that they form a whole. . and which have a particular kind of relationship with each other -- namely that they are mutually influencing ('interdependent') and they tend to maintain the relationship they have with each other over time ('equilibrium').

increased by at least a factor of two the stress on their own subsystem, and only after the threat to their interests as political authorities generally was eliminated did the court system begin to respond in a homeostatic fashion. The assumption of homeostasis, then, obscures the crucial fact that it was precisely actions undertaken by court authorities themselves which were a major determinant of systemic stress and dysequilibrium.

The effort to "rescue" the assumption of homeostasis, moreover, by suggesting that the court authorities' response, while dysequilibrating for the court subsystem, was nevertheless equilibrating for the local political system as a whole, is ultimately unsuccessful, since it entails the critical problem of non-falsifiability. The argument that the assumption of homeostasis should apply only to whole systems rather than to subsystems overlooks the crucial fact that virtually any "whole" system may be considered a subsystem of another system; e.g., that the local political system can be conceptualized as a subsystem of the national political system, the national political system a subsystem of the international political system, etc. If, notwithstanding the postulate of homeostasis, virtually any system can act in a non-homeostatic fashion (because it is labelled a subsystem) yet not contradict the theory, then a theory based on the assumption of homeostasis carries with it no explanatory power and is in fact non-falsifiable. The point is that if a theory is analytic -- if the model on which it is based can be applied to any concrete system or subsystem because all are presumed to share certain regularities by virtue of the common label "system" -- then a postulate which applies to one of them must apply to all of them.

Our analysis suggests, then, that the assumption of homeostasis is not a fruitful heuristic assumption on which to base a general theory of political behavior. In so far as the assumption of homeostasis is a defining assumption of systems analysis -- in so far as "system" is understood to imply that the interactions among the parts contribute under "normal" conditions to the equilibrium of the whole -- then systems analysis must be complemented by a theory of those interests or imperatives within social systems which produce non-homeostatic responses, which lead actors to seek the destruction rather than the maintenance or restoration of the equilibrium of their system. A comprehensive theory, in other words, must be based on the assumption that every social system contains within itself both forces for its own preservation <u>and</u> the seeds of its own destruction.

REFERENCES

1. Dallin H. Oaks and Warren Lehman, A Criminal Justice System and the Indigent (Chicago: The University of Chicago Press, 1968), passim.
2. Robert I. Mendelsohn and James R. Klonoski, "The Allocation of Justice: A Political Approach", Journal of Public Law, XIV, No. 2 (1965), 211-226.
3. James Q. Wilson, ed., City Politics and Public Policy (New York: John Wiley & Sons, 1968), passim.
4. Grant McConnell, Private Power and American Democracy New York: Alfred A. Knopf, 1966), passim.
5. J. David Greenstone and Paul E. Peterson, "Reformers, Machines, and the War on Poverty", in Wilson, ed., City Politics and Public Policy, pp. 267-292.
6. City Politics (Cambridge, Massachusetts: Harvard University Press and the M.I.T. Press, 1963), p. 186.
7. John C. Leggett, Class, Race and Labor: Working Class Consciousness in Detroit (New York: Oxford University Press, 1968), passim.
8. Isaac D. Balbus, Rebellion and Response: A Comparative Study of the Administration of Justice Following Urban Ghetto Revolts in Three American Cities (Chicago: unpublished doctoral dissertation, 1970.)
9. Donald J. McIntyre, "A Study of Judicial Dominance of the Charging Process", Journal of Criminal Law, Criminology, and Police Science, LIX, No. 4 (December 1968), 463-490.
10. Philip Colista and George Domonkos, "Bail and Civil Disorder", Journal of Urban Law. XLV (Spring-Summer 1968), pp. 841-848.
11. "Criminal Justice in Extremis: Administration of Justice During the April 1968 Chicago disorder", The University of Chicago Law Review, XXXVI, No. 3 (Spring 1969), pp. 455-613.
12. Lee Silverstein, "Bail in the State Courts", Minnesota Law Review, L, No. 4 (March 1966), pp. 621-652.
13. Ralph Nutter, "The Quality of Justice in Misdemeanor Arraignment Courts", Journal of Criminal Law, Criminology, and Police Science, LIII, No. 2 (June 1962), 215.
14. Abraham Blumberg, Criminal Justice (Chicago: Quadrangle Books, 1967), p. 51.
15. Revolutionary Change (Boston: Little, Brown and Company, 1966), p. 40.

Chapter 14

A COMPUTER BASED SYSTEM FOR FORMING EFFICIENT ELECTION DISTRICTS

E. S. Savas

The Office of the Mayor of the City of New York, with the cooperation of the Board of Elections, has completed a pilot project that demonstrated that a computerized system for drawing election district lines would arrest the persistent decline in the efficiency with which voting machines are utilized, and could therefore be used to: (1) reduce voter delays at polling places; (2) equalize delays among election districts; (3) reduce the vulnerability of election districts to voting machine breakdowns; (4) reduce the cost of conducting elections by up to one million dollars annually; (5) permit convenient redistricting whenever required by changes in political boundaries, election laws, or population shifts; (6) facilitate analysis of voter registration. However, the value of this system is even broader than these points indicate. Its capability to manipulate geographic data is an important element of a generalized management information, mapping, and analysis system that the city is developing.

The author is First Deputy City Administrator, City of New York.

After every election, complaints are heard about long lines at some polling places and about delays caused by breakdowns of voting machines. It appears that these problems in many areas could be alleviated by more efficient allocation of voting machines.

Under the New York State Election Law, an election district must have one machine for every 750 voters or fewer and two voting machines where there are more than 750 voters, except when the total lies between 750 and 800. In the latter case, the New York City Board of Elections shall decide whether an additional machine is to be used. It has been the policy of the Board to set the upper limit of a two-machine district at 1200. The great majority of the election districts in the city have one machine.

The Board of Elections is responsible for drawing election district lines before each election. These lines must accommodate changes in political boundaries and changes in the election laws. However, there is generally insufficient time to make extensive readjustments in election district lines by present manual methods, and, as a result, there has been a steady decline of efficiency in the use of voting machines.

Figure 1 shows the voter registration for the City of New York from 1949 to 1966. This figure indicates a marked four-year cycle with a peak in voter registration occurring every four years because of interest in presidential elections. However, the long-term trend in the registration shows no apparent increase over the years.

An appropriate measure of the effectiveness of various election-districting methods for a given geographical area is the efficiency percentage,[1] defined as 100 times the ratio of the average number of voters per voting machine to the maximum allowable number of voters per voting machine. Figure 2 depicts the efficiency percentage for New York City for the time period covered by Fig. 1. Prior to the implementation of permanent registration in 1957, the Board of Elections redrew the election district lines each year. Since the advent of permanent personal registration in 1957, election district lines have been redrawn only when required. Because of the lack of adequate time to redistrict, the Board of Elections has often added a second voting machine to a district rather than change existing election-district lines. As a result, since 1960 (the first presidential election under the permanent registration system) there has been a steady decline of efficiency at an average rate of 2.5 per cent per year, to a current value of 61 per cent. This decrease in efficiency is manifested in the increased costs of conducting elections.

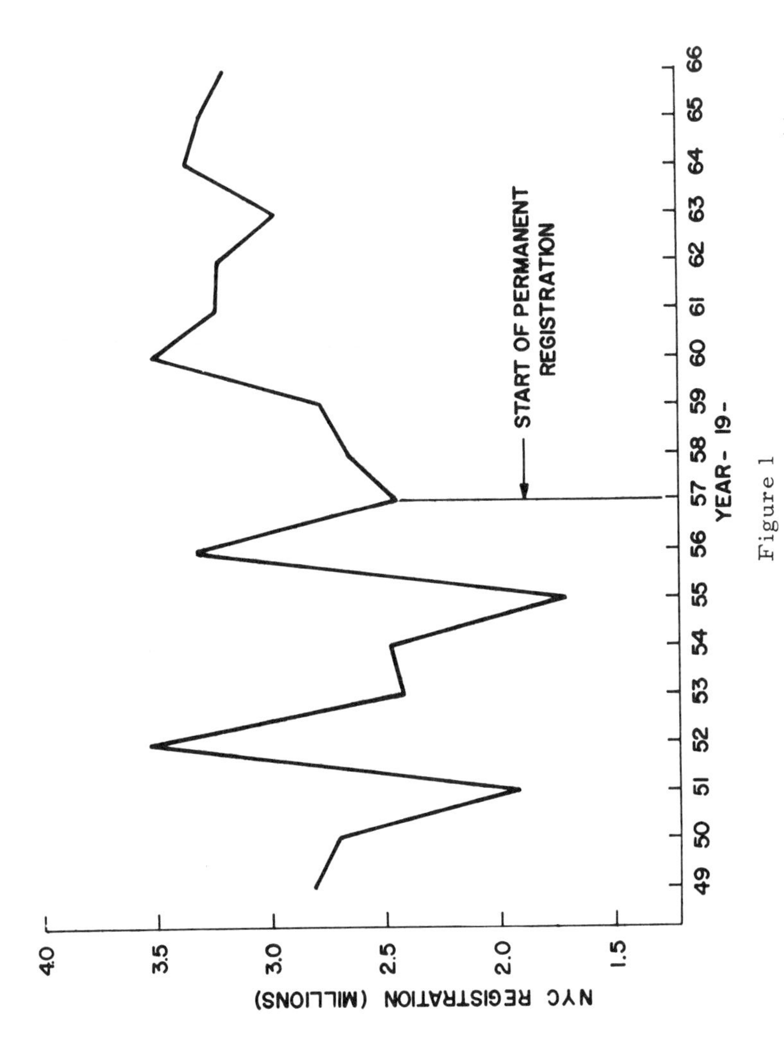

Figure 1
Voter Registration for the City of New York from 1949 to 1966.

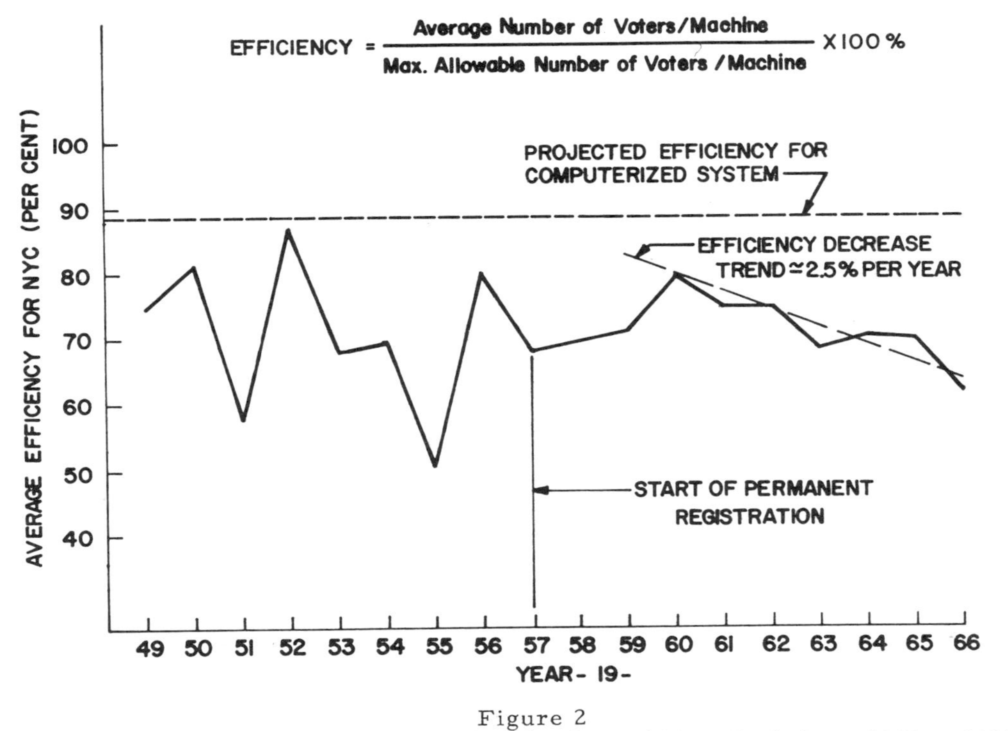

Figure 2
City-Wide Districting Efficiency for the City of New York from 1949 to 1966.

The Office of the Mayor proposed that a computer-based system be developed to reduce drastically the effort needed to redistrict the City. The system would therefore allow the Board of Elections to make a large initial improvement in efficiency and then to retain a high level of efficiency by enabling it to redistrict with relative ease whenever warranted by changes in political boundaries, election laws, or population. The projected efficiency line in Fig. 2 is the estimated efficiency that would be achieved with the proposed computerized districting system, based upon the actual efficiency achieved in a pilot study, reported below.

14-1 OUTLINE OF THE SYSTEM

There are five major technical steps in developing the computerized districting system:

14-1.1 Division into Common Lots

The city must be partitioned into common lots. A common lot is an area that has all political boundaries in common (i. e., is in the same congressional, senatorial, assembly, councilmanic, and judicial districts). For the pilot study, the common lot chosen was an area in the upper west side of Manhattan, the 71st Assembly District, which contains many of the anticipated difficulties (i. e., large housing projects, curved and irregular blocks, bridge approaches, etc.).

14-1.2 Creation of Geographic Master File

For the selected common-lot area, a geographic master file was created. This file is a prototype for the block-face data file that is at the heart of the city's geographic information system, [Ref. 1] which is under development. Among other things, the block-face file contains the house-number range of each blockface and the geographic (X, Y) coordinates of the ends of each blockface. (A brief description of the block-face file appears in Appendix A.)

14-1.3 Merger of Registered Voter File with Geographic Master File

The voter data file was merged with the geographic master file so that geographical coordinates could be associated with each voter's residence for subsequent aggregation, districting, and mapping. Although simple in concept, both this and the preceding step were very difficult to execute in practice, because of the large number of ambiguous addresses, inconsistencies in nomenclature, errors, and incompatible data formats.

14-1.4 Formation of Districts

A procedure was developed [Ref. 2] that permits a computer to search the merged data files and to form compact and contiguous election districts that satisfy all criteria of the Board of Elections (details appear in Appendix B). The computer program can readily be set to any upper limit on the number of voters per election district; for example, trials were conducted with limits of 600, 680, and 750.

14-1.5 Mapping by Computer

A set of official maps must ultimately be produced to satisfy legal requirements. These maps must clearly show the low and high house numbers that straddle each district boundary. With the new system, the computer generates a map overlay showing the number of registered voters in each voter residence; a sample of this map appears in Fig. 3. Next, the computer produces an overlay map that displays the recommended district lines and the house numbers of each voter residence in the district; a sample of this map appears in Fig. 4 (although the district lines are omitted in this rendition). The latter map can then readily be used by a draftsman to produce a finished map, shown in part in Fig. 5.

14-2 RESULTS

The system for computerized redistricting was applied to a trial area, the 71st Assembly District in Manhattan; the results are shown in Table I. Efficiency is boosted from 69 per cent to more than 90 per cent, and the number of voting machines required (equal to the number of districts) is reduced by as much as 25 per cent.

Splitting of blockfaces, which is not done now, is both legal and acceptable to the Board of Elections. However, because it is a break from past practice, both procedures were tried, and the results are shown separately in Table I. If blockface splitting is permitted, the efficiency of computerized redistricting rises from 88 per cent to 93 per cent. On the basis of the pilot study, it appears that, even in the most densely populated areas, where splitting of a blockface between two districts would be most advantageous, less than 1 per cent of all blockfaces would be split.

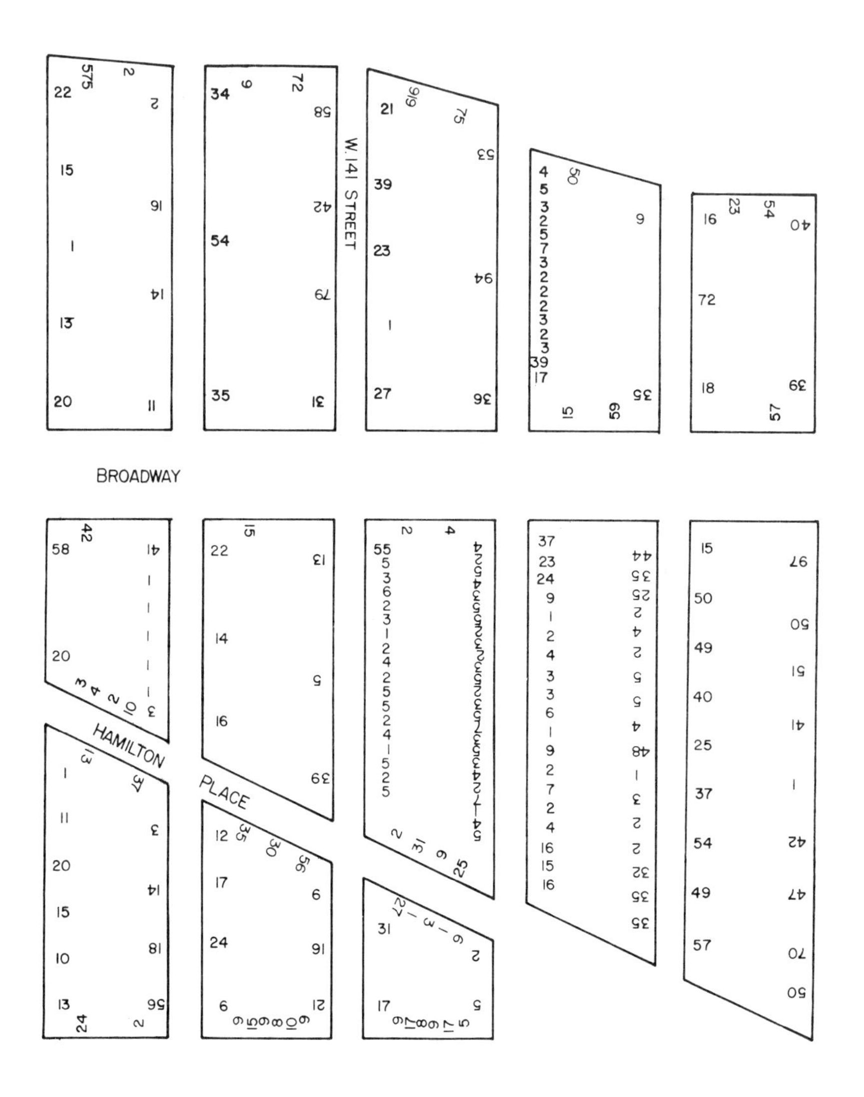

Figure 3
Computer-generated map overlay showing the number of registered voters by block location.

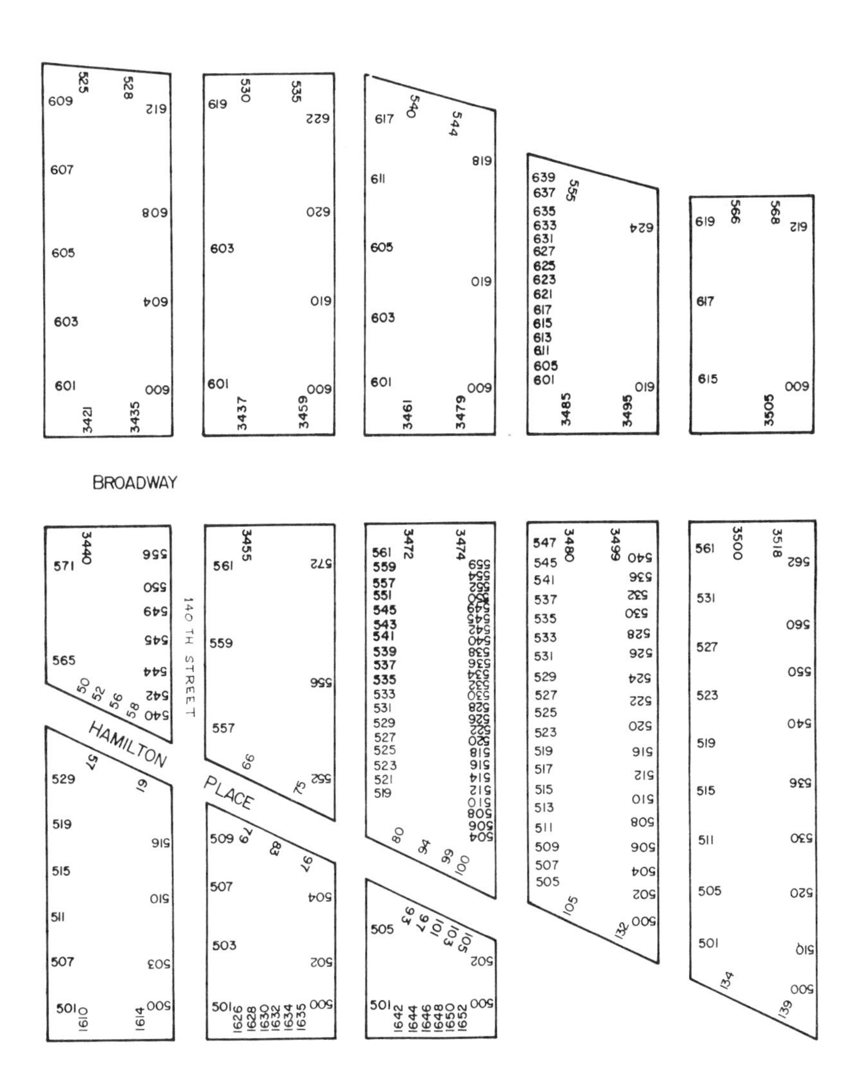

Figure 4
Computer-generated map showing building addresses.
(District lines have not yet been drawn on
this map by the computer.)

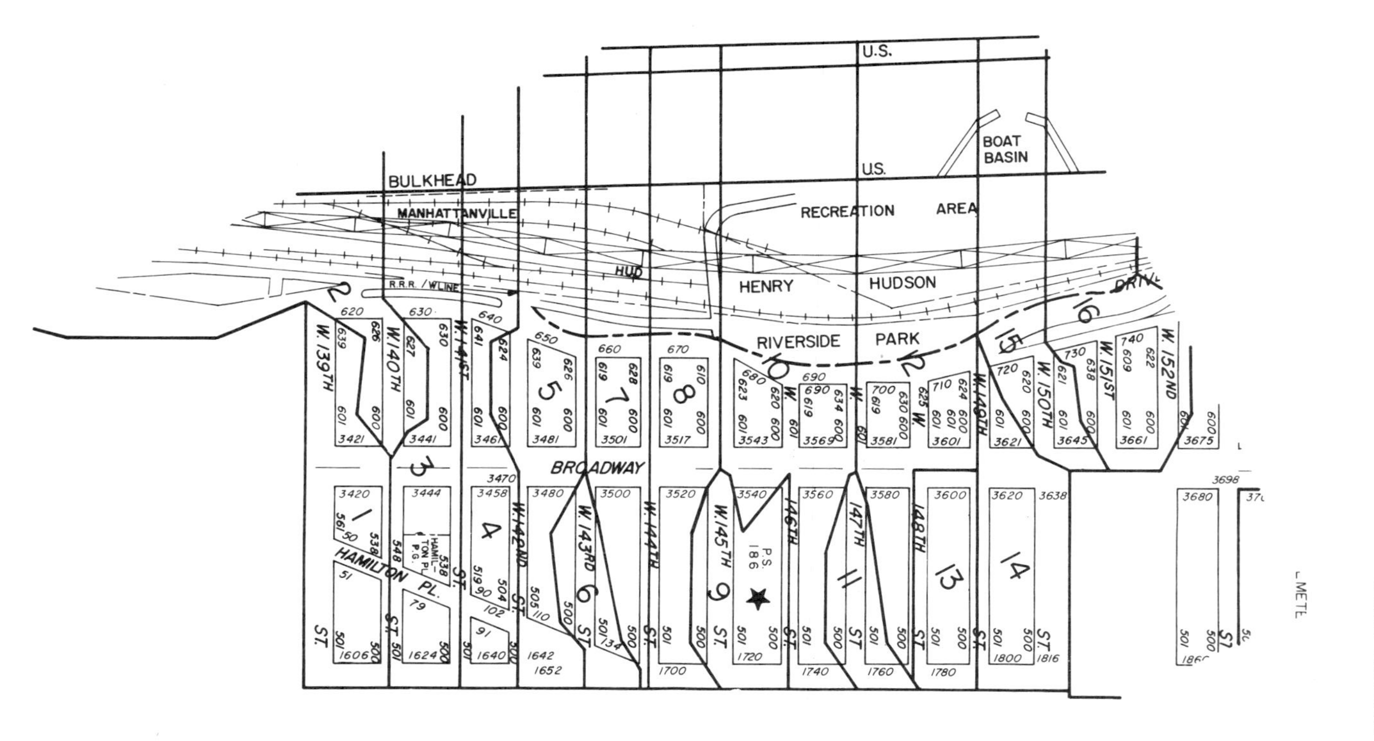

Figure 5
Example of a Finished Map Prepared by an Artist from a Computer-Generated Map

Table 1

RESULTS OF COMPUTERIZED DISTRICTING IN A TRIAL AREA

	Max. no. of voters/ machine	Avg. no. of voters/ machine	No. of machines	Percent of machines saved	Effi- ciency[a] (%)
Existing districts in 71st AD	750	529	85	--	69
Computerized redistricting without splitting blockfaces	750	653	68	20	88
Computerized redistricting with splitting of blockfaces	750 680 600	697 617 573	64 70 78	25 17 9	93 93 95

[a] Efficiency equals the average number of voters per machine divided by the maximum allowable number of voters per machine, multiplied by 100 per cent.

The underlying reason for the increase in efficiency is that the computer designs the election districts more equitably as far as the voters are concerned, and thereby produces a more reasonable apportionment of voting machines to voters. This is shown clearly in Fig. 6. At present, one-quarter of the districts contain more than 640 registered voters, while another quarter contain fewer than 450. (One odd district in the city has only twelve voters!) This situation leads to long lines in some districts and no waiting at all in others. The computer-generated districts are much more uniform in size, with 75 per cent of them containing 700 to 750 registered voters. When the upper limit is reduced to 600 (not shown in Fig. 6), 80 per cent of the districts contain 550 to 600 voters.

It should be emphasized that the location of polling places is generally not affected by this procedure; only the district lines are altered, and voters generally can continue to vote at their accustomed locations with minimum changes, particularly in areas where many districts share a common polling place.

14-3 CONCLUSIONS

14-3.1 Reduced Costs

The increased efficiency achievable by the computer-based system can be translated directly into cost savings. Appendix C examines these in detail and shows that gross savings of more than one million dollars can be realized for each election. (An additional $150,000 saving would be obtained if the existing Board of Election policy were to be modified to permit the splitting of blockfaces. However, this added saving would be somewhat offset by an additional mapping cost incurred by having to indicate the addresses where election-district lines split blockfaces.)

The gross savings are partially offset by the cost of changing poll binders, enrollment files, central records, and computer files in response to election district changes. In addition, the computer-related cost of redistricting the entire city at any one time, after the system is developed, will be $12,000 to $15,000. The total cost of these changes for the whole city is estimated to be approximately 40 per cent of the gross savings. However, a portion of these expenditures will be incurred even under the present procedure because of routine shifts in political boundaries and population. When there are major changes in election laws,

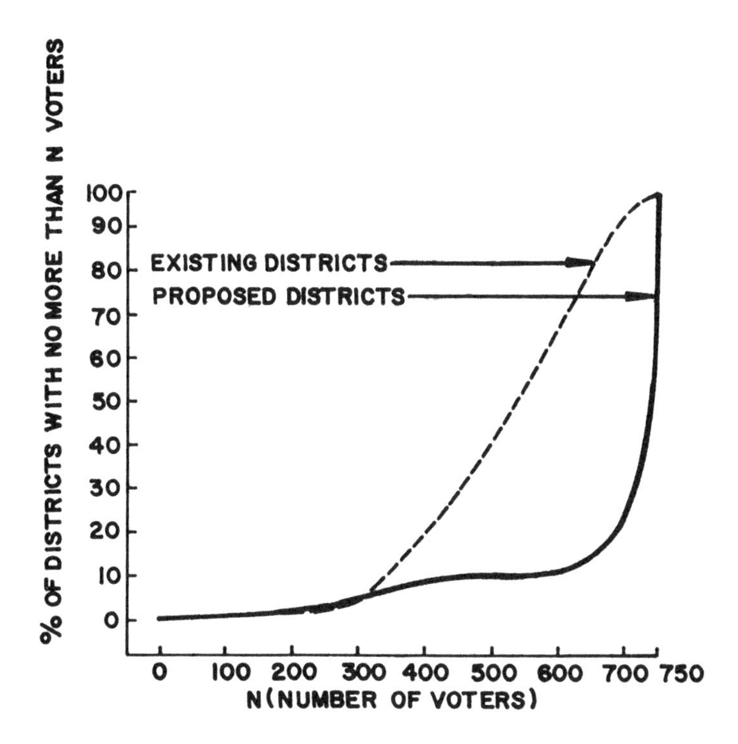

Figure 6
Cumulative distribution of the number of voters per voting machine in the 71st Aseembly District in Manhattan.

for example, resulting from the 1970 census or judicial reapportionments, all of these latter costs would be incurred with the present system. Furthermore, without computerization, such major changes would present formidable problems of crisis proportions to the Board of Elections. Consequently, the net annual savings are estimated to exceed $750,000.

To be conservative, no credit is claimed for the capital savings that would result from the fact that fewer machines and less storage space for them would be needed. Nevertheless, the acquisition of new machines at $1,823 each could confidently be terminated and it might even be possible to sell up to 2,000 machines that would become excess.

Another advantage of increased efficiency is that it would alleviate the problem of finding and training sufficient workers to staff the polls during elections. This is becoming a nuisance in some parts of the city.

14-3.2 Improved Service

Under the computerized districting system, election districts will have approximately equal numbers of registered voters. This should provide more uniform service to the public and reduce the incidence of long waiting lines at some voting machines while other voting machines, sometimes in the same polling place, are underutilized. Furthermore, some of the cost savings achieved by the computerized system can be traded for improved service. Thus, instead of taking full advantage of the State's legal maximum of 750 voters per voting machine, it is recommended that the limit be reduced to 600 and that, in densely populated areas, two-machine districts of 1200 voters become the norm, thus reducing the vulnerability of these districts to machine breakdowns. The decrease below the legal maximum, when coupled with computerized districting, will reduce delays and equalize delays among election districts, and will still produce annual savings of the magnitude indicated above.

14-3.3 Improved Analytical Capability

As an important by-product, the computerized election-districting system will make the voter data of the Board of Elections more accessible. For instance, the mapping, tabulation, and analysis of voter distributions and voting patterns for any segment of the city will be an easy task. The system will be able to provide a variety of listing, large-scale local maps, and other aids to facilitate the orderly and efficient conduct of elections. It will also enable the Board to assess rapidly the effects of various proposed policies and changes in election laws.

More generally, the capability of this system in manipulating, analyzing, and mapping geographic data is a vital element in the city's emerging management information systems. More specifically, the ability to produce a variety of maps and to create equal-sized districts readily according to any relevant variable could be applied to the districting problems affecting many agencies, including sanitation, police, hospitals, and schools.

APPENDIX A

Data Elements in the Block-Face File

Identification Data

- Borough
- Tax Block No.
- Tax Block suffix
- Street name
- Street suffix
- Street ending
- City Sectional Map No.
- Census Tract
- Census Block
- Census Block Suffix
- Type I Adjacent Block sides
- Type II Adjacent Block sides
- Type III Adjacent Block sides

Districts

- Health Area
- Police Precinct
- Fire Company
- Sanitation District
- Hospital District

Location

- Street Segment Identification
- Number of Block sides on Block
- Block Side No.
- High House No.
- Low House No.
- High Intersecting Street
- Low Intersecting Street

High Block side No.
Low Block side No.
Geographic coordinates:
High Corner: X_H
Y_H
Low Corner: X_L
Y_L

APPENDIX B

Districting Algorithm

In order for a computer to be able to aggregate blockfaces into compact, contiguous, and efficient election districts; it is necessary to define the physical relation among nearby blockfaces. This is accomplished by defining three kinds of blockface relations (see Fig. 7):

1. First-order adjacency--blockfaces that share a corner of the same block.
2. Second-order adjacency--blockfaces directly across the street (i. e., sharing the same two intersecting streets).
3. Third-order adjacency--blockfaces along the same side of the same street but across an intersecting street.

For each blockface, the blockfaces that bear a first-, second-, or third-order adjacency with respect to it are identified. A computer program can be written to determine adjacencies automatically.

The districting procedure uses a geographical-preference concept to allocate voters to election districts. In essence, the procedure tends to aggregate blockfaces by traveling back and forth between east and west, starting from the southernmost street and proceeding northward. Wherever there is a possibility that excessively elongated districts will be formed, pseudo-barriers are specified beforehand as input to the computer. These serve the same role as common-lot boundaries, i. e., blockfaces bordering on these pseudobarriers are treated as having no second-order adjacencies.

The districting program is divided into a 'geographic' section and a 'districting' section. A simplified flowchart of the geographic section of the program is shown in Fig. 8. The procedure starts by summing voters starting in the southeast corner of the common lot. It then checks for the number of first-order adjacencies available to the initial blockface. If there is one

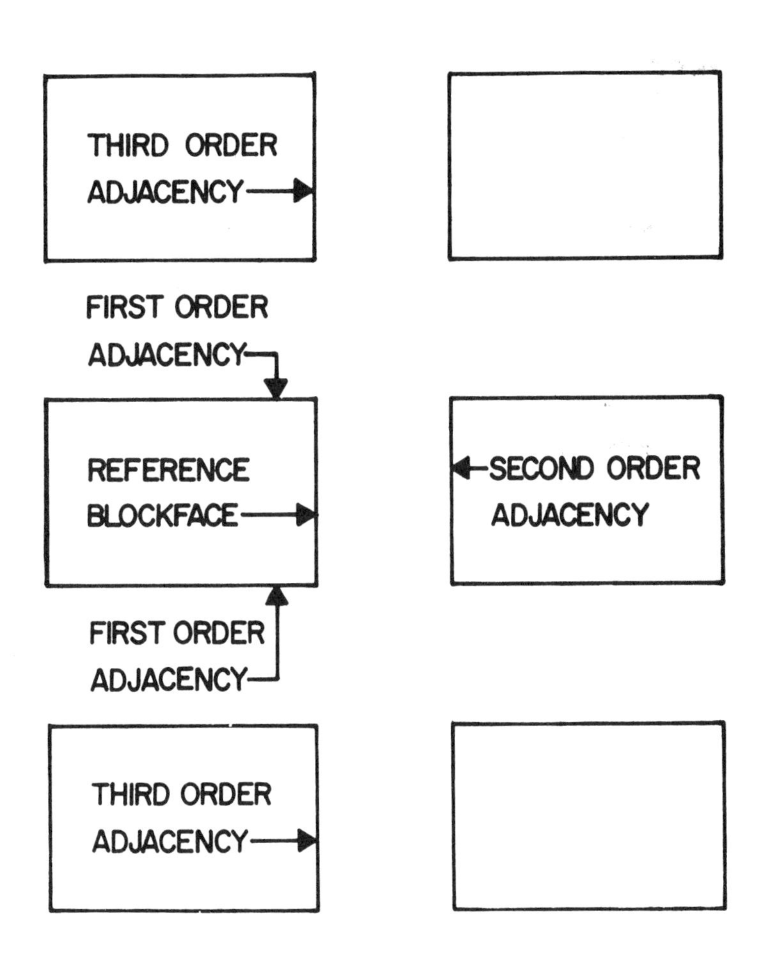

Figure 7
Blockface Adjacencies Diagram

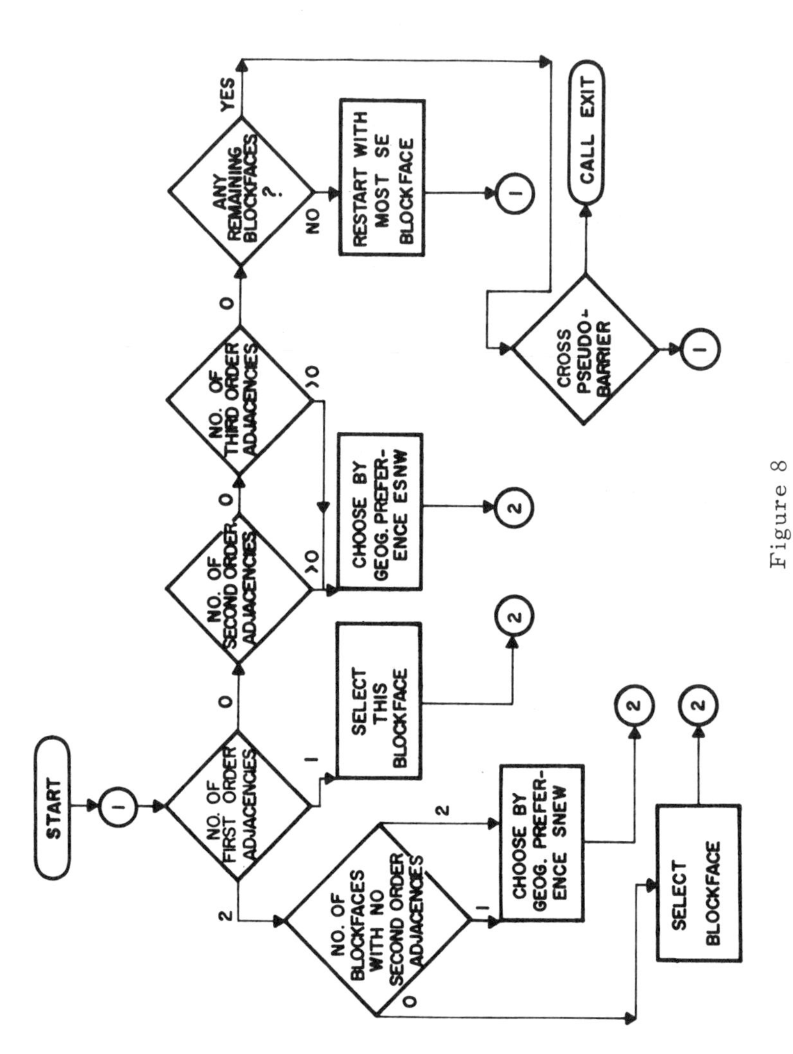

Figure 8
Simplified Flow Chart of Geographic Section of the Districting Program

first-order adjacency, it will select that blockface and proceed to the districting section of the program. If more than one first-order adjacency is available, the procedure selects the blockface with no second-order adjacencies. If there is more than one blockface with no second-order adjacency, a choice is made on the basis of geographic preference (south, north, east, west). If there are no first-order adjacencies, then a second-order adjacency (if available) is selected in the preferential sequence east-south-west-north. If there are no first- or second-order adjacencies, then a third-order adjacency is selected on a geographical preference basis as above. The program then proceeds to the districting section. Any blockface that is not completed will be the starting point for the next district.

A flowchart of the districting section of the program is shown in Fig. 9. The number of voters on each newly selected blockface is tentatively added to the previously accumulated sum, and a test is made to determine whether the new total exceeds the limit. If the limit is exceeded, then a decision must be made whether or not to split this blockface. The criterion is to split a blockface only if this action increases the efficiency of that district by more than some fixed percentage. If it is decided not to split the blockface, the next election district will start with the newly selected blockface. If the blockface is split, the next election district will start with the remaining voters of the split blockface.

The computerized program is expected to district efficiently approximately 95 per cent of all election districts. There will be a few exceptions resulting from erratically oriented blocks that can cause the computer procedure to form 'pockets.' The search procedure developed appears to minimize the number of pockets formed even in areas of very erratic geography. A capability can be provided for mapping the voter distribution for areas where a substantial number of pockets should be formed. This map, together with appropriate listings (which are also generated by the program), would permit manual adjustment of districts by means of punched cards. It is anticipated that such adjustments will not be necessary in many cases.

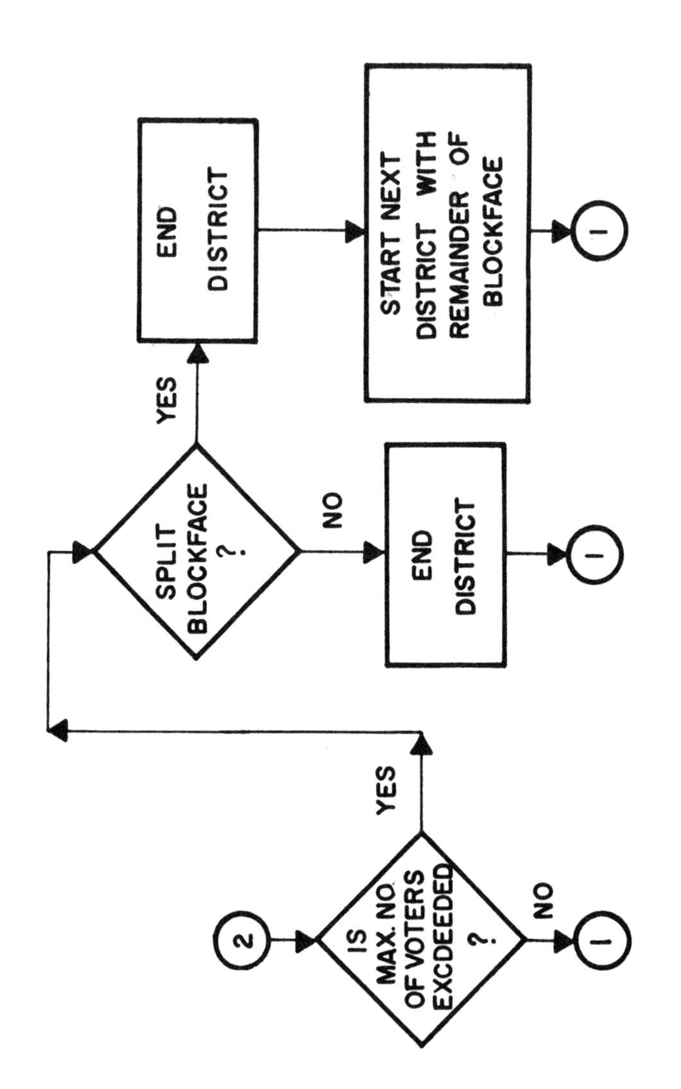

Figure 9
Flow chart of the districting section of the districting program.

APPENDIX C

Analysis of Costs and Savings

The direct cost of conducting annual elections depends upon the operating cost of an election district and the number of election districts (ED). In turn, the number of election districts depends upon the following factors:

Number of voters.

Legal limit on the number of voters in a one-machine or two-machine election district.

Efficiency of election districts, i.e., the ratio of actual voters per ED to maximum permissible number of voters per ED.

Ratio of one-machine to two-machine districts, which is a policy matter.

Each of the factors will be explored in the subsequent discussion.

A. Operating Costs of an Election District

For the purposes of this discussion, the relevant costs that must be considered are the capital costs of voting machines and the direct operating costs associated with election districts.

Capital Costs. The purchase cost of a voting machine is $1,823. It has a useful life of 30 years.

Operating Costs. The operating costs of a one-machine election district are displayed in Table II. This table shows both the costs that appear in the budget of the Board of Elections and the costs that appear in the budget of the Police Department (for police time). It also shows the annual custodial and warehousing costs for that voting machine, as well as the operating costs incurred for a primary election, registration, general election, and School Board election.

It is important to focus on the displaceable costs, which are $600 annually without the School-Board elections and $680 if the latter are included. It is only these displaceable costs that are eliminated or added as a result of eliminating or adding an election district. The corresponding displaceable costs of a 2-machine election district are $680 and $780. These costs, which will be featured prominently in the subsequent discussion, are summarized in Table III.

Table II

AVERAGE ANNUAL OPERATING COSTS FOR A ONE-MACHINE ELECTION DISTRICT

	1	2	3	4	5	6	7	8	9
	Pri-mary	Registration			General election	Ann. cost (1+4+5)	School elec-tion	Avg. ann. School el. (c)	Ann. total (6+8)
		Pres. yr.	Other yrs.	Avg. ann.					
Inspectors salaries	$49	$175	$105	$122	$105	$276	$49	$36	$312
Repairmen	5				5	10	0	0	10
Trucking	40				40	80	6	4.50	84.50
Printing	55	10	10	10	30	95	50	37.50	132.50
Packing supplies	2.50	2.50	2.50	2.50	2.50	7.50	2.50	2.00	9.50
Advertising	3	3	3	3	4	10	3	2.00	12.00
Rent (or equiv.)	20	20	20	20	40	80	10	7.50	87.50
Police expenses	2	2	2	2	6	10	2.50	2.00	12
Police time	25	25	25	25	50	100	25	19	119
Custodial cost(a)						50			50
Warehouse cost(b)						50			50
Total cost						$768.50		$110.50	$879
Displaceable cost						$600			$680

(a) Annual custodial cost of $50 per machine, including maintenance. One custodian is required for every 100 machines.

(b) Annual warehousing cost is $1 per sq. ft. One machine occupies 50 sq. ft.

(c) Three school-board elections will be held every four years.

Table III

DISPLACEABLE COSTS

	1-machine district	2-machine district
Without school-board elections	$600	$700
With school-board elections	$680	$780

Table IV

DISTRIBUTION OF ED's PER POLLING PLACE: QUEENS

No. of ED's per polling place[(a)]	No. of polling places	No. of ED's	Per Cent of ED's total
1	115	115	8.3
2	73	146	10.6
3	58	174	12.6
4	53	212	15.4
5	40	200	14.5
6	26	156	11.3
7	19	133	9.6
8	10	80	5.8
9	11	99	7.2
10	4	40	2.9
13	2	26	1.9
Total	411	1381	100

[(a)] Average number of ED's per polling place: 3.36; based on 1969 general election.

B. Maximum Number of Voters per Election District

According to current State law, the upper limit on the number of voters per ED is 750 for one-machine districts and 1050 for two-machine districts. Taken together, this is a strange pair of limits. It is much more logical that the latter be twice the former, which would tend to equalize the delays for all voters, regardless of whether they are assigned to one-machine or two-machine districts. Accordingly, one outgrowth of this project was a recommendation that the limits be changed in simple fashion to 600 and 1200 voters respectively, a change which requires action by the State legislature. (Actually, queuing theory suggests that a two-machine district can accommodate even more than twice as many voters than a one-machine district, with equal waiting times.)

C. Efficiency of Election Districts

It has been shown above that computerized formation of election districts produces ED's whose efficiency is 93 per cent or 88 per cent depending on whether block-face splitting is or is not employed, in contrast to the gross City-wide efficiency of 61 per cent (see Fig. 2). An examination of the ED's in one county shows that one-machine ED's have an efficiency of 69 per cent while two-machine ED's have an efficiency of 82 per cent; these conservatively high figures are used in the analysis below.

The maximum efficiency that can be achieved by manual redistricting can be calculated both historically and experimentally. Historically, manual redistricting produced efficiencies of 80, 85, 78, and 77 per cent in 1950, 1952, 1956, and 1960 respectively (see Fig. 2). A recent experimental redistricting by the Board of Elections, aimed at increasing the number of two-machine ED's, led to efficiencies of 76 and 86 per cent for one-machine and two-machine ED's respectively. Thus, the experimental evidence corroborates the historical facts, and for the purposes of the subsequent analysis, it is assumed that manual redistricting will result in efficiencies of 76 and 86 per cent for one-machine and two-machine districts, compared to computer-generated districts with efficiencies of 88 per cent (for either one-machine or two-machine districts) if no blockfaces are split, or 93 per cent if blockfaces are split.

D. Fraction of Districts with Two Machines

An election district with two machines is much less vulnerable to breakdown of a voting machine than is a district with only one machine, because the probability of two machines in the same ED breaking down on the same day are slim indeed.

Furthermore, because a two-machine district has an operating cost only $100 more than a one-machine district, there is substantial room for monetary savings as well.

The term 'policy ratio' will be used to refer to the fraction of all ED's that have two machines. As of 1968, only 30 per cent of the ED's in the City had two machines; in Manhattan, the figure was only 18 per cent. The recent experiment by the Board of Elections, referred to in the preceding section, was able to raise this figure from 18 to 47 per cent without much effort.

One possible argument against extensive doubling up is that the resultant election districts would be too large in area and that voters would be forced to walk too far to their polling places. Another related argument might be that redistricting in this fashion would require wholesale changes in polling-place assignments, and that voters would be confused by having to break old habits and appear at different, unaccustomed, polling places.

These points were examined by studying the relation between election districts and polling places. Two boroughs were selected, Manhattan because it is the most densely populated, and Queens because it is the least densely populated of the four boroughs, which together comprise 96 per cent of the city's electorate.

Tables IV and V show the tabulations for these boroughs. Queens has an average of 3.4 ED's per polling place and has 92 per cent of its ED's located in polling places that serve two or more ED's. Manhattan averages 2.7 ED's per polling place and has 86 per cent of its ED's located in polling places that serve two or more ED's. Therefore, one can confidently conclude that the formation of efficient two-machine ED's can be carried out without inconveniencing the electorate. The vast majority of voters would be unaffected and would continue to vote at their accustomed polling places. Only the internal administrative apparatus of the Board of Elections would be altered by such a policy.

E. Effect of the Various Factors on Costs

The total operating cost of conducting an election can be represented by the formula

$$C = c_1x_1 + c_2x_2 \quad (1)$$

Table V

DISTRIBUTION OF ED's PER POLLING PLACE, MANHATTAN

No. of ED's per polling place[a]	No. of polling places	No. of Ed's	Per Cent of ED's to total
1	143	143	13.4
2	102	204	19.1
3	45	135	12.6
4	38	152	14.2
5	20	100	9.4
6	18	108	10.1
7	16	112	10.5
8	6	48	4.5
9	3	27	2.5
10	4	40	3.7
Total	395	1069	100

[a] Average number of ED's per polling place: 2.71; based on 1969 general election.

where C = total operating cost,
c_1 = cost of a 1-machine ED,
x_1 = number of 1-machine ED's,
c_2 = cost of a 2-machine ED,
x_2 = number of 2-machine ED's.

Additional applicable formulas are the following:

$$x_1E_1L_1 + x_2E_2L_2 = N \,, \tag{2}$$

$$R = x_2/(x_1 + x_2) \,, \tag{3}$$

where E_1 = efficiency of 1-machine ED's,
E_2 = efficiency of 2-machine ED's,
L_1 = upper limit on size of 1-machine ED,
L_2 = upper limit on size of 2-machine ED,
N = total number of registered voters,
R = the policy ratio; ratio of 2-machine ED's to total ED's.

Given the values of the cost parameters c_1 and c_2, the policy-determined value of R, the values of E_1 and E_2 (which are determined by the mode of redistricting--manual or computerized), the legally determined values of L_1 and L_2, and N, the number of voters in the City, the above three equations can be solved for the three unknowns: x_1, x_2 and C; that is, the number of one-machine and two-machine districts and the total cost of conducting annual elections.

Using this method, Table VI was constructed to show how different policy ratios, efficiencies, and upper limits affect the cost of elections. The current condition is also shown for purposes of comparison.

The relation among these factors is clarified when displayed graphically, as in Fig. 10.

ACKNOWLEDGEMENTS

The cooperation of Commissioner Maurice J. O'Rourke and Alexander Bassett of the Board of Elections of the City of New York is gratefully acknowledged. The author is indebted to Eric Brodheim, Charles Wang, and Ivor Herzer of the Riverside Research Institute for developing the districting approach utilized here.

Table VI

NUMBER AND COST OF ELECTION DISTRICTS[a]

	R	Efficiency (%)		Max. voters per ED		Number of ED's			No. of machines	Annual operating cost[b] of ED's (millions of $)					Annual savings[b] (millions of $)		
		E_1	E_2	1-M	2-M	1-M	2-M	Total		1-M	2-M	Total	B/E	PD	B/E	PD	Total
Current system[a]	.30	69	82	750	1050	3656	1531	5187	6718	$2.19	$1.07	$3.26	$2.74	.52	---	---	---
Manual redistricting	.30	80	82	750	1050	3276	1404	4680	6084	1.97	.98	2.95	2.48	.47	.26	.05	.31
	.47	76	86	750	1050	2340	2035	4375	6410	1.40	1.42	2.82	2.38	.44	.36	.08	.44
Computerized redistricting (without splitting)	.30	88	88	750	1050	3003	1287	4290	5577	1.80	.90	2.70	2.27	.43	.47	.09	.56
	.50	88	88	750	1050	2000	2000	4000	6000	1.20	1.40	2.60	2.20	.40	.54	.12	.66
	.75	88	88	750	1050	923	2770	3693	6463	.55	1.94	2.49	2.12	.37	.62	.15	.77
Computerized redistricting (with splitting)	.30	93	93	750	1050	2850	1220	4070	5290	1.71	.85	2.56	2.15	.41	.59	.11	.70
	.50	93	93	750	1050	1900	1900	3800	5700	1.14	1.33	2.47	2.09	.38	.65	.14	.79
	.75	93	93	750	1050	874	2620	3494	6114	.52	1.83	2.35	2.00	.35	.74	.17	.91
Computerized redistricting (without splitting)	.30	88	88	600	1200	3235	1388	4623	6011	1.94	.97	2.91	2.45	.46	.29	.06	.35
	.50	88	88	600	1200	2005	2005	4010	6015	1.20	1.40	2.60	2.20	.40	.54	.12	.66
	.75	88	88	600	1200	858	2575	3433	6008	.51	1.80	2.31	1.97	.34	.77	.18	.95

(a) Based on 3,174,096 registered voters in 1966.

(b) Without school-board elections; based on displaceable costs per 1-M and 2-M ED of $600 and $700 respectively, of which $100 is Police Department cost in each case.

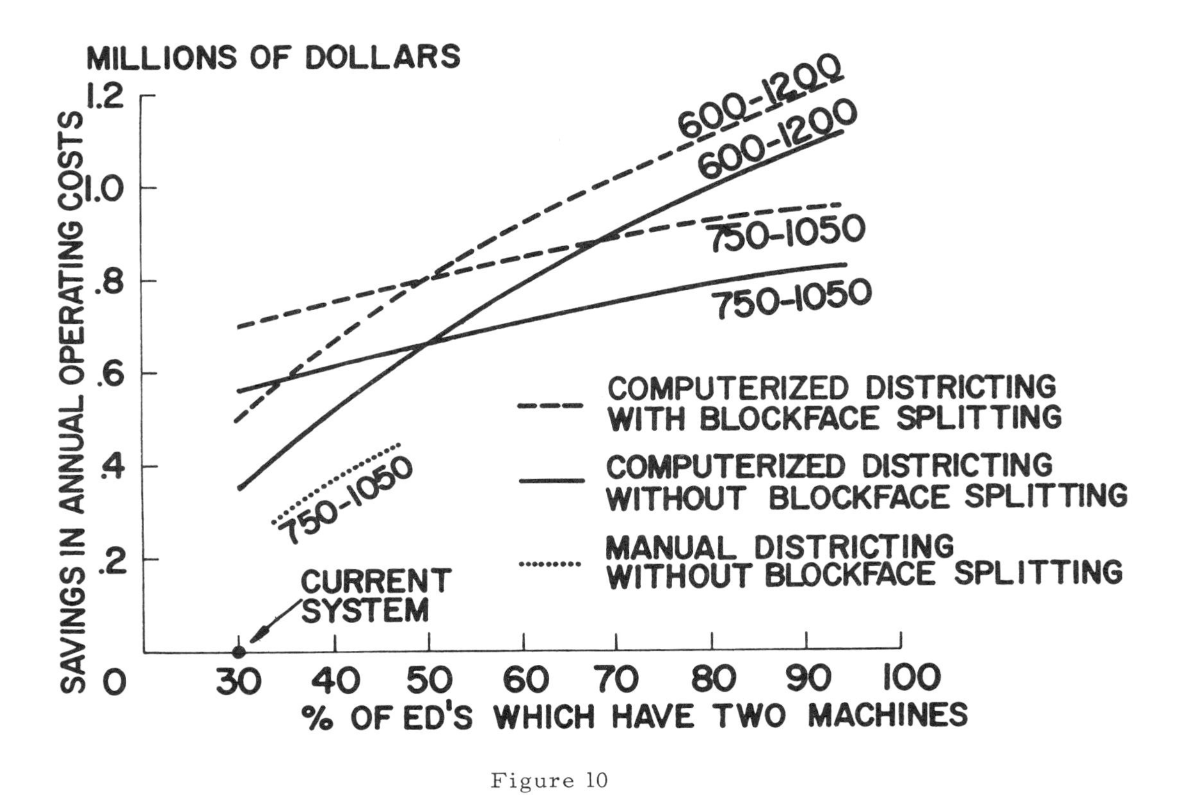

Figure 10

The Effects of Policy Ratio, Efficiency, and Upper Limits on Cost Savings

REFERENCES

1. E.S. Savas, R. Amsterdam, and E. Brodheim, "Creation of a Geographic Information System," Proc., 4th Annual Symposium on the Application of Computers to the Problems of Urban Society, Assoc. for Computing Machinery, New York, 1969.
2. E. Brodheim, C. Wang, and I. Herzer, "A Computerized System for Forming Efficient Election Districts," Proc., 3rd Annual Symposium on the Application of Computers to the Problems of Urban Society, Assoc. for Computing Machinery, New York, 1968.

Chapter 15

SYSTEMS EVALUATION OF SOCIAL AGENCIES

A. L. Service S. J. Mantel, Jr., A. Reisman

ACKNOWLEDGEMENTS

As is indicated in the text, this work is the product of a genuinely interdisciplinary research effort. Therefore, although this report was prepared by a particular subset of the research team, the contributions of numerous other individuals should not be minimized. Current members of the project team include: Mrs. Miriam Jaffee, Dr. Howard Reiger, Mr. Richard Ronis, and Mr. Judah Rubenstein of the Jewish Community Federation of Cleveland; Dr. Arthur Blum and Prof. Raymond Koleski of the CWRU School of Applied Social Sciences; and Dr. Burton Dean and Mr. Robert Reich of the CWRU Department of Operations Research. Furthermore, in view of the highly evolutionary nature of this work, it is also important to acknowledge the influence of former members of the research team; namely, Mr. Alan Beckman (Benjamin Rose Institute), Mr. Norman Eisenberg (City of New York), Mr. Ralph Gregory (Welfare Federation of Cleveland), Dr. Elliot Markus (Israel Institute of Applied Social Research), and Dr. Bernard Olshansky (Combined Jewish Philanthropies of Greater Boston). This work was performed under a grant from the Cleveland Foundation.

The authors are respectively, Instructor, Department of Operations Research and the School of Applied and Social Science Case Western Reserve University; Professor, College of Business Administration, University of Cincinnati; Professor, Department of Operations Research, Case Western Reserve University, Cleveland, Ohio.

15-1 A SOCIAL WELFARE SYSTEM - THE JEWISH COMMUNITY FEDERATION OF CLEVELAND

15-1.1 Organization of the Federation

The Jewish Community Federation of Cleveland (JCFC) is a voluntary association of individual contributors and private social agencies that seeks to serve the needs of the Jewish Community at home and overseas through organized, cooperative action and leadership. Locally, its member agencies include several old age homes, several religious educational institutions ranging all the way from primary grades to higher education, and individual, family, and vocational counselling agencies. These agencies serve the health, welfare, and educational needs of a Jewish Community numbering approximately 80,000 people and utilize a total annual operating budget of approximately nineteen million dollars, toward which the Federation allocates about four and one-third million dollars.

In addition to its institutional members, the JCFC has a large number of individual members; namely, all those members of the Jewish community who have made a contribution of ten dollars or more to the Jewish Welfare Fund Appeal. This latter group is essential to the productivity of the federated system, as is clearly demonstrated by the majority representation of the "community-at-large" on the Board of Trustees, the central administrative body of the Federation. The Board also includes representatives of the various local member agencies and of the Delegate Assembly, a standing forum for discussion of matters relating to the JCFC and the Cleveland Jewish community as a whole. Operationally, the Board of Trustees directs the functioning of the Federation through a staff of social welfare professionals and a network of standing and special committees of laymen.

Three of these committees are especially related to the work described in this report: the Welfare Fund Planning Committee, the Community Services Planning Committee, and the Budget Committee. The Welfare Fund Planning Committee is charged with organizing and administering the annual Jewish Welfare Fund campaign and also acts as the social planning arm of the Federation in relation to national and overseas beneficiary agencies. Local planning is the responsibility of the

Community Services Planning Committee, a body established to "study the needs of the Jewish community in the fields of health, welfare, and education; observe and evaluate the functioning of the agencies operating in such fields;...seek improved ways and means of meeting such community needs, both currently and prospectively; and make recommendations to the Board of Trustees with reference thereto."[1] (The CSPC includes agency executive directors, i.e. "professionals," among its members.) Even though the member agencies do retain a significant degree of autonomy, some central consideration of the allocation of the Federation's budget is nevertheless necessary. This is the province of the Budget Committee, established to "consider budgetary requests made of Federation and to recommend appropriations directly to the Board of Trustees; to keep informed on the financial status of beneficiary agencies; to maintain budget controls; to prepare recommendations to the Board of Trustees regarding the financial operations of Federation; and to consider the appropriate matters relating to the finances of Federation."[2] These three committees, working closely with the Board of Trustees and the local member agencies, thus constitute the core of the Federation's organizational structure.

Before turning to the topic of decision-making *per se*, it is pertinent to note that not all of the JCFC's local member agencies are considered in this systems analysis. Several agencies, not engaged in direct social services to clients, were ignored for purposes of this analysis as were two agencies whose budgets constituted only a miniscule part of the total allocation. The sixteen remaining local member agencies are listed in Table 1.

---

[1] By-Laws of the Jewish Community Federation of Cleveland, Article I, Section A-12.

[2] By-Laws of the JCFC, Article I, Section A-1.

TABLE 1

AGENCIES INCORPORATED IN SYSTEMS ANALYSIS
OF JEWISH COMMUNITY FEDERATION OF CLEVELAND

1. Akiva High School
2. Bellefaire[1]
3. B'nai B'rith Hillel Foundation
4. Cleveland College of Jewish Studies
5. Cleveland Hebrew Schools
6. Hebrew Academy
7. Jewish Children's Bureau-Jewish Day Nursery
8. Jewish Community Center
9. Jewish Convalescent and Rehabilitation Center
10. Jewish Family Service Association
11. Jewish Vocational Service
12. Menorah Park-Jewish Home for Aged
13. Montefiore Home-Home for the Aged
14. United Jewish Religious Schools
15. Workmen's Circle School
16. Yeshivath Adath B'nai Israel[2]

15-1.2 Decision Making in the Federation

It should be clear from the foregoing description that decision-making in the federated system is a complex process involving a multiplicity of individuals, groups, and organizations. Laymen are involved through membership on the Board and the various committees at both the Federation and the member agency levels. Social welfare professionals function not only in fund-raising, budgeting, and planning roles as members of the Federation staff, but also as the providers of direct service to clients in the context of specific agencies. The community-at-large includes two other groups of decision-makers: clients, who must make decisions about the extent to which the federated system can satisfy their needs; and contributors, whose decisions concerning financial support are a key determinant of the Federation's ability to survive and grow. Finally, it should never be forgotten that undertaking research of the sort described herein constitutes the introduction of a new group of decision-makers into the system; namely, the research team. The analyst who ignores the latter reality runs a serious risk of seeing his work badly mauled by the political characteristics of the social system.

[1] A resident school for emotionally disturbed children.
[2] A secondary school emphasizing the study of the Scriptures.

Even though simultaneous and detailed consideration of all these aspects of the decision-making process may represent a desirable ideal, some narrower focus seems a matter of practical necessity given the current state of our techniques for analyzing social systems. This necessity, coupled with the operations researcher's natural propensity to gravitate toward budgeting decisions, has led this research to a primary concern with the Budget Committee's decisions about the allocation of the Federation's limited fiscal resources to the programs and services of member agencies. Of course, no sensible analysis could completely ignore the other elements and types of decision making, and we do not mean to imply that such has been the case here. In fact, some of the elements have been explicitly incorporated in the model developed and no part of the analysis has been divorced from the context of the network of decision-makers. The basic objective of the research, however, has been the development of a tool to be used by the Budget Committee to "improve" its decisions. Since budgeting and planning are so closely linked, both administratively and conceptually, it is entirely appropriate to take a somewhat larger view and regard the systems analysis as an attempt to provide a more systematic framework for the Federation's planning.

When viewed in this light, this work is a natural outgrowth of various recent efforts to improve planning for social welfare services. These have included attempts in Cleveland and other Jewish communities to undertake more scientific studies of population characteristics and to use the accumulated information for systematic planning purposes. [Ref. 1, 2] The incorporation of this broader base of information about the total community into planning is a necessary first step toward utilizing more objectivity than was previously possible in judging the effect of alternative courses of action upon the system as a whole.

This trend toward comprehensive planning and the use of more objective data is reflected in the Cleveland Jewish Federation's "Projection of Future Plans" [Ref. 3]. The projection was an attempt to develop coordinated program plans for the local member agencies and to provide an overview of services as an aid in determining specific directions for the future. The data included the purposes served by the agencies; the community needs addressed; the programs with which the agencies proposed to meet these needs; the estimated costs of such programs, and the agencies' views regarding financing and future capital requirements. Information was provided on personnel and facilities involved, guidelines were developed for future planning decisions, and specific programs were evaluated and priorities attached.

While the study represented a major advance in comprehensive planning for the community, it contained deficiencies, not the least of which was the inadequacy of information available. It was found that the information was difficult to convert into comparable form, that the reliability of some information was questionable for decision-making purposes, and therefore, that there was no mechanism for using this information to assess Federation planning and budgeting decisions. It was as a result of this planning effort and the recognition of its deficiencies that the Jewish Community Federation of Cleveland was stimulated to initiate the study reported in this paper.

15-1.3 The Research

In mid-1968, the Federation began to assemble a team whose function was defined as "exploring the feasibility of applying Operations Research principles and techniques to the Federation system." If the team could demonstrate the feasibility of this objective, it would then move to construct a model of the relationships between agency and Federation inputs (needs, personnel, funds, facilities) and outputs (services, movement toward community goals). Such a model would attempt to increase the ability of Federation planners to utilize the diverse resources of the federated system in a manner consistent with the needs and goals of the community.

The first step in the study was the selection of a research team. It was obvious that to deal with the complex problems of the federated system, a variety of skills and backgrounds would be required. Individuals with interdisciplinary backgrounds would be especially valuable. From among its own staff, the Federation selected a community planner, a researcher, and a budgeting executive. Faculty and doctoral students from the department of Operations Research and the School of Applied Social Sciences of Case Western Reserve University brought competence and experience in systems analysis, operations research, economics, community planning, casework, and group work. Among the several members with skills or knowledge in more than one area were several professionals who also functioned as active laymen within the Federation system.

The assignment of the team was to explore the feasibility of constructing a model to be used in planning for the local community. It seemed clear that an adequate model would have to incorporate not only objective measures of performance but also value judgments as to where community priorities lay. Accordingly, both of these areas were examined and the team addressed itself to a variety of questions:

. What variables should be measured?
. What role do these variables play in the overall model?
. How can these variables be measured?
. How can the various measures be appropriately aggregated?
. What are the commitments and desires of the community?
. Who should make the required judgments?
. How could the value judgments best be obtained?

The team saw its responsibility as including the development of methods for assisting in making such decisions as well as the identification of the system level at which the decision would most appropriately be made. For example, the quality of a given service can probably be best evaluated by the givers and users of that service and, perhaps, by some unbiased outside experts. On the other hand, determination of priorities between aged and youth services involves clarification of broad community objectives and is best left to lay community leaders.

Because of the importance of the linkages between the team and the Federation, the team never regarded itself as separate from the system. Ongoing involvement of Federation leaders was insured through a special committee of laymen charged with overall project direction. Interim reports, the testing of the methods and concepts of the analysis, and repeated discussion of the concepts and the problems involved the Federation laymen in the research process.

Results of the feasibility study [Ref. 4] were encouraging enough to convince the JCFC and the Cleveland Associated Foundations to continue through the testing and implementation stages. That portion of the work is still underway and so the model described here is still somewhat fluid even though the basic concepts are reasonably well developed. Indeed, the dynamic nature of the product of systems analysis here may be a distinguishing characteristic of useful work in this context.

15-1.4 Goals of the System

As is the case with any respectable system analysis, one of the initial tasks attempted by the research team was the definition of the goals of the system. An early problem that had to be resolved concerned the definition of the group of decision makers whose goals were to be defined and used. An iterative procedure eventually developed involving the research team, the Federation staff, and the members of the lay Operations Research Committee. First, a Federation committee whose members were representative of the various elements of the Jewish community was asked for opinions on community goals. These opinions, together with material taken from Federation documents and reports, were edited by the research team into a unified statement of community goals. This version was then resubmitted to the O. R. Committee for review and approval. The resultant "apple pie and motherhood" type of goal statement was acceptable to all and seemed to be an adequate representation of the community's goals. Some questions arose as to whether this process was either necessary or useful. It may even have been somewhat dysfunctional in that it initially sought agreement on specifics in areas where agreement on general principles would have sufficed. There may be merit in revising this goal setting process but for now the Statement of Goals that evolved is being used. It is not possible to say, however, to what extent this statement has in any way influenced the further course of the measurement model or the parameters within that model. In fact, the study of that influence might well be an interesting research topic but such a study is well beyond the scope of this work.

The goal statement produced is given in Table 2. Given these goals, the objective of the research was the development of a rational and consistent method of relating agency activities to the goals and then defining methods of incorporating that relationship in planning and budgeting decisions. The bulk of the research effort has been devoted to the development of a technique for consistently measuring the activities of the Federation's local member agencies. Time has also been devoted to the specification of an allocation model tailored to this particular measurement system.

TABLE 2

STATEMENT OF COMMUNITY GOALS

The primary goals of the Jewish community of Cleveland, are to promote the general welfare of the Jewish community, to serve the needs of its members (as noted below), and to insure the creative survival and continuity of the Jewish people. It seeks, within the framework of American democracy:

1. To develop and maintain a communal system of organizations and a network of agencies to serve the religious, cultural, educational, physical and social needs of Jews as individuals and as a community.
2. To maintain and strengthen the ties which exist between the Jewish community, locally and nationally, other segments of the larger community, and the American society as a whole; and to maintain and strengthen ties between the local Jewish community, the national Jewish community, and Jews throughout the world.

15-1.5 Structuring the System: Service - Client Packages

One of the first contributions of the systems analysis to the federated system was to reveal clearly the need for some uniform unit of account that could serve as a base point.[3] Further examination demonstrated that this fundamental structural element has three critical dimensions: the nature of the specific service delivered, the person to whom the service is delivered, and the agency that provides the service. To deal with the first dimension, a set of 30 services was developed.[4] Tests were made to insure that every agency could decompose its activities into some subset of these thirty categories. A set of "descriptors" for each service category was provided. It should be

---

[3] Current social welfare administrative practices tend to use services given in problem areas, e.g., refugees, drugs, etc. or agencies as units of account. These categories were felt to be too broad and/or overlapping to have much operational meaning in a study such as this.

[4] See Table 3.

noted that developing these descriptions was a non-trivial task. Attempts to find "definitions" for each service that would be applicable across all agency contexts were unsuccessful. It was possible, however, to describe the services by finding "discriminators" that would allow an agency to categorize services by discriminating them from other services. The agencies found this solution satisfactory. The set of 30 services that is to be used is closely compatible with the functional categories developed for the Federation's concurrent implementation of a functional accounting system. Without such compatibility, the systems analysis and the resultant models would be under a severe handicap since data collection costs would then have been prohibitive.

TABLE 3

AGENCY SERVICES

1. INDIVIDUAL COUNSELLING (Vocational-Educational)
2. GROUP COUNSELLING (Vocational-Educational)
3. INDIVIDUAL COUNSELLING (Psycho-social)
4. GROUP COUNSELLING (Psycho-social)
5. FAMILY COUNSELLING (Psycho-social)
6. FAMILY LIFE EDUCATION
7. INTAKE AND REFERRAL
8. FORMAL EDUCATION (Primary Jewish Focus)
9. JEWISH DAY SCHOOL EDUCATION
10. REGISTERED ACTIVITIES (Primary Jewish Focus)
11. REGISTERED ACTIVITIES
12. UNREGISTERED ACTIVITIES (Primary Jewish Focus)
13. UNREGISTERED ACTIVITIES
14. SOCIAL ACTION
15. ADOPTIVE SERVICE
16. FOSTER CARE
17. GROUP FOSTER CARE
18. INSTITUTIONAL HEALTH AND LIVING
19. TEMPORARY SHELTER
20. HOMEBOUND SERVICES
21. DAY CARE
22. DAY CAMP
23. RESIDENTIAL CAMP
24. RESETTLEMENT
25. JOB PLACEMENT
26. RELIGIOUS ACTIVITIES

27. INSTITUTIONAL HEALTH SERVICES (Short-Term)
28. INSTITUTIONAL HEALTH SERVICES (long-term)
29. NON-INSTITUTIONAL HEALTH SERVICES
30. SHELTERED WORKSHOP

The second and third dimensions of the structural unit presented less problem. Clients are identified as members of six age groups,[5] and the member agencies themselves provide a set of identifiers for the "server" dimension. The basic unit of account of the model is, thus, a service-client-agency package, P_{ijk}, where:

i = the service being offered,

$$i = 1, \ldots, 30$$

j = the client group receiving the service,

$$j = 1, \ldots, 6$$

k = the agency offering this service,

$$k = 1, \ldots, 16$$

TABLE 4

CLIENT GROUPS AND DESCRIPTIONS

GROUP	DESCRIPTION[6]
1. Pre-School Age	In general, includes all clients five years old or younger. Does not include any children participating in programs that are part of a formal school system.

[5] See Table 4

[6] Note that the first determinant of a particular person's client group is the orientation or focus of the service that individual is receiving, as determined by the serving agency. Age guidelines and other descriptors are meant to supplement this primary determinant.

2. Elementary School Age	Includes kindergarten through grade six level clients. These will be predominently ages six through twelve.
3. Junior High School - High School Age	Includes grade seven through grade twelve level clients. These will be predominantly ages thirteen through eighteen.
4. Young Adult	Includes all clients between ages eighteen and twenty-two.
5. Adult	Includes all clients between ages twenty-two and sixty-five who are not involved in a program aimed primarily at one of the other client groups (College Age or Aged).
6. Aged	All clients involved in programs geared to the aged. This will typically be clients over sixty-five years of age but may include some younger clients (e.g., some women in the sixty-two to sixty-five range).

15-2 MODELING THE SYSTEM: MEASUREMENT OF OUTPUT

15-2.1 The Concept

Basically, the objective of the research was a cost-effectiveness type of model to aid in the allocation of scarce community resources to local social services for the community. An essential prerequisite to such a model is an operational definition of "output" that can be applied with equal validity to each of the agencies considered in the systems analysis. Accordingly, the major thrust of the research effort to date has been the development of a mechanism for measuring the output associated with each of the service-client-agency packages in the system.

The approach to the problem, utilized throughout the generation of the model, has been a continuing attempt to separate concepts into their component parts. Thus even a brief examination of the notion of output or system preformance reveals three inherent questions that merit consideration: What is the system doing? How well is the system doing this? How important is it for the system to be doing this? The measurement model seeks to answer these questions in a uniform and quantitative manner.

The three questions define the parameters of interest. We will attempt to measure "what the system is doing" by measuring its physical throughput. To answer "how well" requires a measure of the quality of the service. The "how important" question requires an assessment of the relative value of each of the client-service packages. Each of these parameters can be further subdivided, as will be seen in what follows.

It is immediately clear that two distinct types of input information are required. Objective data, sufficient to quantify the above concepts completely, simply do not exist and so subjective information must play a vital role in the measurement model. Further, the distinction between subjective and objective is at times blurry, e.g., when "objective data" is estimated because it is too difficult (or expensive) to obtain directly. Thus some measurements are objective, some completely subjective, and some mixed. In general, the physical throughputs are measured objectively, the relative package values are entirely subjective, and the quality measures involve both types of information.

15-2.2 The Model Components

Physical throughput, the first of the parameters of interest, is quantified through observation of two sets of variables:

N = the number of different individuals served in a particular service-client-agency package during a given period of time.

STR = the amount of direct contact service time received by individuals in a particular service-client-agency package during a given period of time.

These are intended to be objective measures although some estimation or approximation may be necessary. It should be noted that the problems of collecting this information are eased in those cases where the specific throughputs can be linked to the functional accounting system currently being installed within the Federation system.

In contrast to the objectively measured physical throughput, the measurement of package quality involves both objective and subjective information. Since no "true" measures of quality could be obtained, the concept of quality was split into seven elements or quality criteria. Each of these elements was further decomposed into sub-elements in order to identify as explicitly as possible information that would serve as indicators of relative levels of package and/or agency quality. The seven criteria, together with their sub-elements, are given below:

ACCESSIBILITY - The degree to which service-client packages[7] are judged to be easily reached by segments of the community.

Judgments obtained from panel of "gatekeepers" concerning:

1. Physical accessibility
2. Economic accessibility
3. Program accessibility (includes timing and content)

CAPABILITY - The degree to which service-client packages satisfy expressed needs.

EFFECTIVENESS - The degree to which service-client packages achieve client-oriented goals

Judgments obtained by questionnaire and/or panel from:

1. Local experts external to JCFC system
2. Clients
3. Service providers

---

7 Criterion element ratings will be obtained with respect to service-client-agency packages whenever possible. In certain areas, the information available will not permit this. In these cases, ratings will be obtained with respect to groups of packages. These groups might result, for example, from the aggregating of all client groups for a particular service or the aggregating of all packages for a particular agency. Ratings would then be obtained with respect to services and agencies, respectively.

EFFICIENCY - The number of clients and client hours (throughput) per unit of input resources.

1. $\frac{\text{Cumulative hours of service received by all clients in this service}}{\text{Cost of professional staff assigned to this service}}$

2. $\frac{\text{Hours of direct staff contact with clients in this service}}{\text{Total cost of providing this service (including share of overhead)}}$

INTERRELATION - The degree of cooperation among agencies in the Federation

1. The amount of time spent by staff of one agency in providing service under the auspices of another agency or in acting as host for another agency.

2. The amount of time spent by staff of one agency engaging in planning and/or providing direct service jointly with another agency.

SYSTEM CONTRIBUTION - The degree to which service-client packages achieve community oriented goals.

1. Number of administrative and policy volunteers
2. Number of service volunteers
3. Hours of direct service given by service volunteers
4. Judgment on agency contribution to Jewish community's ability to raise funds
5. Judgment on agency contribution to the maintenance of the Jewish communal system
6. Judgment on agency contribution to the maintenance of good relations with the non-Jewish community.

UNIQUENESS - The degree to which service-client packages are available elsewhere in the community.

1. Number of Federation agencies offering this service-client package.

Two other concepts supplement this set of quality criteria and complete the specification of the quality component of the measurement model. These concepts are similar to ideas developed and tested in the context of personnel evaluation [Ref. 5] and program evaluation [Ref. 6]. Recognizing that the criteria were not necessarily equally important parts of service quality, provision was made for the generation of a set of relative weights. These weights would be designed to account for the differential significance of the criteria. In a similar vein, a device was included for translating the "raw" criterion ratings onto a uniform scale measuring intrinsic worth or "utility" to the system as a whole. It is important to note that this translation is not necessarily linear.

The final parameter requires a measure of the importance of value of client-service packages, a completely subjective piece of information. Aside from the actual measurement problem, which is serious enough, a basic question is raised by the decision to integrate "value" in the measurement model. Whose values will be integrated? Which set of decision-makers is to be considered paramount?

It is not clear that there is a single set of values that may properly be characterized as the opinion of the Jewish community of Cleveland. Furthermore, even if such a set of values does exist, it is not at all evident how one would go about determining them.

Several possible approaches were considered. At least four candidate groups were identified: representative members of the community at large, representative community leaders, selected Federation and agency professionals, and the research team itself. The most appealing concept involved convening a special panel of laymen experienced in community service and representative of the various interest groups within the Cleveland Jewish community. Previous experience indicated that a panel of approximately fifteen people could adequately represent all segments of the community, that a group of that size would be operationally manageable, and that the value position of such a panel would be acceptable to the general community and so the research moved in that direction.

15-2.3 The Output Measurement Function

The set of components described in the previous section are to be the basis for measuring the "output" of the system. Recalling that p_{ijk} refers to the package consisting of service, i, delivered to client group, j, by agency, k, we define:

$N_{ijk} \equiv$ number of different individuals served in p_{ijk}

$STR_{ijk} \equiv$ total amount of direct service time received by individuals in p_{ijk}

$Q_{ijk} \equiv$ quality of p_{ijk}

$V_{ij} \equiv$ value of p_{ijk}[8]

One of these terms, Q_{ijk}, merits further explanation. Based fundamentally upon the various element data, the quality score results from the following process, a direct analogue of the methods described in [Ref. 5, 6].

1) raw scores for the elements of each quality criterion are scaled and averaged to obtain a rating for the p_{ijk} with respect to each of the seven criteria. Let

$n_{ijke} \equiv$ rating of p_{ijk} with respect to criterion e,

$$e = 1, \ldots, 7.$$

2) appropriate utility transformations are used to find the utility score associated with each rating. Let:

$u_e(\cdot) =$ utility transformation for criterion e.

3) utility scores are weighted with the appropriate criterion weights where:

$w_e \equiv$ weight assigned to criterion e.

---

8 The value of a package is not affected by the serving agency and hence V does not require an agency index.

4) the resultant weighted utility scores are aggregated by addition[9] to yield a single number for each p_{ijk},

$$Q_{ijk} = \sum_{e=1}^{7} w_e \cdot u_e (n_{ijke})$$

Having identified the four basic measurements, it is now possible to state, at least conceptually, that:

$$O_{ijk} = f(N_{ijk}, STR_{ijk}, Q_{ijk}, V_{ij})$$

where $O_{ijk} \equiv$ output associated with p_{ijk}.

The specific form of the functional relationship between these parameters is as yet unspecified. In fact, it is not even clear whether a single indicator number or a vector-valued measure of output appears to be most desirable and current efforts are focused on identifying and preparing to test various alternative forms of such a measure. At this point in time a vector-valued measure would admittedly diminish the value of the model as an allocation tool, but the disparate nature of the parameters may force that sacrifice if validity is to be maintained.

15-3 CHANGING THE SYSTEM: IMPLEMENTATION OF THE MEASUREMENT MODEL

15-3.1 Service Statistics

Operationalizing the model whose components are described in the previous section may be viewed as a three-part process:

1) the establishment of a mechanism for collecting certain specific pieces of objective data, referred to as "service statistics";

9

It should be noted that addition of weighted utility scores presumes independence among the criteria. Even though this was explicitly considered in selecting the criteria, system realities did not permit complete satisfaction of the requirement. Thus the additive relation should be treated as a reasonable first approximation.

2) the implementation of processes or instruments for generating certain subjective information of an evaluative nature, e.g. client opinions about service effectiveness
3) the implementation of devices (similar to those employed in 2) for generating certain subjective information of a valuative nature, e.g. judgments concerning the relative importance of service-client packages.

Service statistics necessary for the measurement model include not only the basic throughput parameters, N and STR, but also much of the data used to quantify quality criterion elements. Examination of the set of quality criteria reveals a need for data on: clients who do not receive service, time spent by staff in contact with clients, cost of professional staff, total service costs, extent of cooperative activities, and volunteer activities. It is no exaggeration to say that the "success" of the systems analysis will depend in large part upon the ability of member agencies to provide the requisite service statistics. Accordingly, considerable attention and effort has been devoted to the construction of a reliable and workable information system. Extensive field testing of concepts and approaches, close liaison between agencies and the research team, and a continuing conscientious effort to utilize current data collection procedures wherever possible, have all played a role in this effort. One especially important aspect has been work aimed at insuring the compatibility of the JCFC's new functional accounting system with the proposed measurement model. Thus, for example, great care was taken to maintain consistency between each agency's set of service-client packages and its functions as defined for accounting purposes. Some service statistics, primarily cost data for the criterion of Efficiency, will be taken directly from functional accounting reports. The collection of other data, e.g. service time received, will also be closely tied to functional accounting since both systems will utilize the same random sample of days for amassing data related to staff time usage.

Agencies are now beginning to implement the actual collection system. A data collection manual has been developed to provide general instructions, definitions of data items sought, timing of the data collection, and forms for accomplishing the collection. Members of the research team are now working closely with agency executives and staff members to produce the cooperation and understanding necessary. Only time will tell whether current optimism in this area is justified; but, at this point, it seems entirely reasonable to suppose that the information system will provide a legitimate set of service statistics for use in the measurement model.

15-3.2 Evaluative Information

Three of the seven elements of quality (see Section 2. 2) require the evaluation of certain system characteristics. These elements are Effectiveness, System Contribution, and Accessibility.

Effectiveness is concerned with the degree to which a service-client package meets the client's goals. The sources of such information include the clients who receive a service, individuals who provide the service, and knowledgeable outside experts. Clients and direct servers have a similar source of knowledge, immediate experience with individual service-client-agency packages, but they have different points of view. Outside experts must base their judgments on less intimate knowledge, but they can be expected to have yet another point of view. Finally, service-client packages are delivered through agencies, and the service cannot be divorced from the agency that offers it. If we assume, temporarily, that qualitative differences among agencies probably dominate the qualitative differences among the services offered within an agency, we can obtain and "average" rating for each agency and assign that rating to each package delivered by that agency unless exceptional packages are identified.

Through short questionnaires, a random sample of an agency's clients are asked to characterize the service they received as "excellent," "good," "satisfactory," "poor," or "very poor." After identifying the specific service-client packages involved, clients are asked to indicate which packages, if any, differ from the general rating by noting that a specific service was "much better," "better," "worse," or "much worse" than the agency average.

Direct servers also complete a short questionnaire in which they rate the agency's general ability to satisfy client needs and then follow such a rating with exceptions as above.

Outside experts, professional social workers who are highly knowledgeable about specific JCFC agencies, are asked to rate overall agency effectiveness on a 0-10 scale. Among other things, such a rating might include consideration of agency goals, programs, facilities, equipment, staff, available resources, organization, and clientele. The outside expert is left completely free to base his rating on any factors he wishes.

The three subjective ratings are weighted and combined to form one rating. While the relative weights for different classes of raters have not been set at this writing, pre-testing of the client and server questionnaires indicates that raters understand the instruments and use them as predicted.

Gathering evaluative data on System Contribution and Accessibility is somewhat simpler. For the quality element System Contribution, three basic evaluative questions must be answered:

1) To what extent does a given agency contribute to a "sense of community" or to community "goodwill" within the local Jewish community?
2) To what extent does a given agency contribute to the JCFC's ability to raise funds within the community:
3) To what extent does a given agency contribute to good relations with the non-Jewish community?

The ideal set of respondents to these questions would combine a high level of experience in fund raising in the community with extensive observation of community decisions in which individual agencies were dealt with at least partially according to the system in which they were held by the community. Various Federation committees were selected as respondents in this area. For example, members of the Budget Committee were asked to respond to the second question noted above for each of the system's agencies. The form of the answer required each agency to be placed on a five-point scale, the end points of which were labeled "very much" and "not at all." The responses obtained are given in Table 5, as a sample of the kind of data obtained via this process.

TABLE 5

SYSTEM CONTRIBUTION SCORES[10]

Agency	very much				not at all
Akiva H. S.	0	1	3	4	3
Bellefaire	4	5	0	2	0

10 Entries in the table represent the number of raters placing an agency at each point in the scale.

B'nai B'rith Hillel	1	2	3	4	1
CCJS	1	1	6	2	1
CHS	2	2	3	3	1
HA	0	3	3	4	1
JCB-JDN	3	5	2	1	0
JCC	2	6	3	3	0
JCRC	1	5	3	2	0
JFSA	5	3	3	0	0
JVS	1	6	2	2	0
Menorah Park	7	4	0	0	0
Montefiore	4	6	1	0	0
UJRS	0	1	4	4	2
WCS	0	0	1	7	3
Yeshivath Adath	0	2	3	3	3

The quality criterion Accessibility is composed of three sub-elements; physical, economic, and program accessibility. The evaluative ratings for each agency--and for service-client packages within agencies--are made by a panel of "gatekeepers."

"Gatekeepers" are individuals who exert some control or influence over the entry of a client into an agency. For example, doctors, lawyers, clergymen, and teachers are among the "gatekeepers" for a family counseling agency. They make judgments about the physical, economic, and programmatic accessibility of an agency and make recommendations to individual clients based on these judgments and on knowledge of the client's needs. Some "gatekeepers" are inside the system of social agencies, social workers concentrating on intake and referral, for example.

The representative panel will rank each agency on each of the three facets of accessibility. If specific service-client packages are considered to be exceptions to the average agency ranking, a form is provided to note exceptions. Otherwise, all p_{ijk} in a given agency share the overall agency rating. All three aspects of accessibility are equally weighted in determining the final rating.

15-3.3 Valuative Information

Valuative data is required for three parts of the output measurement process: determination of the value (importance) of the service-client packages; determination of the relative weights of the seven quality criteria; and, construction of the curves which relate raw scores on the quality elements to their "utility" equivalents. Various modifications of the "Delphi Method" were used to get this valuative data.

The Delphi Method, developed at the Rand Corporation, is a technique for combining a group of opinions into a simple composite opinion. It has three important characteristics:

1) Anonymity - Effected by the use of written responses or other formal communication channels.
2) Controlled feedback - Effected by conducting the exercise in a sequence of rounds between which a summary of the results of the previous round and justification of extreme positions are communicated to the participants.
3) Statistical definition of group responses - Use of a statistical measure, usually the median, assures that the opinion of every member of the group is represented in the final response.

The use of Delphi in estimating subjective parameters is well established [Ref. 7, 8, 9 and 10] but its use in the context of this research has been of an exploratory nature. The underlying principles of the technique have been maintained but various modifications in operating procedures have been developed to elicit the several different types of subjective information required by the measurement model. Details of these modifications and explanations of all procedures used are available elsewhere [Ref. 4]. This same approach to the generation of subjective information has also been employed in other contexts [Ref. 4, 11].

During the discussion of the logistics of this approach, a serious difficulty became evident -- the time required for the lay panel to evaluate the large number of service-client packages in depth would be absolutely prohibitive. The research team was thus faced with a choice between allowing the panel of laymen to perform only part of the task or allowing them to perform the entire task superficially. Recognizing that the former approach, though not perfect, was still to be preferred to the latter, the team proceeded to develop a three part evaluation process. As a first step, certain members of the research team evaluated all the service-client packages in depth. The use of the research team for this purpose was not suggested by mere convenience. The team included three Federation professionals, four lay members of the Cleveland Jewish community, and two Jews who were not a part of the Cleveland community (out of a total of thirteen team members). While the Jewish members of the team were not a perfect surrogate for this community, they did represent a variety of viewpoints about the relative importance of different areas of social service and produced a reasonable set of values for use as a base point.

Following this step, a group of key members of the Jewish Community Federation's staff estimated values for a large random sample of service-client packages. Scores were generated from the raters opinions via the use of simple weighted averages and a statistical comparison between the two sets of values was made. Reasonable consistency existed, as hypothesized, and the base set of values generated by the research team continued to be a candidate for incorporation in the measurement model. The lay panel was then convened and asked to address itself to a second random sample of the service-client packages and was once again examined and demonstrated. Some typical service-client package values are given in Table 6 and the data on correlations among the three sets of values is shown in Table 7.

The entire process was a most useful exercise. It involved a relatively large group of people in the determination of community priorities and there are significant indications that the acceptance of this research was thereby greatly enhanced. More importantly, it permitted the development of a complete set of numerical value ratings for service-client packages.

TABLE 6

TYPICAL SERVICE-CLIENT PACKAGE VALUE RATINGS[11]

Service-Client Package	Panel Distribution[12]					Score
	VL	L	M	H	VH	
Adoptive Service for Elementary School Age	0	0	1	4	2	29
Job Placement for Adults	0	0	3	4	0	25
Formal Education (Primary Jewish Focus) for Young Adults	0	1	1	4	1	26
Institutional Health and Living for Aged	0	0	0	3	4	32
Day Care for Pre-School Age	0	0	0	7	0	28
Individual Counselling (Psycho-social) for Jr. H.S. - H.S. Age	0	0	2	4	1	27

---

11 Selected from among those generated by research team.

12 Entries indicate number of raters utilizing each of five categories of importance: Very low (VL), Low (L), Moderate (M), High (H), and Very high (VH).

TABLE 7

VALUE RATING CORRELATIONS

	Sample Size	Correlation[14]	
Consensus[13] items (both panels)	60	0.61	.001
All items	83[14]	0.69	.001

Research Team Scores Versus Federation Staff Scores

	Sample Size	Correlation	
Consensus items (both panels)	50	0.69	.001
All items	76	0.71	.001

Research Team Scores Versus Lay Panel Scores

Section 2.2 describes the elements of the quality measure. Several steps are required to make the concept operationally useful. First, for all quality elements composed of two or more sub-elements, the sub-elements must be weighted and aggregated. Second, the resulting "raw scores" for each of the seven elements must be reduced to a common dimension and numerical scale. Third, the resulting "utility scores" must be weighted and aggregated. Each of these three steps require valuative information.

The Operations Research Team, with both professional social workers and lay members of the community, developed weights for the sub-elements within a given quality criterion. The reduction of the raw scores for each quality element to a common measure was accomplished by developing a set of quality criterion-utility relationships, or "utility curves." The procedure is explained in more detail in [Ref. 4], but amounted to a modification of the Delphi Method in which respondants actually drew curves relating the raw scores for a quality

---

13 Pearson Product Moment Correlation Coefficient

14 Any package for which approximately 80% of the panelists' votes were in two contiguous levels of importance was regarded as a "consensus" item.

element to "utility." A typical curve is shown in Figure 1.

To carry out the third step, a group of laymen used the Delphi Method to determine a set of weights for the seven basic elements of quality. The resulting weights are given in Table 8.

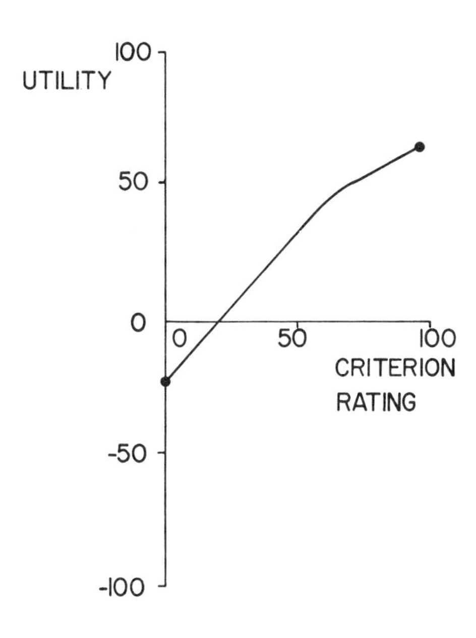

Figure 1

Utility Curve for Criterion Accessibility

TABLE 8

QUALITY CRITERIA WEIGHTS

Criterion	Weight
Accessibility	6.0
Capability	8.5
Effectiveness	9.5
Efficiency	8.0
Interrelation	5.0
System Contribution	7.0
Uniqueness	5.0

As indicated in Section 2.3, the technique of aggregating across quality criteria is to calculate the weighted sum of the utilities associated with each criterion, a technique that assumes independence among the criteria. (A similar assumption is, of course, made when the elements for each component are summed to calculate the component score.) It is highly probable that this assumption is not strictly true. For example, it seems likely that a sizable increase in "Effectiveness" would cause an increase in "System Contribution." Similarly, a decrease in the "Uniqueness" of a service might result in an increase of the "Effectiveness" with which the service is offered because of the competition between agencies.

Fortunately, the system has characteristics which allow us to ignore these interdependencies in practice. First, social systems exhibit considerable inertia. Sub-element scores within each quality criterion tend to shift slowly over time. Certainly, scores which depend on the subjective perception of raters (individuals or groups) will change slowly. Second, periodic re-evaluation of quality criterion scores will reflect updated sets of ratings, which will automatically take into account changes in a given quality-element score "caused by" a change in another element score. As a result, simple summation of the weighted scores should not introduce major errors in the quality measure.

With the exception noted above, the basic general assumptions common to utility measures apply here. The relationship between a quality-element score and the decision-makers preferences is stated. There are finite upper and lower bounds on the related utility score associated with each component score. Step functions are allowable (though they did not appear here).

It is obvious that the quantification techniques employed in measuring value and quality are subject to easy criticism. Neither the research team nor the professional and lay community people involved suffer the delusion that the various quality criteria are uniquely measured by the sub-elements contained in each. All groups, however, feel that criteria scores and the associated utilities are highly indicative of the "quality" of a service-client-agency package. Perhaps, just as important, both the methodology and the outcomes are acceptable to the parties involved.

15-3.4 Using the Model

Once the output measure model is developed, its use as a budgeting and planning tool must be delineated. As indicated in an earlier section, budgeting is the primary focus of this research, but it is inappropriate, if not conceptually impossible, to separate budgeting and planning. Both orientations call for some type of resource allocation model.

The problem as seen by the Budget Committee requires that a limited budget be allocated across the member agencies in such a way as to maximize system output. The allocation decision, however, is subject to physical, policy, and political constraints. While the specifications of these constraints is a non-trivial problem, it is by no means as difficult as it appears at first glance. For example, such limits on agency capacity as are operationally meaningful are known; the community will insist on maintaining some residual capacity in certain areas, even if no need currently exists in the system or is foreseen; the model will not allow any agency to be given an allocation that is smaller than last year's allocation unless the Budget Committee specifically requests that this constraint be removed.

It must be recognized that allocations are made for the future but are based on system output in the past. We could, of course, state a decision rule that relates current allocation to past system performance, but this does not appear to be either desirable or necessary. It was noted above that one of the political constraints on the system would require that an agency's current allocation be no smaller than the allocation for the preceding period. If this constraint holds for all agencies, it follows that the Budget Committee does not have access to the entire budget, but must exercise its allocative wisdom on the difference between this year's available funds and last year's. Hopefully, this difference will be positive.

In any event, even if this political constraint were not present, allocational flexibility would probably still focus on the incrementally available funds. Systems as large and complex as the JCFC cannot even afford the psychic threat of "zero base budgeting," let alone the actuality.

If it is the incrementally available funds that will be allocated, one step toward a realistic allocation procedure is to utilize estimates of the changes in output associated with a change in allocation to a particular service-client-agency package. To be consistent with this kind of model, an agency would be well advised to justify any requests for an increased budget by indicating the probable consequences in the form of increased agency output. These and other models are discussed in some detail in [Ref. 12].

While the Budget Committee will tend to be interested in an allocation model, the Community Services Planning Committee can use the output measurement model directly in its decision process. The CSPC is directed to "study the needs" of the community, to "evaluate" the agencies, and to "seek improved ways and means" of meeting the community's needs.

The "needs" of the community express themselves through changes in the usage rate of service-client packages. Agencies offering services are sensitive to the changing needs of their potential clientele. As the agencies approach the CSPC with creative plans for dealing with "new problems," the CSPC fulfills its directive to "study the needs" of the community as a part of the process of interacting directly with a service agency. The scope and urgency of the need is reflected in its value, V_{ij}. It is important that this statement of priorities be kept current through periodic reconsideration and re-valuing of the V_{ij}. Finally, the quality measure should assist the CSPC rather directly to accomplish its assignment to "evaluate" the agencies and, somewhat less directly, to meet its commitment to "seek improved ways and means of meeting community needs."

In addition to affecting the decision processes of the Budget Committee and the Community Services Planning Committee, it seems apparent that the use of the output measurement and allocation models will also have an impact on the Federation's member agencies. Aside from its obvious use as a basic planning tool, the output measurement model can serve as an internal control mechanism for an agency. A large part of the information an agency must feed into the information system serving these models and the functional accounting system is also useful to an agency that wishes to exert more control over its own activities. Several agencies have recognized this possibility and are considering the collection of additional information useful internally, but not demanded by the centralized information system. Above all, during the course of this research, the individual agencies have grown much more conscious of their roles as "producers" of services in a federated system of social service.

It would be most unfortunate if the existence of the measurement and allocation models led agencies to make opportunistic, short-run decisions. In view of the high level of technical skill and dedication found in the agency executives, such an event is highly unlikely. It is, nonetheless, critically important that both the measurement and the allocation models are kept in proper perspective. These models are <u>planning aids</u> and whatever value they may have will arise from their ability to <u>assist</u> human decision-makers in social welfare planning. The recognition of the dominant role of the human element in these kinds of systems by no means obviates the need for procedures of the type described here. On the contrary, the ever-increasing complexity and expanding scope of social planning problems makes it imperative that new and better "tools" be placed at the disposal of the human decision-maker. Although these models are technically capable of generating resource allocations, it must be made crystal clear that no automatic decisions should result, that the final decision ought to rest in the hands of responsible community leaders and not be, in any sense, mechanistic or predetermined. This entire model building process is an attempt to provide decision-makers with a planning tool that allows them to analyze a large number of variables in a more rational way and to study the implications of policy alternatives in a systematic and consistent manner.

With Heckman and Taylor [Ref. 13],

> "Our view is that the role of the management scientist is not to provide one optimal plan under a given set of assumptions, but to provide a set of efficient options or alternatives together with some measure of their feasibility or cost."

REFERENCES

1. Axehod, J., F. J. Fouler, Jr., and A. Guren, A Community Survey for Long Range Planning, Boston, Mass., Combined Jewish Philanthropies of Greater Boston, 1967.
2. Goldstein, S., The Greater Providence Jewish Community, Providence Rhode Island: The General Jewish Community of Providence, Inc., 1964.
3. The Jewish Community Federation of Cleveland, Projection of Future Plans, Cleveland, Ohio: JCFC, 1967.
4. Reisman, A., S. J. Mantel, Jr., B. V. Dean, N. C. Eisenberg, E. Markus, and A. L. Service, "Measurement of Output in a System of Social Agencies", CWRU Department of Operations Research, Technical Memorandum No. 188, June, 1970.
5. Reisman, A., and M. I. Taft, "On a Computer-Aided Systems Approach to Personnel Administration," The American Society of Mechanical Engineers, Winter Annual Meeting, 1968, preprint 68-WA/MGT-7, Forthcoming in Socio Economic Planning Sciences.

6. Reisman, A., Managerial and Engineering Economics: A Unified Approach, Allyn and Bacon, Inc., Boston, 1971.
7. Brown, B., S. Cochran, and N.C. Dalkey, The Delphi Method II: Structure of Experiments, Santa Monica, California: The RAND Corporation, RM-5957, June, 1969.
8. Dalkey, N.C., The Delphi Method: An Experimental Study of Group Opinion, Santa Monica, California, The RAND Corporation, RM-5888-PR, April, 1969.
9. Pill, J., "The Delphi Method: Substance, Context, A Critique, and an Annotated Bibliography," Socio Economic Planning Sciences, Vol. 5, pp. 57-71, 1971.
10. Esbeck, E.S., N.M. Balwally, S. Srinivasan, A. Reisman, B.V. Dean, and S.S. Rao, Toward a R&D Laboratory Evaluation Methodology, CWRU Department of Operations Research, Technical Memorandum No. 212, Jan. 1971.
11. Reisman, A., J. Herling, G. Kaminski, and M.G. Fancher, "Timeliness of Library Materials Delivery: A Set of Priorities", forthcoming in Socio Economic Planning Sciences.
12. Reisman, A., S.J. Mantel, Jr., B.V. Dean, N.C. Eisenberg and A.L. Service, "Evaluation and Budgeting Model for a System of Social Agencies", CWRU Department of Operations Research, Technical Memorandum No. 167, November, 1969.
13. Heckman, L.B., and H.M. Taylor, "School Rezoning to Achieve Racial Balance: A Linear Programming Approach," Socio-Economic Planning Sciences, Vol. 3, 2, pp. 127-133, 1969.

Chapter 16

DILEMMAS FOR SYSTEMS ANALYSIS OF URBAN PUBLIC PROGRAMS

S. Goldstone

16-1 INTRODUCTION

In the public-policy economics group at Battelle we were frequently involved in studies aimed at evaluating and recommending how public programs might be improved. Preparing for this symposium has given me an opportunity to sit back and see if there were any lessons I could draw from this experience that pertain to the question of the systems approach and the city. The main lesson, which I am sure will be of little surprise to those of you who have been involved in such studies, is that applying the systems approach in this context is fraught with certain basic difficulties. The purpose of this paper is to explore some of the reasons underlying these difficulties in hopes that some positive contribution may thereby be made towards their alleviation.

The author is Senior Economist, Battelle Memorial Institute, Columbus, Ohio.

At the root of the difficulties is the principle, enunciated by Churchman, that every system is embedded in a larger system [Ref. 1]. When it comes to the provision of public services, the city may be thought of as a political system embedded in a hierarchy of higher level systems represented by county, state and federal governments. Two extremely important facts about cities are their economic and political interdependence with these higher level governments. These interdependencies, moreover, characterize a political-economic system in which many smaller systems, commonly referred to as public programs, are embedded. I shall try to show that viewing public programs from this perspective enhances our understanding of the difficulties inherent in applying the systems approach to their improvement.

16-2 THE LARGER SYSTEM

As indicated the larger system may be characterized by looking at the economic and political interdependencies between cities and higher level governments.

16-2.1 Economic Interdependence

Casual examination of available data is sufficient to reveal the high degree of economic dependence that cities have on higher level government for financing public services. In fiscal year 1969, city governments on the average acquired over 30 percent of their general revenues (i.e., 7.3 out of 24.3 billion dollars) from county, state and federal levels of government [Ref. 2, Table 1]. The degree of dependence, interestingly, increases with city size, with the smallest cities (population less than 50,000) having a dependence rate of 21% and the largest cities (population over 1 million) a dependence rate almost double that of 41% [Ref. 2, Table 3]. These figures, moreover, understate the degree of economic dependence because they do not include money spent on programs that benefit city residents but that are administered by higher level governments. Most important in this respect is the Federal government which has dramatically increased its expenditures on domestic aid programs-over tenfold within the last two decades. In fiscal year 1971, the level of federal grants-in-aid will reach over 27 billion dollars of which approximately 19 billion will be spent in urban areas. For urban areas this represents a very sharp increase of 5 billion dollars in the short span of only three years [Ref. 3].

16-2.2 Political Interdependence

While the economic dependence of cities is reflected by the large amount of funds they receive from higher level governments, their political dependence is reflected by the fact that nearly all these funds have strings attached. Federal grant funds, for example, are channeled to cities and other constituencies by well over 400 minutely specified authorizations which cover virtually every important public service activity [Ref. 4].

Of central importance is the fact that any particular public program in order to be authorized must have sufficient political support from the various constituencies involved in the decision process. In the case of higher level political units, such as state and federal governments, this means support must typically be obtained from a variety of different kinds of constituencies, including not only cities but suburban and rural areas as well. Because these different kinds of constituencies have so many different kinds of problems and priorities, a particular public program is bound to benefit some constituencies, and disregard others. It also means that programs tailored to meet the particular problems of a particular type of constituency, such as large cities, will have difficulty in obtaining the necessary political support. This would be especially true if political support for any given program were based entirely on the direct economic benefits it generated. The general implication of this argument is that if the public decision process depended entirely on the mechanism of direct economic benefits to obtain political support for proposed programs then many "worthwhile" public programs would never exist [Ref. 5].

Examination of how the political process actually works reveals several types of mechanisms that are used to overcome the problem of inadequate political support for specific programs. Two particularly important such mechanisms are what we might call (1) economic side payments, and (2) legitimacy appeal[1]. The

[1]Other important political support mechanisms not discussed in this paper because of the difficulty in relating them to program specific consequences involve the role of the political party in packaging groups of programs. Through a complex of party related sanctions and rewards support for a particular program can be uncoupled to some extent from benefits received from that program. One obvious example, is using a program that benefits one type of constituency as a payment to obtain the support of its representatives on other programs that benefit other types of constituencies.

former typically involves extension of a program's scope or coverage to enable side payments to be made beyond those necessary to meet the needs which initially motivated the program. Log-rolling as traditionally used to obtain support for "pork barrel" projects is one example of this type of mechanism. A major drawback to this mechanism, of course, is that its use is limited by budgetary considerations.

By contrast, gaining support for a program on the basis its inherent legitimacy makes no excess budgetary demands. If a program is perceived as doing things that people feel are legitimate, i.e., ought to be done, they will tend to support it despite the lack of direct or immediate economic benefits to themselves. We may think of this mechanism as involving the substitution of symbolic gratification or what we might call psychic benefits for material economic benefits [Ref. 6]. Identifying a program with the objective of eliminating poverty serves to gratify the non-poor who will not benefit directly but who favor poverty reduction. To the extent that the poverty program is inadequate with respect to achieving its objective this tactic also substitutes the symbolic promise of future benefits for actual benefits in the present, i.e., the so called "foot-in-the-door" tactic.

16-3 SOME IMPLICATIONS FOR PUBLIC PROGRAMS

Use of these two political support mechanisms have several, what we might call "undesired" consequences for the character of public programs. Programs supported through use of economic side payments will be characterized by spending that is unrelated or at best only weakly, i.e., inefficiently, related to a program's objectives. These inefficiencies will be paid for by some combination of excessive total program funding and dilution of effort in relation to intended objectives. To the extent the former is used the program will be extremely vulnerable during times of budgetary stress. To the extent the latter is used the program's effectiveness will be jeopardized.

Programs aimed at providing federal aid to depressed areas are a classic example of what can happen under these latter circumstances. When initially conceived during the 1950's, the federal area redevelopment program was a modest proposal to aid a small number, certainly no more than one or two hundred of depressed areas [Ref. 7]. By the time the program had gone through the legislative mill and completed two years of operation over 1,000 counties, more than 1/3 of all the counties in the U.S., had been designated as eligible to receive aid. Various devices such as broadening the eligibility criteria for depressed counties and "rounding out" areas to include non-depressed counties next to depressed counties permitted this dramatic expansion of eligible areas while the total level of funding remained virtually unchanged. The inevitable result of this tremendous overextension of coverage was a program that was extremely ineffective in relation to its originally intended objective.

Although not quite so obvious, certain characteristic undesirable consequences are also likely to be associated with programs whose support is based on their legitimacy appeal. One might expect that programs with high legitimacy appeal would be able to obtain adequate funding. And indeed this is true in those cases where the perceived legitimacy is strongly felt and widely shared. Programs with such high degrees of perceived legitimacy, however, are extremely rare, for the most part being limited to programs developed in response to crises and fundamental threats to society such as major shooting wars. More typically, programs of this kind are likely to be inadequately funded relative to their objectives because prospective beneficiaries and others favoring the program can be paid off with psychic rather than with material economic benefits.

Besides low funding, another reason for ineffectiveness of highly legitimate programs is the pressures that their legitimacy produce for premature expansion. Initially, for example, Headstart was intended to be an experimental program to be tried on a limited number of children. But the idea turned out to be too good. It became a showcase symbol for the war on poverty. Under this atmosphere, it became expanded well beyond the initially intended coverage, before it could be adequately designed and evaluated [Ref. 8].

Another "undesirable" consequence frequently associated with high legitimacy programs is perverse distributional patterns. By focusing on their highly legitimate objectives, attention is distracted from the more mundane considerations of how a program's economic benefits are distributed. This in effect enhances the political feasibility of a program having a pattern of benefit distribution that contradicts prevailing equalitarian norms, e.g., favoring well-to-do suburban areas over more needy inner city neighborhoods.

For example, we recently examined a county park program aimed at conserving scenic naturally wooded areas [Ref. 9]. Because virtually all such areas have been located in the outer parts of the county, this program has developed a beautiful system of parks which are comparatively inaccessible to black inner city neighborhoods. A study of participation patterns showed, in fact, that these inner city neighborhoods participate on the average less than half as much as suburban neighborhoods located nearer the parks. Despite the skewed distribution of benefits, the high perceived legitimacy of the program has resulted in consistent support, registered in periodic votes on tax levies, by all sectors of the community. Even more to the point of how perceived legitimacy can uncouple political support from the direct receipt of benefits is the detailed voting pattern on tax levies. I am told that while the average support for the park program in the county as a whole tends to be about 55 percent, inner city black wards have typically given supporting votes on the order of 85 to 90 percent[2].

16-4 IMPLICATIONS FOR APPLIED SYSTEMS ANALYSIS

In general, these examples serve to illustrate how a particular aspect of a public program may take on fundamentally different interpretations depending on which system level they are viewed from. What from the micro-system viewpoint is an undesired consequence to be avoided is, from the higher level macro-system viewpoint, a rational and desirable solution to the problem of developing political support in a democratic government with heterogeneous constituencies. This fundamental duality pervades all attempts to apply systems analysis to improving public programs. It means that applied micro-systems analysis is inherently in opposition to the mechanisms of political support. Whereas these mechanisms are aimed at making

---

2 Private conversation with official involved in program. Partly, of course, this high support could be due to the fact that these wards have a higher percentage of renters, who, since they aren't directly responsible for paying property taxes, don't perceive the tax levy as a burden.

programs politically feasible, applied systems analysis will tend to jeopardize that feasibility by attacking the basis for political support.

According to the conventional wisdom the systems approach is a normative-instrumental aid to public decision makers. It involves analyzing the relationship between a program's cost and its effectiveness with respect to achieving its objectives. The purpose of applied systems studies is to both evaluate and determine ways of improving either the efficiency or effectiveness with which programs achieve their objectives. The logical starting point for this endeavor is an attempt to pin down exactly what a program's objectives are. And it is at this point that the analyst runs into his first conflict with the mechanisms of political support. In his attempt to specify objectives the analyst will typically review the relevant legislative documents and interview decision makers involved in the program. Invariably this yields a glowing account of high minded vague sounding objectives which, more than anything else reflect the program's need for legitimacy appeal. The need to maintain as broad an appeal as possible dictates against the kind of precise specification and weighing of different, but potentially conflicting, objectives that systems analysts would ideally like to work with. It is true that a number of techniques have been developed to help analysts move from the standard vague statement of objectives to more precise quantitative objective functions. But for a number of reasons these techniques are either not particularly useful or simply not practicable in the highly politicalized environment surrounding most public programs. For all practical purposes, therefore, the analyst must rely on some relatively imprecise interpretation of a program's stated objectives.

Using stated objectives as a reference point the application of systems analysis will tend to jeopardize the basis for political support in a number of different ways depending on which of the three types of "undesired" consequences - (1) ineffectiveness, (2) inefficiency, and (3) perverse benefit distribution - are involved.

16-4.1 Ineffectiveness

For a variety of reasons we have seen that the use of standard political support mechanisms lead to programs which are less that perfectly effective with respect to their stated objectives. In those cases where the actual effectiveness is very low and where the techniques for measuring effects are relatively crude, systems evaluation studies are likely to find that a program is not producing noticeable effects. Such findings could, of course, be used as a spur to increased effort. They may also, however, be used as ammunition by opponents of the program to undermine support, thereby insuring that an effective level of support will never be achieved. Because of this possible dysfunctional aspect, such studies are themselves likely to be attacked and become a matter of considerable controversy. The controversy surrounding the Westinghouse study which found that by the time children, who had attended Head Start, reached the early elementary grades their intellectual achievement was not significantly different from comparable children who had not attended Head Start, is a case in point[3].

16-4.2 Inefficiency

In programs where side payments are made which are unrelated or only weakly related to objectives, systems evaluation studies are able to find a substantial degree of inefficiency. From the conventional micro systems analysis viewpoint, side payments, since they do not contribute their share toward the achievement of objectives, are expendable parts of a program. Clearly, however, if there is any validity to the macro analysis, cutting off these side payments may threaten a program's political viability. Possibly under these circumstances the analyst might want to consider the existence of side payments as a fixed political constraint and focus his analysis elsewhere. In general, he faces a dilemma because there is no way of knowing for sure whether or not the existing level of side payments is, in fact, essential to a program's viability. If the side payments are not essential then systems evaluation can effectively contribute to their reduction without jeopardizing the program's viability. On the other hand, if the side payments are essential and the analyst chooses to focus his analysis on exposing them it is difficult to imagine any productive result.

---

[3] For an informative account of this controversy see [Ref. 8].

A recent Battelle study of an urban fire department illustrates this kind of dilemma [Ref. 10]. The study was intended as a preliminary investigation into the feasibility of applying systems analysis to improving the department's operations. At an early stage the analysts were led into focusing an important part of their efforts on the possibility of improving the efficiency of man-power utilization. Partly this was because salaries and wages were found to represent over 90 percent of the departments total costs. Partly it was due to the prodding of the city manager, who felt that such a focus might prove useful to him in upcoming contract negotiations with the fireman's union.

As is apparently typical of many fire departments, the studied department made a practice of assigning firemen on 24 hour shifts. This meant that the number of firemen on duty was necessarily the same over all 24 hours of the day. The analysis revealed, however, that fires are not evenly distributed over the day. Rather they are heavily concentrated during the last half of the day, between the hours of 2 p.m. and midnight. A fairly obvious conclusion is that efficiency can be improved by varying the daily pattern of fire department staffing to more closely match the pattern of fire occurrence.

Implementing this recommendation would, however, mean doing away with the 24 hour shift. And as most firemen will tell you, the 24 hour shift has important advantages. It means that the fireman, through use of a "24 hour-on-24 hour-off" rule, can get in 48 hours of work, a normal weeks worth, in an elapsed time span of only three days. This leaves four days of the week which he can use to supplement his income by taking a second job. In effect the 24 hour shift may be viewed as an economic side payment to the fireman. It is unlikely, therefore, that the firemen would sit idly by and allow this benefit to be taken away.

Needless to say the study was welcomed by the city manager, who viewed it as a bargaining counter he could use in the upcoming contract negotiations. The fire chief, possibly anticipating negative feelings on the part of the firemen, responded very unenthusiastically being quoted in a local newspaper as saying that "he didn't learn anything from the study he didn't already know". Subsequent discussion with the city manager's staff revealed that the study had indeed produced considerable negative response from the firemen's union and that for the time being they had decided not to push for either any implementation or further analysis. There are a number of things one might conclude from this example. Most important it illustrates how the systems analyst, whether he chooses to do so or not, cannot avoid becoming involved in the larger political system of which the studied program is merely a subsystem. In this instance, the analysts threatened what now, in retrospect, appears to have been a rather essential economic side payment. Their study, consequently, did not have any productive real world impact.

16-4.3 Perverse Distributional Consequences

In the case of perverse distributional consequences the difficulties associated with applied systems analysis are likely to be somewhat different. To the extent that the analysis is tied to a program's stated objectives there will be a tendency to ignore distributional consequences unless these consequences are somewhat related to stated objectives. In the case of the county park program, for example, there would be no reason for a systems study to consider the question of benefit distribution. The stated objective of this program is conservation of naturally scenic land wherever it might be located in the district. The spatial distribution of sites, which is the factor most directly related to benefit distribution in this case, would logically be considered immaterial to an evaluation of the program's effectiveness. Systems analysis in this instance would serve to reinforce the device of legitimacy appeal by leaving such distribution inequities, which could alienate support from the program if they were known, unexposed.

Frequently, however, the situation is not quite so clear cut. Though a program's primary objectives may be unrelated to distribution patterns, there may be secondary considerations derived from the broader context in which the program takes place, which make distributional consequences germane to an evaluation of effectiveness.

A good example is a program that we evaluated several years ago, for providing state financial aid to local school districts in Ohio for the construction of school buildings [Ref. 11]. The need for this program arose because of Ohio's legal limit on the amount of bonded indebtedness as a percentage of their tax bases that local school districts can incur. Politically this law draws considerable support on the basis of the prevailing norm against "mortgaging ourselves to future generations" through deficit financing. The debt ceiling also directly contributes to efficiency by maintaining high quality ratings on school bonds which in turn insures favorable interest rates. The law creates practical difficulties, however, because under certain circumstances school districts run into their debt ceilings before they can meet all their school building needs. The specific purpose of the state aid program under investigation was to provide the means whereby such districts could meet their school building needs without violating their legal debt limits.

Implementation of the program produced perverse distributional consequences because only certain types of districts have ever and are ever likely to become eligible for aid. These are small poor rural districts and well-to-do, but rapidly growing, suburban districts. The poor rural district is likely to qualify because its tax base is so low that even the financing of one new school building might force it over its debt limit. For these districts the problem is a permanent one. For the qualifying suburban districts, however, the problem is temporary. It is due to the combination of (1) a temporarily low tax base and (2) unusually large start-up school building costs that are associated with the early stage of suburban development.

Probably the most perverse distributional consequence of the program is the fact that no city school district, despite the fact that such districts typically have more than their share of very old and dilapidated school facilities, is ever likely to qualify for aid. Because many of their school buildings are older, but not yet ready for replacement, a much higher percent of city school district's facility costs are for maintenance and repair rather than new building construction. Since maintenance and repair costs are considered part of operating expenses, they are not covered in the state school facility aid program.

The pattern of aid distribution was germane in this particular study because the program is administered by the Ohio Department of Education. And even though distributional matters are not explicitly included in the stated objectives of the particular program, they are part of the department's broader objectives which include the encouragement of equality of educational opportunity throughout the state. A program which subsidizes comparatively well-to-do suburbs but gives nothing to the substantially more needy large city school districts is obviously inconsistent with this broader objective.

One recommendation coming out of this study, which I now understand has actually been implemented, was to allow local school districts to utilize future bonding capacity. This has the effect of allowing growing suburbs to finance their own school building needs without any adverse affect on interest costs. This revision, thus, reduces the amount of subsidization to well-to-do school districts which don't really need it, without any reduction in overall efficiency.

This revision in the debt ceiling provision was adopted about a year ago. Currently, according to my most recent conversation with the administrator of this program, the program is in great danger of not receiving any new funds. In general there appears to be little interest in or support for the program, except for about forty, mostly poor rural, districts who have joined together to lobby for its continuation. I don't know what the eventual outcome will be. However, it is interesting to note in retrospect that during the course of the systems evaluation study, the question of how the reduction in benefits to suburban districts would effect the political viability of the program was never seriously raised. From a micro systems analysis viewpoint the recommended revision that has been adopted improves the program's effectiveness with respect to its objectives. Yet we are now confronted with the real, though unintended, possibility that this revision will also contribute to the program's demise.

16-5 SUMMARY AND CONCLUSIONS

In summary, I have argued that application of the systems approach to improving public programs, especially the kind typically related to the city, involves a fundamental dilemma. This dilemma occurs in various forms. It occurs in connection with specifying program objectives; it occurs in connection with specifying constraints; and ultimately it occurs in the translation of findings into recommendations. The dilemma is rooted in the inherent hierarchy of systems, the fact that every system is embedded in a larger system. The analyst plays the role of detached scientific observer with respect to the micro-system he is analyzing. At the same time he also, whether he consciously chooses to do so or not, plays the role of participant in the larger political system of which the studied program is a subsystem. These two roles are invariably in conflict because normative systems analysis inherently conflicts with the standard democratic mechanisms for obtaining political support. I pose this as a dilemma because I don't know the resolution, nor do I believe that in general there is a resolution. Instead, I believe each systems analyst must resolve the dilemma for himself in light of his own situation. At one extreme he may choose to single mindedly pursue the goal of program efficiency and effectiveness, essentially ignoring the political context. At the other extreme, he may choose to narrow his analysis by taking the existing political context as a fixed constraint. Or finally, he may try to steer a middle course, in which he recognizes the potential contribution of his analysis to changing some but not all of the existing political context and in which he attempts to develop a creative synthesis that compromises between the conflicting demands of the two systems.

REFERENCES

1. Churchman, C. West, THE SYSTEMS APPROACH, Dell Publishing Co., New York, New York, 1968.
2. City Government Finances, 1968-69, U.S. Department of Commerce, Bureau of Census, U.S. Government Printing Office, Washington, D.C., August, 1970, Table 1.
3. Special Analyses, Budget of the United States, Fiscal Year 1971, Special Analyses O, p. 1219, U.S. Government Printing Office, Washington, D.C., 1970.
4. Freeman, Robert A., "What Federal Fiscal System", Federal-State-Local Fiscal Relationships Symposium conducted by Tax Institute of America, November 29-30 - December 1, 1967, Princeton, Tax Institute of America, 1968, p. 67.
5. Easton, David, A Systems Analysis of Political Life, John Wiley & Sons, Inc., New York, 1965, for an excellent general treatment of this question.
6. Merelman, Richard M., "Learning and Legitimacy", The American Political Science Review, September, 1966, No. 3, Vol. LX, pp. 548-561.
7. Levitan, Sar A., Federal Aid to Depressed Areas, An Evaluation of the Area Redevelopment Administration, The Johns Hopkins Press, Baltimore, 1964.
8. Williams, Walter and Evans, John, "The Politics of Evaluation: The Case of Head Start", Evaluating the War on Poverty, the Annals of the American Academy of Political and Social Science, September, 1969.

9. Whitman, Ira, Davis, D., Goldstone, S., Molholm, L., Dee, N., Evaluating Urban Core Usage of Waterways and Shorelines, report to Office of Water Resources, Department of Interior, from Battelle Memorial Institute, to be published January, 1971.
10. Gordon, B., Drozda, W., Stacey, G., Cost-Effectiveness in Fire Protection, Working Paper, Urban Studies Center, Battelle Memorial Institute, December 31, 1969.
11. Brown, L., Goldstone, S., Hoovey, H., Rosenberg, H. Programs To Finance Ohio's School Facilities, Ohio Department of Education, November 15, 1968, Research Report from Battelle Memorial Institute.

Chapter 17

THE EXPERIENCE OF LIVING IN CITIES

S. Milgram

17-1 ASPECTS OF URBAN EXPERIENCE

Adaptations to urban overload create characteristic qualities of city life that can be measured.

> "When I first came to New York it seemed like a nightmare. As soon as I got off the train at Grand Central I was caught up in pushing, shoving crowds on 42nd Street. Sometimes people bumped into me without apology; what really frightened me was to see two people literally engaged in combat for possession of a cab. Why were they so rushed? Even drunks on the street were bypassed without a glance. People didn't seem to care about each other at all."

This statement represents a common reaction to a great city, but it does not tell the whole story. Obviously cities have great appeal because of their variety, eventfulness, possibility

The author is Professor, Department of Psychology, The City University of New York, New York.

of choice, and the stimulation of an intense atmosphere that many individuals find a desirable background to their lives. Where face to face contacts are important, the city is unparalled in its possibilities. It has been calculated by the Regional Plan Association [Ref. 1][1] that in Nassau county, a suburb of New York City, an individual can meet 11,000 others within ten minutes of his office by foot or car. In Newark, a moderate sized city, he could see more than 20,000 persons. But in mid-town Manhattan an office worker can meet 220,000 persons within 10 minutes of his desk. So there is an order of magnitude increment in the communication possibilities offered by a great city. That is one of the bases of its appeal, and indeed, of its functional necessity. The city provides options that no other social arrangement permits. But there is a negative side also as we shall see.

Granted that cities are indispensable in complex society, we may still ask what contribution psychology can make to understanding the experience of living in them. What theories are relevant? How can we extend our knowledge of the psychological aspects of life in cities through empirical inquiry? If empirical inquiry is possible, along what lines should it proceed? In short, where do we start in the construction of urban theory and in laying out lines of research.

Observation is the indispensable starting point. Any observer in the streets of mid-town Manhattan will see: 1) large numbers of people, 2) high density, 3) heterogeneity of population. These three factors need to be at the root of any socio-psychological theory of city life, for they condition all aspects of our experience in the metropolis. Louis Wirth [Ref. 2], if not the first to point to these factors, is nonetheless the sociologist who relied most heavily on them in his analysis of the

---

1

This article is based on an Invited Address to Division of General Psychology of the American Psychological Association delivered by the author in Washington, D. C. on September 2, 1969. Barbara Bengen worked closely with the author in preparing the present version of this paper, and its expository values reflect her skill. The author thanks Dr. Gary Winkel, editor of Environment and Behavior, for useful suggestions and advice.

city. Yet, for a psychologist there is something unsatisfactory about Wirth's theoretical variables. Numbers, density and heterogeneity are demographic facts but they are not yet psychological facts. They are external to the individual. Psychology needs an idea that links the individual's experience to the demographic circumstances of urban life.

One link is provided by the concept of overload. This term, drawn from systems analysis, refers to the inability of a system to process inputs from the environment because there are too many inputs for the system to cope with, or because successive inputs come so fast that input A cannot be processed when input B is presented. When overload is present adaptations occur. The system must set priorities and make choices. A may be processed first while B is kept in abeyance, or one input may be sacrificed altogether. City life, as we experience it, constitutes a continuous set of encounters with adaptations to overload. Overload characteristically deforms daily life on several levels, impinging on role performance, evolution of social norms, cognitive functioning, and the use of facilities.

The concept has been implicit in several theories of urban experience. In 1903 George Simmel [Ref. 3] pointed out that since urban dwellers come into contact with vast numbers of people each day, they conserve psychic energy by becoming acquainted with a far smaller proportion of people than their rural counterparts and maintaining more superficial relationships even with these acquaintances. Wirth (2) points specifically to "the superficiality, the anonymity, and the transitory character of urban social relations."

One adaptive response to overload, therefore, is that less time is given to each input. A second adaptive mechanism is that low priority inputs are disregarded. Principles of selectivity are formulated so that investment of time and energy are reserved for carefully defined inputs (the urbanite disregards the drunk sick on the street as he purposefully navigates through the crowd). Third, boundaries are redrawn in certain social transactions so that the overloaded system can shift the burden to the other party in the exchange; thus, harried New York bus drivers once made change for customers, but now this responsibility has been shifted to the client who must have exact fares ready[2]. Fourth, reception is blocked off prior to entering a system; city dwellers increasingly use unlisted telephone numbers to prevent individuals from calling them, and a small but growing number resort to keeping the telephone off the hook to prevent incoming calls. More subtly, one blocks inputs by assuming an unfriendly countenance, which discourages others from initiating contact. Additionally, social screening devices are interposed between the individual and environmental inputs (in a town of 5,000 anyone can drop in to chat with the Mayor, but in the metropolis organizational screening devices deflect inputs to other destinations). Fifth, the intensity of inputs is diminished by filtering devices, so that only weak and relatively superficial forms of involvement with others are allowed. Sixth, specialized institutions are created to absorb inputs that would otherwise swamp the individual (welfare departments handle the financial needs of a million individuals in New York City, who would otherwise create an army of mendicants continuously importuning the pedestrian). The interposition of institutions between the individual and the social world, a characteristic of all modern society, and most acutely present in the large metropolis, has its negative side. It deprives the individual of a sense of direct contact and spontaneous integration in the life around him. It simultaneously protects and estranges the individual from his social environment.

Many of these adaptive mechanisms apply not only to individuals, but to institutional systems as well, as Meier [Ref. 4] has so brilliantly shown in connection with the library and the stock exchange.

In sum, the observed behavior of the urbanite in a wide range of situations appears to be determined largely by a variety of adaptations to overload. We shall now deal with several specific consequences of responses to overload, which come to create a different tone to city and town.

---

2 Considerations of the safety of drivers and of money collected in fares were of course instrumental in the decision to make this particular change.

17-1.1 Social Responsibility

The principal point of interest for a social psychology of the city is that moral and social involvement with individuals is necessarily restricted. This is a direct and necessary function of excess of input over capacity to process. Restriction of involvement runs a broad spectrum from refusal to become involved in the needs of another person, even when the person desperately needs assistance (as in the Kitty Genovese case) through refusal to do favors, to the simple withdrawal of courtesies (such as offering a lady a seat, or saying "sorry" when a pedestrian collision occurs.) In any transaction more and more details need to be dropped as the total number of units to be processed increases and assaults an instrument of limited processing capacity.

The ultimate adaption to an overloaded social environment is to totally disregard the needs, interests, and demands of those whom one does not define as relevant to personal need satisfaction, and to develop optimally efficient perceptual means of identifying whether an individual falls into the category of friend or stranger. The disparity in treatment of friends and strangers ought to be greater in cities than towns; the time allotment and willingness to become involved with those who can make no personal claim on one's time will be less in cities than in towns.

17-1.2 Bystander Intervention in Crisis

The most striking deficiencies in urban social responsibility occur in crisis situations, such as the Genovese murder in Queens. As is well known, in 1964 Catherine Genovese, coming home from a night job in the early hours of an April morning, was stabbed repeatedly and over an extended period of time. Thirty-eight residents of a respectable New York City neighborhood admit to having witnessed at least a part of the attack but none went to her aid or called the police until after she was dead. Milgram and Hollander [Ref. 5] writing in The Nation analyzed the event in these terms: "Urban friendships and associations are not primarily formed on the basis of physical proximity. A person with numerous close friends in different parts of the city may not know the occupant of an adjacent apartment. This does not mean that a city dweller has fewer friends than does a villager, or knows fewer persons who will come to his aid; however, it does mean that his allies are not constantly at hand. Miss Genovese required immediate aid from those physically present. There is no evidence that the city had deprived Miss Genovese of human associations, but the friends who might have rushed to her side were miles from the scene of her tragedy."

"Further, it is known that her cries for help were not directed to a specific person; they were general. But only individuals can act, and as the cries were not specifically directed, no particular person felt a special responsibility. The crime and the failure of community response seem absurd to us. At the time, it may well have seemed equally absurd to the Kew Gardens residents that not one of the neighbors would have called the police. A collective paralysis may have developed from the belief of each of the witnesses that someone else must surely have taken that obvious step."

Latane and Darley [Ref. 6] have reported laboratory approaches to the study of bystander intervention and established experimentally the principle that the larger the number of bystanders the less likely is it that any one of them will intervene in an emergency. Gaertner and Bickman [Ref. 7] of The City University of New York have extended the bystander studies to an examination of help across ethnic lines. Blacks and whites, with clearly identifiable accents, called strangers (through an apparent error in telephone dialing), gave them a plausible story of being stranded on an outlying highway without more dimes and asked the stranger to call a garage. The experimenters found that white-accented callers had a significantly better chance of obtaining assistance than black callers. This suggests that ethnic allegiance may well be another vehicle for coping with overload: the city inhabitant can reduce excessive demands and screen out urban heterogencity by responding along ethnic lines; overload is made more manageable by limiting the "span of sympathy".

In any quantitative characterization of the social texture of city life a first order of business is the application of these experimental methods to field situations set in large cities and small towns. Theorists argue that the indifference shown in the Genovese case would not be present in the small town, but in the absence of solid experimental findings the question remains an open one.

More than just callousness prevents bystanders from participating in altercations between people. A rule of urban life is respect for other people's emotional and social privacy, perhaps because physical privacy is so hard to achieve. And in situations for which the standards are heterogeneous, it is much harder to know whether taking an active role is unwarranted meddling or an appropriate response to a critical situation. If a husband and wife are quarreling in public, at which point should a bystander step in? On the one hand, the heterogeneity of the city produces substantially greater tolerance about behavior, dress, codes of ethics, than does the small town, but this diversity also encourages people to withhold aid for fear of antagonizing the participants or crossing an inappropriate and difficult to define line.

Moreover, the frequency of demands present in the city gives rise to norms of non-involvement. There are practical limitations to the Samaritan impulse in a major city. If a citizen attended to every needy person, if he were sensitive to and acted on every altruistic impulse that was evoked in the city, he could scarcely keep his own affairs in order.

17-1.3 Favor Doing on Trust

We may now move away from crisis situations to less urgent examples of social responsibility. For it is not only in situations of dramatic need, but in the ordinary, everyday willingness to lend a hand that the city dweller is said to be deficient relative to his small town cousin. The comparative method must be employed in any empirical examination of this question. A commonplace social situation is staged in an urban setting and a small town, a situation to which a subject can respond either by extending help or withholding it. The response in town and city are compared.

One factor in the purported unwillingness of urbanites to extend themselves to strangers may well be their heightened sense of physical (and emotional) vulnerability -- a feeling that is supported by urban crime statistics. A key test for distinguishing between city and town behavior, therefore, is how city dwellers compare with town dwellers in offering aid that increases their personal vulnerability and requires some trust of strangers. Altman, Levine, Nadien, and Villena [Ref. 8] of The City University of New York devised a study to compare city and town dwellers in this respect. The criterion used in this study was the willingness of householders to allow strangers to enter their home to use the telephone. The investigators (two males and two females) completed a total of 100 requests-for-entry in the city and 60 in the small towns. The results gleaned from middle income housing developments in Manhattan were compared with data gathered in several small towns in Rockland County, outside of New York City (Stony Point, Spring Valley, Ramapo, Nyack, New City, and West Clarkstown).

Table 1

Percentage of Entries by Investigators for City and Town Homes

Experimenter	Location and Percent of Entries	
	City (n=100)	Small Town (n=60)
Male		
1	16	40
2	12	60
Female		
3	40	87
4	40	100

As Table 1 shows, in all cases there was a sharp increase in the proportion of entries gained by an experimenter when he moved from city to small town. In the most extreme case the experimenter was five times more likely to gain admission to a home in a small town than in Manhattan. Although the female experimenters had noticeably higher levels of entry in both cities and towns than the male experimenters, all four students did at least twice as well in gaining access to town homes as they did in the city, suggesting that the city-town distinction overrides even the predictably greater fear of male strangers than of female ones.

The lower level of helpfulness by city-dwellers seems due in part to recognition of the dangers of Manhattan living, rather than to mere indifference or coldness. It is significant that 75 percent of all city respondents received and answered messages either by shouting through closed doors or by peering out through peepholes; in the towns, by contrast, about 75 percent of the respondents opened the doors, with no barriers between them and the experimenter.

Supporting the experimenters' quantitative results was their general observation that the town-dwellers were noticeably more friendly and less suspicious than the city-dwellers. In seeking to explain the sense of psychological vulnerability city-dwellers feel, above and beyond differences in actual crime statistics, Villena [Ref. 8] points out that for a village resident, if a crime is committed in a neighboring village, he may not perceive it as personally relevant, though the geographic distance may be small. But a criminal act committed anywhere in the city, though miles from the city-dweller's home is still verbally located within the city, "therefore...the inhabitant of the city possesses a larger vulnerable space."

17-1.4 Civilities

Even at the most superficial level of involvement -- the exercise of everyday civilities, urbanites are reputedly deficient. Persons bump into each other and frequently do not apologize. They knock over another person's packages, and as often as not, proceed on their way with a grump, rather than taking the time to help the victim. Such behavior, which many visitors to great cities find distasteful, is less common, we are told, in smaller communities where traditional courtesies are more likely to be maintained.

In some instances it is not simply that in the city traditional courtesies are violated; rather, the cities develop new norms of non-involvement. They are so well defined and so deeply a part of city life, that these constitute the norms people are reluctant to violate. Men are actually embarrassed to give up a seat on the subway for an old woman; they will mumble "I was getting off anyway" instead of making the gesture in a straightforward and gracious way. These norms develop because everyone realizes in situations of high density, people cannot implicate themselves in each others' affairs, for to do so, would create conditions of continual distraction which would frustrate purposeful action.

The effects of overload do not imply that at every instant the city dweller is bombarded with an unmanageable number of inputs, and that his responses are determined by the input excess at any given instant. Rather, adaptation occurs in the form of the gradual evolution of norms of behavior. Norms are created in response to frequent discrete experiences of overload; they persist and become generalized modes of responding.

17-1.5 Overload on Cognitive Capacities Anonymity

It is a truism that we respond differently toward those whom we know and those who are strangers to us. An eager patron aggressively cuts in front of someone in a long movie line to save time only to confront a friend; he then behaves sheepishly. A man gets into an automobile accident caused by another driver, emerges from his car shouting in rage, then moderates his behavior on discovering a friend driving the other car. The city dweller, when moving through the midtown streets, is in a state of continual anonymity viz a viz the other pedestrians.

Anonymity is part of a continuous spectrum ranging from total anonymity at one end to full acquaintance at the other, and it may well be that measurement of the precise degrees of anonymity in cities and towns would help to explain important distinctions between the quality of life in each. Conditions of full acquaintance, for example, offer security and familiarity but they may also be stifling, because the inhabitant is caught in a web of established relationships. Conditions of complete anonymity, by contrast, provide freedom from routinized social ties, but they may also create feelings of alienation and detachment.

Empirically one could investigate the proportion of activities in which the city-dweller and town-dweller are known by others at given times in their daily lives, and the proportion of those the urbanite or town-dweller interacts with who know him. At his job, for instance, the city-dweller may be known to as few people as his rural counterpart. While not fulfilling his occupational role, however -- say, in merely traveling about the city -- the urbanite is doubtless more anonymous than his rural counterpart.

Limited empirical work on anonymity has begun. Zimbardo [Ref. 9] has tested whether the social anonymity and impersonability of the big city encourages greater valdalism than do small towns. Zimbardo arranged for one car to be left for 64 hours near the New York University campus in the Bronx (New York) and a counterpart to be left near Stanford University in Palo Alto. The license plates on both cars were removed and the hoods opened, to provide "releaser cues" for potential vandals. The New York car was stripped of all moveable parts within the first 24 hours, and was left a hunk of metal rubble by the end of 3 days. Unexpectedly, however, most destruction occurred during daylight hours usually under scrutiny by observers, and was led by well-dressed, white adults. The Palo Alto car was left untouched.

Zimbardo attributes the difference in the treatment accorded the two cars to the "acquired feelings of social anonymity provided by life in a city like New York," and he supports his study with several other anecdotes illustrating casual, wanton vandalism in the city. Any study comparing the effects of anonymity in city and town, however, must satisfactorily control for other confounding factors: the large number of drug addicts in New York, the higher proportion of slum-dwellers in the city, etc.

Another direction for empirical study is the investigation of the beneficial effects of anonymity. Impersonality of city life breeds its own tolerance for the private lives of inhabitants. Individuality and even eccentricity we may assume can flourish more readily in the metropolis than the small town. Stigmatized persons may find it easier to lead comfortable lives without the constant scrutiny of neighbors. To what degree can this assumed difference between city and town be shown empirically? Judith Waters [Ref. 10] at The City University hypothesized that avowed homosexuals would be more likely to be accepted as tenants in a large city than in small towns, and she dispatched letters from homosexuals and normals to real estate agents to cities and towns across the country. The results of her study were inconclusive. But the general idea of examining the protective benefits of city life to the stimatized ought to be pursued.

17-1.6 Role Behavior in Cities and Towns

Another product of urban "overload" is the adjustment in roles made by urbanites in daily interactions. As Wirth has said: "Urbanites meet one another in highly segmental roles...

They are less dependent upon particular persons, and their dependence upon others is confined to a highly fractionalized aspect of the other's round of activity." [Ref. 2] This tendency is particularly noticeable in transactions between customers and those offering professional or sales services: the owner of a country store has time to become well-acquainted with his dozen-or-so daily customers; but the girl at the checkout counter of a busy A & P, handling hundreds of customers a day, barely has time to toss the green stamps into one customer's shopping bag before the next customer has confronted her with his pile of groceries.

Meier's stimulating analysis of the city [Ref. 4] discusses several adaptations a system may make when confronted by inputs that exceed its capacity to process them. Specifically, Meier argues (see Figure 1) that according to the principle of competition for scarce resources, the scope and time of the transaction shrinks as customer volume and daily turnover rise. This, in fact, is what is meant by the brusque quality of city life. New standards have developed in cities about what level of services are appropriate in business transactions.

McKenna and Morgenthau [Ref. 11], in a seminar at the City University of New York, devised a study (a) to compare the willingness of city-dwellers and small-towners to do favors for strangers that entailed a small amount of time and inconvenience but no personal vulnerability and (b) to determine whether the more compartmentalized, transitory relationships of the city would make urban salesgirls less likely than small-town salesgirls to carry out tasks not related to their customary roles for strangers.

To test for differences between city dwellers and small towners, a simple experiment was devised in which persons from both settings were asked (by telephone) to perform increasingly onerous favors for anonymous strangers.

Within the cities (Chicago, New York and Philadelphia), half the calls went to housewives, and the other half to salesgirls in women's apparel shops; the same division was made for the 37 small towns, which were in the same states as the cities. Each experimenter represented herself as a long-distance caller who had mistakenly been connected with the respondent by the operator. The experimenter began by asking for simple information about the weather for travel purposes. Next the experimenter excused herself on a pretext (stating "please hold on:), put the phone down for almost a full minute, and then picked it up again and asked the respondent to provide the phone num-

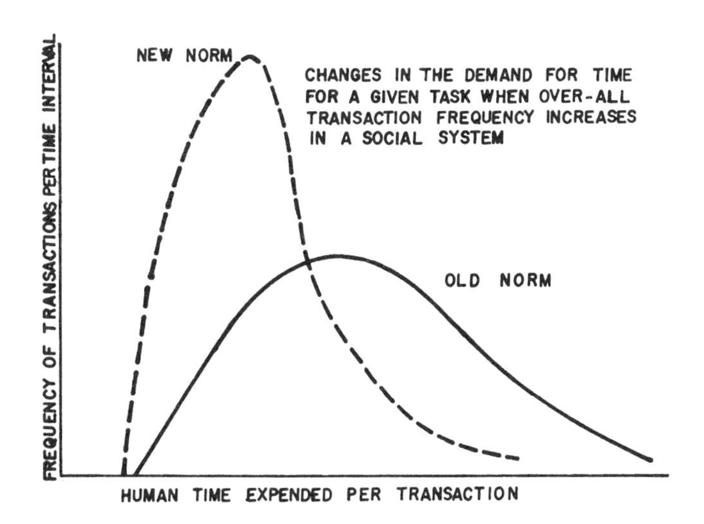

Figure 1
Source: Meier - A communications theory of urban growth.

ber of a hotel or motel in her vicinity at which the experimenter might stay during a forthcoming visit. Scores were assigned to the subjects depending on how helpful they had been. Mckenna summarizes her results in this manner: "People in the city, whether they are engaged in a specific job or not, are less helpful and informative than people in small towns; ... People at home, regardless of where they live, are less helpful and informative than people working in shops." However, the absolute level of cooperativeness for urban subjects was found to be quite high, and does not accord with the stereotype of the urbanite as aloof, self-centered, and willing to help strangers. The quantitative differences obtained by McKenna and Morgenthau are less great than one might have expected. This again points up the need for extensive empirical research in rural-urban differences, research that goes far beyond that provided in the few illustrative pilot studies presented in this paper. At this point we have very limited objective evidence on differences in the quality of social encounters in city and small town.

But the research needs to be guided by unifying theoretical concepts. As this section of the paper has tried to demonstrate, the concept of overload helps to explain a wide variety of contrasts between city and town behavior: (1) the differences in role enactment (the urban tendency to deal with one another in highly segmented, functional terms; the constricted time and services offered customers by sales personnel); (2) the evolution of urban norms quite different from traditional town values (such as the acceptance of non-involvement, impersonality, and aloofness in urban life); (3) the adaptation of the urban-dweller's cognitive processes (his inability to identify most of the people seen daily; his screening of sensory stimuli; his development of blase attitudes toward deviant or bizarre behavior; and his selectivity in responding to human demands); and (4) the far greater competition for scarce facilities in the city (the subway rush; the fight for taxis; traffic jams; standing in line to await services). I would suggest that contrasts between city and rural behavior probably reflect the responses of similar people to very different situations, rather than intrinsic differences between rural personalities and city personalities. The city is a situation to which individuals respond adaptively.

17-2 FURTHER ASPECTS OF URBAN EXPERIENCE

Some features of urban experience do not fit neatly into the system of analysis presented thus far. They are no less important for that reason. The issues about to be raised are difficult to treat in quantitative fashion. Yet I prefer discussing them in a loose way to excluding them because appropriate language and data have not yet been developed. The aim shall be to suggest how phenomena such as "urban atmosphere" can be pinned down through techniques of measurement.

17-2.1 The Atmosphere of Great Cities

The contrast in behavior between city and town has formed a natural starting point for urban social scientists, but even among great cities there are marked differences in 'atmosphere'. The feeling tone, pacing, and texture of social encounters are different in London than in New York, and many persons willingly make financial sacrifices for the privilege of living within a specific urban atmosphere which they find pleasing or stimulating. A second perspective for the study of cities, therefore, is to define exactly what is meant by the atmosphere of a city, and to pinpoint the factors that give rise to it. It may seem that urban atmosphere is too evanescent a quality to reduce to a set of measurable variables, but I do not believe the matter can be judged before substantial effort has been made in this direction. It is obvious that any approach in this vein needs to be comparative. It makes no sense at all to say that New York is "vibrant" and "frenetic" unless a specific comparison city is kept in mind.

In an undergraduate tutorial I conducted at Harvard University some years ago - New York, London, and Paris served as the reference points for attempts at measurement of urban atmosphere. We began with a simple question: Does any consensus exist about the qualities that typify given cities? To answer this question one could undertake a content analysis of travel books, literary descriptions, and journalistic accounts of cities. A second approach, which we adopted, is to ask people to characterize (with descriptive terms and typical experiences) cities they have lived in or visited. Advertisements placed in The New York Times and the Harvard Crimson asked persons to give us accounts of specific incidents in London, Paris or New York that best illuminated the character of each city. Questionnaires were then developed and administered to persons who were familiar with at least two of the three cities.

Some distinctive patterns emerged [Ref. 12]. The distinguishing themes concerning New York, for example, dealt with its diversity, its great size, its pace and level of activity, its cultural and entertainment opportunities, and the heterogeneity and segmentation ("ghettoization") of its population. New York elicited more descriptions in terms of physical qualities, pace, and emotional impact than did Paris or London, suggesting that these are particularly important aspects of New York's ambiance.

A contrasting profile emerges for London, where respondents placed far greater emphasis on their interactions with the inhabitants than on physical surroundings. There was near unanimity on certain themes: i.e. those concerned with London's tolerance and courtesy. One respondent said:

> "When I was 12, my grandfather took me to the British Museum...one day by tube and recited the Aeneid in Latin for my benefit...He is rather deaf, speaks very loudly and it embarrassed the hell out of me, until I realized that nobody was paying any attention. Londoners are extremely worldly and tolerant."

In contrast, respondents describing New York as aloof, cold, and rude referred to such incidents as:

> "I saw a boy of 19 passing out anti-war leaflets to passers-by. When he stopped at a corner, a man dressed in a business suit walked by him at a brisk pace, hit the boy's arm, and scattered the leaflets all over the street. The man kept walking at the same pace down the block."

Many more incidents of this sort need to be obtained, employing careful methods of sampling. By the application of factor analytic techniques, relevant dimensions for each city can be discerned.

Paris evoked responses divided about equally in number in regard to its inhabitants and its physical and sensory attributes. Cafes and parks were mentioned often as contributing to the city's sense of amenities, but there were many complaints that Parisians were inhospitable, nasty, and cold.

We cannot be certain, of course, in what degree these statements reflect actual characteristics of the cities in question or simply tap the respondents' knowledge of widely held pre-conceptions. Indeed, one may point to three factors apart from the actual atmospheres of the cities that determine the subjects' responses:

1. A person's impression of a given city depends on his implicit standard of comparison: a New Yorker who visits Paris may well describe that city as "leisurely", while a compatriot from Richmond, Virginia, may consider Paris too "hectic". Obtaining reciprocal judgments, in which New Yorkers judge Londoners and the reverse, seems a useful way to get at not only the city being judged, but the home city that serves as the visitor's baseline.
2. Perceptions of a city are also affected by whether the observer is a tourist, a newcomer, or a longer-term resident. First, a tourist will be exposed to different features of the city than a long time resident. Second, a prerequisite of adapting to continuing life in a given city seems to be the filtering out of many observations about the city that the newcomer or tourist find particularly salient -- this selective process of observation seems to be part of the long-term resident's mechanism for coping with overload. In the interest of psychic economy, the resident simply learns to tune out many aspects of daily life. One method for studying the specific impact of adaptation on perception of the city is to ask several pairs of newcomers and old-timers to walk down the same city blocks and then report separately what each observed.

Additionally, many persons have noted that when travellers return to New York from an extended sojourn abroad, they are often confronted with "brutal ugliness" [Ref. 13] and a distinctive, frenetic atmosphere whose contributing details are, for a few hours or days, remarkably sharp and clear. This period of fresh perception ought to receive special attention in the study of city atmosphere. For in a few days, details which are initially salient, become less easy to specify. They are assimilated to an increasingly familiar background atmosphere which, though important in setting the tone of things, is difficult to analyze into its constituent parts. There is no better point to begin the study of city atmosphere than at that moment when travellers have just returned from abroad.

3. The popular myths and expectations each visitor brings to the city will also affect how it is perceived [Ref. 14]. Sometimes a person's preconceptions about a city are relatively ac-

curate distillations of its character, but preconceptions of the city to conform to his expectations. Preconceptions affect not only a person's perceptions of a city, but what he will report about it.

The influence of a person's urban baseline on his perceptions of a given city, the differences between the observations of the long-time inhabitant and those of the newcomer, and the filtering effect of personal expectations and stereotypes raise serious questions about the validity of travellers' reports. Moreover, no social psychologist wishes to rely exclusively on verbal accounts if he is attempting an accurate and objective dessciption of the cities' social texture, pacing, and general atmosphere. What he needs to do is to devise means of embedding objective measures in the daily flux of city life, measures that can accurately index qualities of a given urban atmosphere.

17-2.2 Experimental Comparisons of Behavior

Roy Feldman [Ref. 15] incorporated these principles in a comparative study of behavior toward compatriots and foreigners in Paris, Athens, and Boston. Feldman wished to see a) whether absolute levels and patterns of helpfulness varied significantly from city to city, and b) whether inhabitants in each city tended to treat compatriots differently from foreigners. He examined five concrete behavioral episodes, each carried out by a team of native and a team of American experimenters in the three cities. The episodes involved: (1) asking natives of the city for street directions: (2) asking natives to mail a letter for the experimenter; (3) asking natives if they had just dropped a dollar bill (or the Greek and French equivalent) when the money actually came from the experimenter himself; (4) deliberately overpaying for goods in a store to see if the cashier would correct the mistake and return the excess money; and (5) investigating whether taxicab drivers overcharged strangers and whether they took the most direct route available.

Feldman's results suggest some interesting contrasts in the profiles of the three cities. In Paris, for instance, certain stereotypes were borne out: Parisian cab-drivers overcharged foreigners significantly more often than compatriots. But other aspects of Parisians behavior contradict American preconceptions: in mailing a letter for a stranger Parisian treatment of foreigners was significantly better than Athenian or Bostonian behavior toward foreigners, and when asked to mail letters that were already stamped, Parisians actually offered better treatment to foreigners than to compatriots. Similarly, Parisians were significantly more honest than Athenians or Bostonians in resisting false claims to money, and Parisians were the only citizens who treated foreigners better than compatriots in this experiment.

Feldman's studies not only begin to quantify some of the variables that give a city its distinctive texture, but they provide a methodological model for other comparative research. His most important contribution is his successful application of objective, experimental measures to everyday situations, thus providing more pertinent conclusions about urban life than those achieved through laboratory experiments.

17-2.3 Tempo and Pacing

Another important component of a city's atmosphere is its tempo or pace, an attribute frequently remarked on but less often studied. Does a city have a frenetic, hectic quality, or is it easygoing and leisurely? In any empirical treatment of this question, it is best to start in a very simple minded way. Walking speeds of pedestrians ought to be measured from one city to the next, and from city to town. William Berkowitz [Ref. 16] of Lafayette College has undertaken an extensive series of studies of walking speeds in Philadelphia, New York, and Boston, as well as in small and moderate sized towns. Berkowitz writes that "there does appear to be a significant linear relation between walking speed and size of municipality, but the absolute size of the difference varies by less than ten percent."

Perhaps the feeling of rapid tempo is due not so much to absolute pedestrian speeds as the constant need to dodge others in a large city to avoid pedestrian collisions. (One basis for computing the adjustments needed to avoid collisions is to hypothesize a set of mechanical manikins, sent walking along a city street, and to calculate the number of collisions when no adjustments are made. Clearly, the higher the density of manikins the greater the number of collisions per unit of time, or conversely, the greater frequency of adjustments are needed in higher densities to avoid collisions.)

Patterns of automobile traffic contribute to a city's tempo. Driving an automobile provides a direct means of translating feelings about tempo into measurable acceleration, and a city's pace should be particularly evident in vehicular velocities, patterns of acceleration, and response latencies to traffic signals. The inexorable tempo of New York is expressed, further, in the manner in which pedestrians stand at busy intersections, impatiently awaiting a change in traffic light, making tentative excursions into the intersection, and frequently surging into the street even before the green signal light is activated.

17-2.4 Visual Components

Hall has remarked [Ref. 17] the physical layout of the city also affects its atmosphere. A grid-iron pattern of streets gives the visitor a feeling of rationality, orderliness, and predictability but is sometimes monotonous. Winding lanes or streets branching off at strange angles, with many forks (as in Paris or Greenwich Village) create feelings of surprise, and aesthetic pleasure while forcing greater decision-making in plotting one's course. Some would argue that the visual component is all-important -- that the "look" of Paris or New York can almost be equated with its atmosphere. To investigate this hypothesis, we might conduct studies of using only blind or at least blindfolded respondents - we would no doubt discover that the texture of each city is distinctive even without its visual component.

17-2.5 Sources of Ambiance

Thus far we have tried to pinpoint and measure some of the factors that contribute to the distinct atmosphere of a great city. But we may also ask: why do differences in urban atmosphere exist? How did they come about and are they in any way related to the factors of density, large numbers, and heterogeniety discussed earlier in the paper?

First, there is the obvious factor that even among great cities populations and densities differ. The metropolitan areas of New York, London, and Paris, for example, contain 15 million, 12 million, and 8 million persons respectively. London possesses average densities of 43 persons per acre, while Paris is more congested with average densities of 114 persons per acre [Ref. 18]. Whatever characteristics are specifically attributable to density are more likely to be pronounced in Paris than London.

A second factor affecting the atmosphere of cities is the source from which the populations are drawn [Ref. 19]. It is a characteristic of great cities that they do not reproduce their own populations, but that their numbers are constantly maintained and augmented by the influx of residents from other parts of the country. This can have a determining effect on its atmosphere. For example, Oslo is a city in which almost all of the residents are only one or two generations removed from purely rural existence, and this contributes to its almost agricultural norms.

A third source of atmosphere is the general national culture that the city participates in. Paris is a combination of the adaptations to the demography of cities and certain values specific to French culture. New York is an admixture of American values and the peculiar needs of extraordinary high density and large population.

Finally, one could speculate that the atmosphere of one great city relative to another is traceable to the specific historical conditions under which adaptations to urban overload occurred. For example, a city which acquired its mass and density during a period of commercial expansion, will respond to the new demographic facts by adaptations which are designed to serve purely commercial needs. Thus Chicago, which grew and became a great city under a purely commercial stimulus, adapted in a manner that emphasizes business needs. European capitals, on the other hand, incorporate many of the adaptations which were appropriate to the period of their increasing numbers and density. The prevalence of aristocratic values at the time of their growth turned coping mechanisms away from pure considerations of efficiency. Thus manners, norms, and facilities of Paris and Vienna continue to reflect the idealization of leisure and aesthetic values.

17-2.6 Cognitive Maps of Cities

When we speak of "behavioral comparisons" among cities, we must specify which parts of the city are most relevant for sampling purposes. In a sampling of "New Yorkers" should we include residents of Bay Ridge or Flatbush as well as inhabitants of Manhattan? And if so, how should we weight our sample distribution? One approach to defining relevant boundaries in sampling is to determine which areas form the psychological or cognitive core of the city. We weight our samples most heavily in the areas considered by most people to represent the "essence" of the city.

The psychologist is less interested in the geographic layout of a city or its political boundaries than the cognitive representation of it. Hans Blumenfeld [Ref. 20] points out that the perceptual structure of a modern city can be expressed by the "silhouette" of the group of skyscrapers at its center and of smaller such groups at its "subcenters" but that urban areas can no longer, because of their vast extent, be experienced as fully articulated sets of streets, squares, and space.

In Image of the City [Ref. 21] Kevin Lynch created a cognitive map of Boston by interviewing Bostonians. Perhaps his most significant finding was that while certain landmarks, such as Paul Revere's house and the Boston Commons, as well as the paths linking them, are known by almost all Bostonians, vast areas of the city were simply unknown to its inhabitants.

Using Lynch' a technique, Donald Hooper [Ref. 22] created a psychological map of New York from the answers to the questionnaire study on Paris, London, and New York. Hooper's results were similar to those of Lynch: New York appears to have a dense core of well-known landmarks in mid-town Manhattan, surrounded by the vast unknown reaches of Queens, Brooklyn, and the Bronx. Times Square, Rockefeller Center, and the Fifth Avenue department stores alone comprise half the total places specifically cited by respondents as the haunts in which they spent most of their time. Outside the densely packed icons of the mid-town area, however, only scattered landmarks achieved recognition. Another interesting pattern is that even the best-known symbols of New York are relatively self-contained, and the pathways joining them appear to be insignificant on the map.

The psychological map can be used for more than just sampling techniques. Lynch [Ref. 21] argues, for instance, that a good city is highly "imageable" -- with many known symbols jointed by widely known pathways -- while dull cities are gray and nondescript. We might test the relative imagibility of several cities by determining the proportion of residents who recognize sampled geographic points and accompanying pathways.

If we wish to be even more precise, we could construct a cognitive map that would not only show the symbols of the city but would measure the precise degree of cognitive significance of any given point in the city relative to any other. By applying a pattern of points to New York City, for example, and taking photographs from each point, we could determine what proportion of a sample of the city's inhabitants can identify the locale specified by each point on the map.

The most important scientific requirement is that we obtain an impartial and fair sampling of viewing points within the city. We wanted a form of geographic sampling that is objective, readily applied, and could be used not only in New York, but in any city in the world (should comparative studies be attempted). To achieve this we allowed an international coordinate system to determine our sampling points. We used an established grid system based on the 1000 Meter Universal Transverse Mercator grid. Wherever a 1000 meter line of latitude intersected a 1000 meter line of longitude, we took a viewing point for our study (see Figure 2). For purposes of economy, viewing points were systematically thinned out in the Bronx, Brookly, Queens, and Staten Island; the final pattern of viewing points is shown in figure 2. Color slides taken from each of the viewing points were presented to 200 subjects, who were geographically representative of the population distribution of New York City. Subjects were asked to identify the scenes in terms of increasingly more precise criteria of recognition, first placing the scene in its correct borough, then its correct neighborhood, and finally, on its exact street.

Examining the information on neighborhood placement we find that very substantial differences emerge according to borough:

Bronx	5.85%	
Brooklyn	11.42	percentages of placement of scenes into correct neighborhoods
Manhattan	31.98	
Queens	10.76	
St. Island	5.40	

A randomly selected scene in Manhattan is five times more likely to be placed in its correct neighborhood than a randomly selected scene in the Bronx. Manhattan scenes do almost three times as well as scenes in Brooklyn and Queens. What we must keep in mind is that the views were <u>not</u> selected because of their recognizability, but by a mechanical geographic sampling procedure. Figure 3 shows the pattern of neighborhood recognition for 152 viewing points in New York City.

The superior information value of Manhattan becomes even more pronounced when a more exacting criterion of recognition

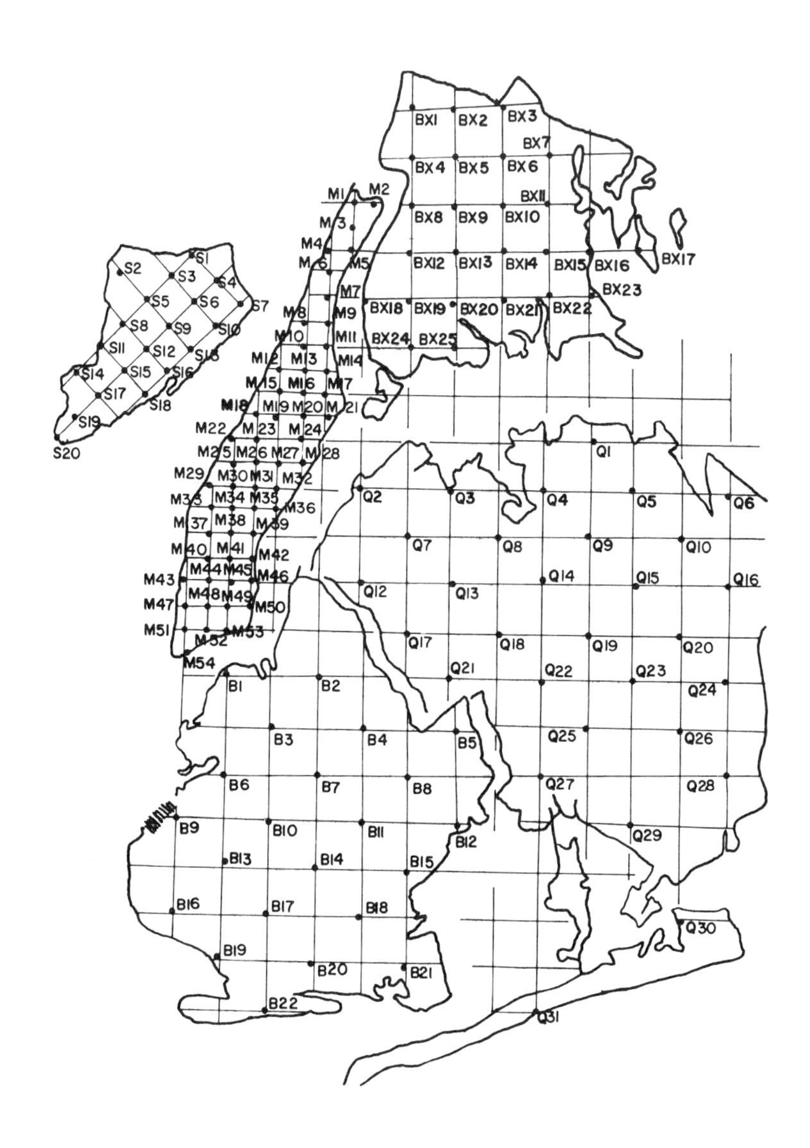

Figure 2
Proposed New York Grid

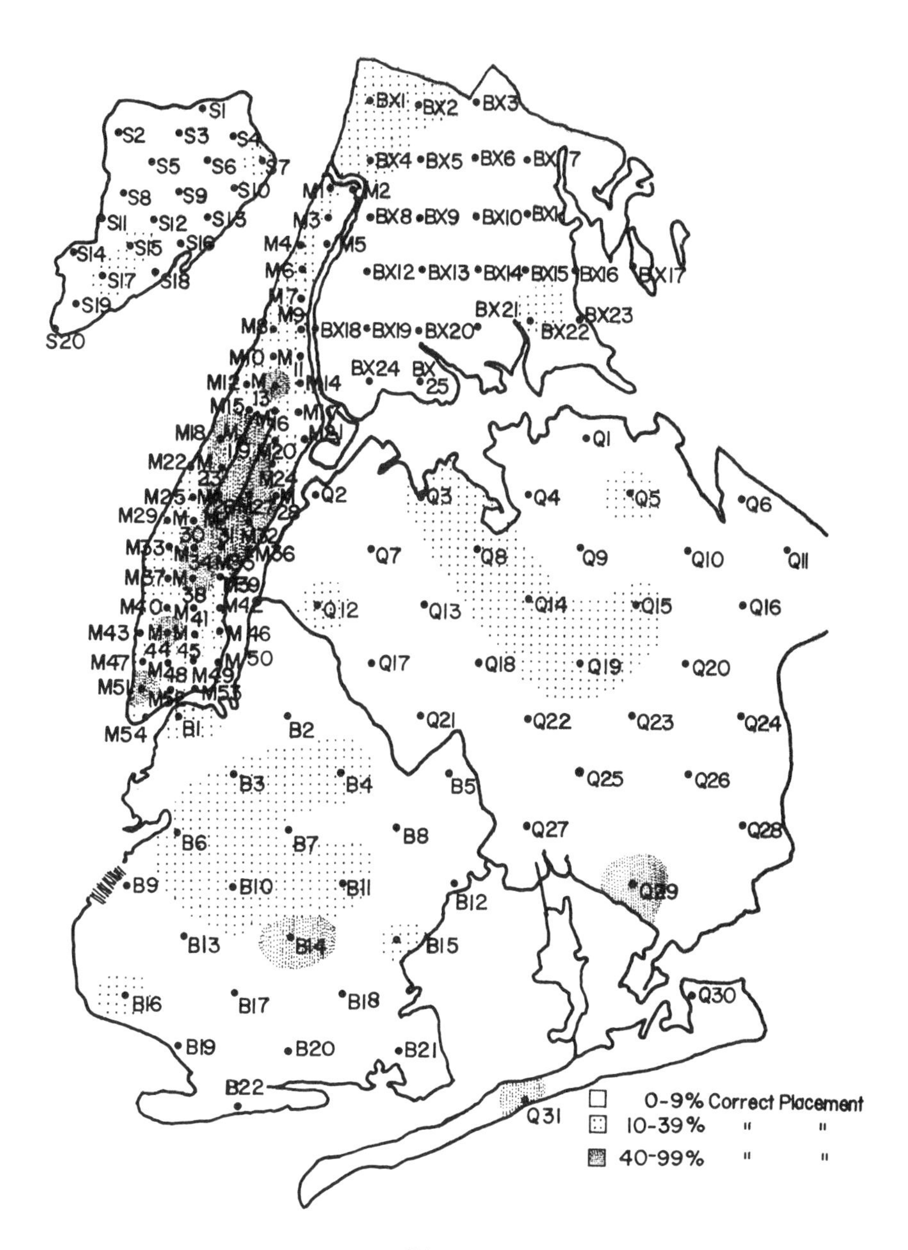
S1
S2
S3
S4
S5
S6
S7
S8
S9
S10
S11
S12
S13
S14
S15
S16
S17
S18
S19
S20
BX1
BX2
BX3
BX4
BX5
BX6
BX7
BX8
BX9
BX10
BX12
BX13
BX14
BX15
BX16
BX17
BX18
BX19
BX20
BX21
BX22
BX23
BX24
BX 25
M1
M2
M3
M4
M5
M6
M7
M8
M9
Q1
Q2
Q3
Q4
Q5
Q6
Q7
Q8
Q9
Q10
Q11
Q12
Q13
Q14
Q15
Q16
Q17
Q18
Q19
Q20
Q21
Q22
Q23
Q24
Q25
Q26
Q27
Q28
Q29
Q30
Q31
B1
B2
B3
B4
B5
B6
B7
B8
B9
B10
B11
B12
B13
B14
B15
B16
B17
B18
B19
B20
B21
B22
0-9% Correct Placement
10-39% " "
40-99% " "

Figure 3

is applied. For when we ask subjects to identify each scene in terms of a street location, we find the following distribution of accurate guesses for each borough:

Bronx	2.56%
Brooklyn	2.83
Manhattan	15.52
Queens	2.21
St. Island	0.6

The reader must again be reminded that these scenes were not selected because of the likelihood they would be recognized, but were mechanically sampled in a manner that was completely independent of their scenic value. The figures presented above are the judgments of 200 individuals who come from each of the five boroughs and their numbers are in proportion to the actual populations of those boroughs. However, this overall picture also holds true no matter which borough the person comes from. A resident of Queens is four times more likely to recognize a street location in Manhattan than in his own borough (3.76% for his home borough of Queens vs. 15.00% for Manhattan). Areas of Queens have often been accused of being non-descript, and taxi drivers are reputed to fear entering Queens lest they never find their way out. And with good reason! Even the people who live in Queens are lost in their home borough compared with the sense of place they experience in Manhattan. So that if we are interested in the psychological representation of the city in the minds of its inhabitants, we find that no matter in which borough the person resides, the most differentiated section of his psychological map is Manhattan, far outstripping even his home borough. In this sense it is correct to say that New York City is not merely culturally, but also imagistically rooted in Manhattan.

What principles account for these findings? First, an area can only be recognized if people are exposed to it. Manhattan's high index of recognizability is due, in important measure, to the fact that as cultural and entertainment core of the city, it attracts persons from all over the city. Even a highly distinctive architectural display will not be widely recognized if it is too far off the beaten path.

The second major factor is the overall architectural or social distinctiveness of the area. Columbus Circle and Rockefeller Center impress themselves because of their unique configuration of spaces and buildings. Chinatown and Little Italy communicate themselves through cultural and racial features of their respective neighborhoods. The degree to which a part of the city will be recognized can be expressed by the simple formula:

$$R = f\left[\begin{array}{l}\text{architectural and} \\ \text{social distinctiveness}\end{array} \times \text{centrality}\right]$$

One might also use psychological maps to gain insight into the differing perceptions of a given city held by members of its cultural subgroups and how their perceptions may change. In the earlier stages of life whites and Negroes probably have only a limited view of the city, centering on the immediate neighborhood in which they are raised. In adolescence, however, the white teenager probably undergoes a rapid enlargement of his field of knowledge, learning of opportunities in mid-town and outlying sections, and comes to see himself as functioning in a larger urban field. But the process of ghettoization, to which the black teenager is subjected, may well hamper the expansion of his sense of the city. These are speculative notions but are readily subject to precise test.

17-3 CONCLUSION

I have tried to indicate some organizing theory that starts with the facts of city life: large numbers, density, and heterogeneity. These are external to the individual. He experiences these factors as overloads at the level of roles, norms, cognitive functions, and facilities. These lead to adaptive mechanisms which create the distinctive tone and behaviors of city life. These notions, of course, are largely speculative, and need to be examined by objective comparative studies of cities and towns.

A second perspective concerns the differing atmosphere of great cities, such as Paris, London, and New York. Each has a distinctive flavor, offers a differentiable quality of experience. More precise knowledge of urban atmosphere seems attainable by applying tools of experimental inquiry.

REFERENCES

1. The New York Times, Sec. 12, June 15, 1969.
2. L. Wirth, Amer. J. Soc. XLIV, 1-24, 1938.
3. G. Simmel, The Sociology of George Simmel, Wolff, Ed., The Free Press: MacMillan, New York, 1950.
4. R. L. Meier, A Communications Theory of Urban Growth, (M. I. T. Press, Cambridge, 1962).
5. S. Milgram and P. Hollander, The Nation, 25, 602-604, 1964.
6. B. Latane and J. Darley, Amer. Sci., 57, 244-268, 1969.
7. S. Gaertner and L. Bickman, The Ethnic Bystander. Unpublished Research. (Graduate Center, The City University of New York, New York, 1968).
8. D. Altman, M. Levine, M. Nadien, and J. Villena, Trust of the Stranger in the City and the Small Town. Unpublished Research. (Graduate Center, The City University of New York, New York, 1969).
9. P. G. Zimbardo, in Neb. Symp. on Motivation. U. S. A., 1969.
10. J. Waters, The Relative Acceptance Accorded a Discreditable Person in Rural and Metropolitan Areas. Unpublished Research. (Graduate Center, The City University of New York, New York, 1969).
11. W. McKenna and S. Morgenthau, Urban-Rural Differences in Social Interaction: A Study of Helping Behavior. Unpublished Research. (Graduate Center, The City University of New York, New York, 1969).
12. N. Abuza, The Paris-London-New York Questionnaires. Unpublished Paper. (Harvard University, Cambridge,1967).
13. P. Abelson, Sci. 165, 853, 1969.
14. A. L. Strauss (Ed.), The American City: A Sourcebook of Urban Imagery (Aldine, Chicago, 1968).
15. R. E. Feldman, J. Pers. Soc. Psy. 10, 202-214, 1968.
16. W. Berkowitz, Personal Communication, 1969.
17. E. T. Hall, The Hidden Dimension (Doubleday, New York, 1966.
18. P. Hall, The World Cities (McGraw Hill, New York, 1966.
19. R. E. Park, E. W. Burgess, and R. D. McKenzie, The City (The University of Chicago Press, Chicago, 1967), pp. 1-45.
20. H. Blumenfeld, in The Quality of Urban Life (Sage Publications, Beverly Hills, 1969).

21. K. Lynch, The Image of the City (M.I.T. and Harvard University Press, Cambridge, 1960).
22. D. Hooper, A Pedestrian's View of New York, London and Paris. Unpublished Paper. (Harvard University, Cambridge, 1967).

Chapter 18

HEALTH SYSTEMS - AN URBAN VIEW

C. Flagle

18-1 INTRODUCTION

I have enjoyed very much the opportunity to prepare this paper for the Fifth Systems Symposium, devoted this time to urban problems. After some years of viewing health services from an institution's point of view, or a patient's point of view, or a physician's, or society at large, it has been quite interesting to take another look, this time from the vantage point of the city and its neighborhoods. Whether a large city or a small city, the focal point of concern must be a population's health and medical care; and all the influences on health, not just facilities and services, must be related to that population.

At the outset the task seemed reasonable. One should be able to trace an evolution of fairly consistent values related to health and health care and from these values develop a set of objectives, measures of effectiveness, alternative courses of

---

The author is Professor, Department of Public Health Administration, School of Hygiene and Public Health, The Johns Hopkins University, Baltimore, Maryland.

actions, and choice of strategy--the familiar process of systems analysis. To help out, some nice urban-based examples were close at hand to illustrate the evolutionary process: one example is a clinic from a poverty area in the old part of a city, another a comprehensive health plan in a fresh, new town, and lastly a pediatric center in an area surrounding the teaching hospital in a city.

In examining the examples, I soon became very uneasy for quite different experiences were developing: obstruction in one case, success and movement in another, and a disturbing kind of indifference in the third. These varying reactions of the population to new medical services were occurring even though the goals and objectives were the same in each example as were the resources and many of the professionals involved. Something was missing then in my abstraction of an urban-based medical system. The difference, it became apparent, was rooted in the beliefs of the populations affected, beliefs about the values of medical care and the intentions of those who provide it. The roots of difference of beliefs seemed to lie in the histories of the different populations. Then I was reminded of Jose Ortega y Gasset's essay, "History as a System." [Ref. 1] His theory is that beliefs form an accumulating hierarchy in a culture, with the result that active responses will be determined not only by new circumstances but by an accumulation of experience. New external circumstances affect only in part how a population or a person will respond to developments; history plays the rest. How, specifically, relevant is this to my argument here? In the case of a new health system directed toward inner-city citizens, where the initial obstructive response came to everyone's disappointment if not surprise, one can find the explanation, in part, in history. Two hundred years of persecution and exploitation or exploitative paternalism have made the population nonreceptive initially to a new idea from the same source. The problem is in part ethnic (related to the surging demand of the black population, much of it recently immigrated, for control of its own affairs) and in part to long-standing community attitudes toward the hospital. Perhaps some description of the particular setting will help sort out the problems and influences to be considered.

18-2 BACKGROUND

Much of this is written in my office in the Johns Hopkins Medical Institutions in Baltimore. From my high window I have a remarkable view of the city, looking south over the historic harbor and west to the rebuilding skyline downtown. Closer at hand the view is not so attractive. A sea of rooftops of slum houses, inner-city decay, pocked by the debris of condemned areas and recently cleared lots, extends in all directions. Here and there, neat new structures are rising in evidence of what the city might be. There are signs of a plan, architecturally; except occasionally for the insensitive (and often bitterly contested) knocking down of antique specimens, one can have hope that someday a better physical plant will stand here than ever existed before.

But what about life and health of those who live here now? We might get some notion by taking another look from the window--or better yet a walk--this time in the direction of the hospital's dispensary and emergency room. Here, the stream of humanity seems endless; the demands are greater than ever anticipated, for the hospital in the past few years has become the family doctor to the community, and it is ill prepared to fill the role.

It was not always like this, and we must know something of the history of the city and its medical care and health problems in order to understand the predicament well enough to seek solutions. It goes without saying that Philadelphia, New York, Washington, Chicago or Los Angeles would fit the story as well.

It is no coincidence that the large teaching institutions are in the decayed areas of the city. Their locations reflect the post-renaissance concept of medicine and hospitals -- a merger of charitable impulses in the community to help the sick poor with the needs of an emerging medical profession for what is still called "teaching material."

For a time--about two decades ago--there was concern that rising affluence and privileges entitling local residents to "private" medical care might deprive the medical school of its supply of teaching cases. Then came the wave of migrations of displaced farm workers. Like Oscar Handlin's uprooted European peasants [Ref. 2] of several centuries ago, the newly uprooted agricultural worker, predominantly southern, flocked to the industrial cities. It is not our purpose here to dwell on all the consequences of this turn of affairs, but some must be noted. There followed an outmigration of older residents, both black and white. A significant effect of this was the destruction of the equilibrium between private medical services in the inner city and the teaching centers; few doctors remained there in private practice and the burden of care fell upon the hospital--or on no one at all.

We find now in the surrounding census tracts a ratio of one physician per 6-7,000 persons, about one-tenth of the availability in the nation or city at large. (The ratio may be expected to worsen since the average age of neighborhood physician is over 60.) As might be expected the mores of medical care reflect the low availability of physicians--a dearth of regular, routine care and a near doubling of emergency department visits. Visualize the population then as one typical of urban poor, about 90% black. The population is predominantly young, almost half under 21, a third under 12. Half the families receive some form of public assistance, a fifth have single parents as head of household. Infant mortality, at 37 deaths per 1,000 live births, is nearly double the rate in affluent sections, with half the utilization of prenatal care or well-baby clinics [Ref. 1].

Two more points need to be made by way of background. The impact of the urban poor on the hospital emergency services revealed a fragmentation of resources on all sides. That the health services themselves were not integrated is well known and publicized, e.g. in the Health Manpower Report of 1967 [Ref. 3].

> "...the word system is a convenient one for our purpose, but we recognize that it is inaccurate if it implies the existence of an organized, coordinated, planned undertaking. Medical care in the United States is more a collection of bits and pieces with overlapping, duplication, great gaps, high costs and wasted efforts. It is more that than an integrated system in which needs and efforts are closely related."

In addition, the sources of support for health care are fragments. As noted by the President of the Johns Hopkins Hospital, nineteen separate agencies must be involved in support of a comprehensive clinic operation. The challenge to management and to system analysis is to develop mechanisms, structure, and organization to bring about rational operation of the many autonomous elements of the urban health care process. Such integration now appears to be coming about, and it is interesting to note that much of the initiative and leadership are coming from the large medical teaching centers. Many factors have been at work to influence the university role: an emerging social conscience and voice among students, an awareness of the European departments of social medicine, with the creation here of departments of community medicine, and federal priorities for primary care to the underprivileged. Finally, researchers of many disciplines have been at work--including operations research and systems analysis--and one may trace their activities through the extensive literature of health services research [Ref. 4]. The following discussion of models gives some picture, both of direction of research and trends in services themselves.

18-3 EVOLUTION OF A HEALTH SERVICE MODEL

Figure 1 shows schematically the major patient care activities of a hospital as they were reflected in the concerns of administrators and researchers a decade ago (2). Inpatient and outpatient elements of a hospital are shown separately with an interconnecting flow of patients. One should note that from a teaching point of view, the inpatient facility was regarded as most important, with much of outpatient care something of a nuisance, existing, in teaching centers, primarily to "find" cases for inpatient care--for surgery or complex diagnosis and medical treatment. An ecological relationship was recognized with a stable proportion of outpatients being admitted as inpatients, thereby permitting a proportionate number of interns and residents to be trained (hence justifying the existence of a related number of professors and administrators). The inpatient areas shown are graded according to intensity of care, reflecting administrative approaches to a number of problems [Ref. 6].

To combat the shortage of nurses, the hospital created an intensive care unit where patients who were critically ill could be brought together and watched over by a small but highly skilled group of nurses with specialized life-saving equipment nearby. At the other extreme it was observed that many patients in hospitals did not need the full array of equipment and service; and so in some hospitals, self-care or light care units were created--sometimes physically separated from the hospital, yet administered by it. These notions of placing patients in facilities related to level of need contained the nucleus of comprehensive systems toward which we have moved slowly since that time.

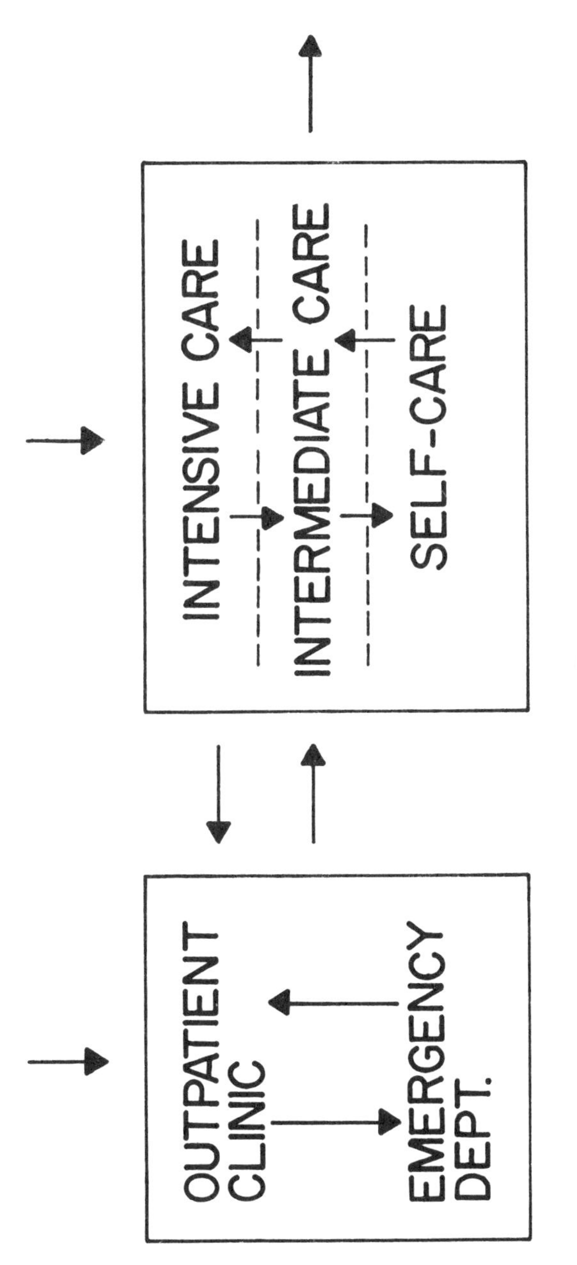

Figure 1
Graded Patient Care In Hospitals

However, one should note the institutional parochialism reflected in the diagram. The arrows indicating patient flow into the system were simply numbers whose rates one kept count of but knew little about in terms of the population served; its demographic description, its socio-economic character, or its alternatives for care. The arrow indicating patient flow out gives no hint of the subsequent experience in the home, or in another institution. The model indicated nothing of external influences that might affect future demands.

Through a painstaking evolution, this view of a health service system, by administrators and analysts, has broadened to the form shown in Figure 2. Here the demographically described population served is added to the model, important for two major reasons. Most illness and accident rates are correlated with age and sex, hence age distribution as well as total size of a population determine today's patterns of health conditions. Further, today's age distribution, immigration and emigration rates, and age specific fertility and mortality rates, determine the makeup of tomorrow's population--a major factor in planning. Significantly, an individual is taken into account, even though he may not be under medical care at a given moment.

Note, also, that what was the hospital alone in our first model has been redefined as a larger system by the inclusion of other ambulatory and long-term care facilities. Arrows from one segment to another represent patient flow rates as before, but they also symbolize the decision process to admit, discharge, and transfer. Formally, these decision processes are physicians' prerogatives, in some cases dominated by medical factors, e.g. admission to intensive care; in others by social or economic factors, e.g. the choice of a long-term care facility.

There are, however, many influences outside the physician/patient decision process. We have attempted imperfectly to represent these as "regulating agencies." Although the model now has an economic appearance with demand and supply linked by intervening regulating agencies, we take these agencies to be any administrative or informational device that influences patient flow within the systems.

An example of a regulating agency, shown in the model, is multi-phasic screening, a process for gathering base line medical data and for uncovering possible disease. Whatever its medical usefulness, there is a far-reaching significance for comprehensive medical care. The pressure of pain or accident no longer becomes the major mechanism for bringing an individual into the medical care system. Just being a recognized

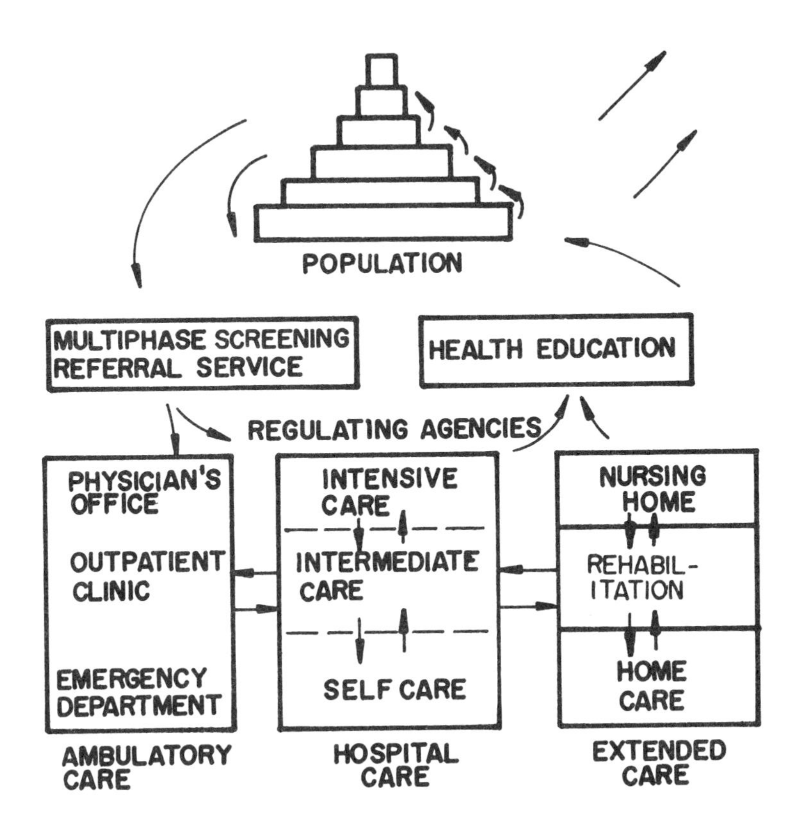

Figure 2
Comprehensive care with regulating agencies.

member of the population covered by the care systems is sufficient to initiate the process of preventive care and to create a continuing medical record.

The model as shown has the shortcoming of being aspatial--it does not show the fact or the effect of population location with respect to services, a factor in a patient's choice of one particular facility over another. Nevertheless, it does represent a system of comprehensive care to an enumerated population, and many of its linkages have in fact been brought into being in a number of settings in recent years.

18-4 DEVELOPMENTS IN URBAN HEALTH SYSTEMS

Three examples were cited earlier of newly developing, urban-based health plans. All have tended to concentrate on primary care and have had the characteristic of reaching out to a geographically defined population nearby. None provides all elements of care to all age groups; thus comprehensiveness is a direction of movement, a goal, rather than an existing state. All have developed under the aegis of a large medical teaching complex.

The first example is the Comprehensive Child Care Clinic based in the Johns Hopkins Medical Institutions. All the children in a neighborhood surrounding the hospital are considered the responsibility of the institution, not just those who become ill and are brought in. Thus the medical care system becomes concerned with the interface between the clinic and the patient family. The reluctance of some families to accept the services as offered, to keep appointments, to follow regimens, has brought into use such terms as "no-show," and "noncompliance." Our system concept must be enlarged to include considerations of individual acceptance or rejection of care and perception of need, to understand the behavioral attributes, the underlying beliefs and conflicts of values of the medical staff and the patients.

The second example, known as the East Baltimore Plan, described in planning stages in [Ref. 3] is just beginning services. It differs from the child care clinic in that the first point of contact is a clinic physically removed from the hospital complex. The key issue is not individual patient acceptance, but community insistence on a role in goverance of the program. From a systems view, the message is that health planning in the community is not the sole prerogative of the surveyors of health services. The potential patient population will have a hand in the planning and management. This is not simply a matter of pride, but is one also of preservation of values. The offering of a service, say family planning, seen as a benefit by the server, who is concerned with population balance, is often seen as a subtle form of genocide by the potential recipient.

In the third example we move from problems of rejection and under-utilization to the opposite, an unusually high demand for physician services. The Johns Hopkins-Columbia medical care plan, offered to residents of the new town of Columbia, Maryland, recently completed its first year of operation. This is a prepaid group practice arrangement with outpatient clinical services in Columbia and, for the time being, hospitalization in Baltimore. By payment of a monthly premium, an individual or family receives all the medical care sought or prescribed. Experience to date for this population, predominantly white middle class, shows seven physician consultations per year, compared to a usual five, and about half the expected hospitalization. This tradeoff of hospital days to physician visits is characteristic of prepaid plans and salaried physicians, but other factors may be present. The city, the population, the medical plan are all new. There is no residual, historic animosity to the medical institution; in fact, there may be a bias in the population toward high utilizers of care because of the existence of the plan. Time will tell, but already obvious is the striking influence of financial incentives and past beliefs and attitudes on the workings of a health system designed to serve a specified population.

18-5 SOME CONCLUSIONS

Three examples of urban-based health care systems in their early stages of development do not provide the broadest base for generalization about system problems. However, if we risk drawing conclusions from what may prove to be only transient experience, a few observations can be made.

18-5.1 Decision-Making and Authority

The medical teaching institution emerges as the innovator and focal point of medical care reform. Its power rests in its supply of physicians, the prestige of faculty appointment and hospital privileges it can offer to participating doctors to compensate for the foregone benefits of conventional practice. On the political side, urban government does not appear as a major factor, but local or neighborhood voluntarism does.

The physician's authority of knowledge appears to be insufficient to produce compliance on the part of patients. It will be interesting to see whether community involvement in planning and operations of clinics will produce social pressures on the population to conform to norms of seeking and accepting medical care.

18-5.2 Complexity

The traditional independence of medical care planning from other urban planning appears more absurd than ever as medicine reaches out to serve a community. Particularly in poverty areas, factors of nutrition, housing, economic opportunity, crime and drug addiction all interlock to influence problems of mental and physical health. An attack on infant mortality--often cited as an embarrassing measure of effectiveness of American medical care--had better include attention to some of the societal factors noted.

It is tempting to draw parallels between urban poverty areas and frontier communities. The newcomers are seekers, sometimes surplus agricultural workers, sometimes the restless and unstable. Eric Hoffer's description of the role of the "undesirable" in society [Ref. 7] could give great hope if the analogy is correct. Arguing for the analogy are problems of law and order, a ruthless provincialism quite separate from existing political subdivisions. Arguing against are the nature of the threats, e.g. of drug addiction, basically an internal problem, and the large residue of indigenous residents, in the main those who have been unable to gain sufficient affluence to move out. We need to know more, demographically and sociologically, of the urban population.

18-5.3 Techniques of System Analysis

The system models described earlier are useful for intra-institutional or communitywide estimates of resource requirements, but being aspatial they fail to treat the problem of facilities utilization as a function of location. Lubin et al. [Ref. 8] have shown that distance, or time of travel, from home to hospital is a major factor in utilization of a health facility. Thus local population density becomes a major variable in patient flow into a system, not only in terms of numbers of people in the "catchment area," but in the fact that density is a correlate of some patterns of morbidity and custom in use of facilities.

Our model, Figure 2, through inclusion of a population pyramid, takes into account population changes as they occur through birth, death, and aging. But instability of population, serious immigration and emigration are characteristic of the city, especially in those areas where health service problems are most severe. Returning to my window, it is apparent that there have been many changes in the scene since this writing began--more houses abandoned or torn down, further progress in a high-rise for the elderly, the stopping of expressway construction pending settlement of disputes over its route. How does one plan in the face of such rapid and radical change! It is natural to seek some alternatives to the formalistic approaches to systems analysis. Boguslaw [Ref. 10] has used the notion of a heuristic approach, in which design is paired upon some principles of operation rather than on a prediction of a future state of affairs, principles that will be valid over a wide range of circumstance. This idea, it seems, is most appropriate to health services in urban setting. Uncertain about the future population, and its needs, about the composition of services and distribution of tasks among evolving personnel types, the most obvious principle of design is one of flexibility in capacity and nature of services. Whatever future demand is expected now, we should be in position to double or halve that level without costly consequence. Many caveats or sub-principles emerge from the first: small isolated clinics of rigid design and organization should be avoided, implying the need for some umbrella organization with the sensitivity and authority to shift resources to match the social patterns as they emerge.

REFERENCES

1. Jose Ortega y Gasset, History as a System, Norton and Co., Inc., New York, 1962.
2. Handlin, O., The Uprooted, Grosset & Dunlap, New York, 1951.
3. National Advisory Commission on Health Manpower, Report, Vol. I, U.S. Government Printing Office, Washington, D.C., 1967.
4. Williamson, John M., Health Services Research Bibliography, National Center for Health Services R & D, HSMHA, Dept. of H.E.W., 1970.
5. "Quantitative Methods Applied to Biology and Medicine," published in Engineering and Medicine, a report of the 1968 Automn Meeting of the National Academy of Engineering, published by the Academy, 1970., pp. 65-72.
6. See for example Weeks, L.E. and Griffith, J.R. (Eds.), "Progressive Patient Care," The University of Michigan Press, Ann Arbor, Mich., 1964.
7. Hoffer, Eric, The Ordeal of Change, Harper & Row, 1967.
8. Lubin, J.W., et al., "How Distance Affects Physician Activity," Chicago, Illinois: The Modern Hospital, 107(1):80, 1965.
9. Boguslaw, Robert, The New Utopians, Prentice-Hall, Inc., Englewood Cliffs, New Jersey, 1965.

Chapter 19

SYSTEMS ANALYSIS OF URBAN AIR POLLUTION

F. F. Gorschboth

Over the last few years the nation has developed an extreme consciousness of its environment. Because the environment affects and is affected by man's numbers, the rate at which he uses up his resources, the products he fashions from their use and the means of their disposal, this consciousness impacts all aspects of modern life. The concerns lately expressed, after an examination of the relationship between modern man and his environment, begin ultimately to question his life style. Modern man has achieved his dominance over his environment and enjoyed affluence and comfort from that dominance, as a result of his development and use of modern technology. In the past, man was quick to point with pride at what he was able to

The author is Manager of Environmental Programs, I. B. M. Corporation, Riverdale, Maryland.

achieve with his technology; today he is wont to indict it for all the changes in his world--changes he found undesirable and now finds unacceptable. In fact, he has come to realize that those changes, wrought by an uncontrolled technology, threaten his survival. But technology is without morality or ethics. It is a tool to be used as man wishes. Man now proposes the question: Can this same technology be used to insure, rather than threaten, his survival?

There are two characteristics of modern technology that are prime considerations in this proposal: its capacity to manage large amounts of data and its ability to approach problems systematically. A very significant aspect of the determination of the environment is the extent of its pollution. Some years ago at IBM, it was suggested that the production of pollutants is similar to other processes. If that be so, it would follow, as seen in Figure 1, that process control techniques would be applicable, and the classical components of such control systems--inputs, outputs, means of measurement and means of control--would be identifiable. This approach was developed over the years--critiqued at various IBM symposia and presented in a paper at the American Society for Cybernetics, entitled, "Ecological Cybernetics." In the course of this development, the urban pollution problem was considered and described in process-control terms. As is seen in Figure 2, the inputs to the pollution producing process in an urban area are the pollution sources (which are controllable) and the natural phenomena affecting them -- in this case principally the weather (which are uncontrollable). The outputs are emissions -- which upon mixing with the atmosphere result in a certain air quality. Traditionally, controls in the urban area have been applied to the sources (the only controllables) in the forms of legal prohibitions and in the form of codes. Code enactment has generally been the result of complaints filed and inspections conducted by urban authorities. Absent from such a system is the means of determining the effectiveness of the measures of their economic and social impact upon the urban area to be controlled.

As a consequence, the control approach was modified at IBM to that depicted in Figure 3. In this new system the air quality is to be monitored by some measuring system (preferably automatic) labelled "aerometric system." Thus, the margin between actual ambient air quality and that desired or tolerated can be determined. However, "air quality" of itself is rather an academic term. The effects of that air quality upon

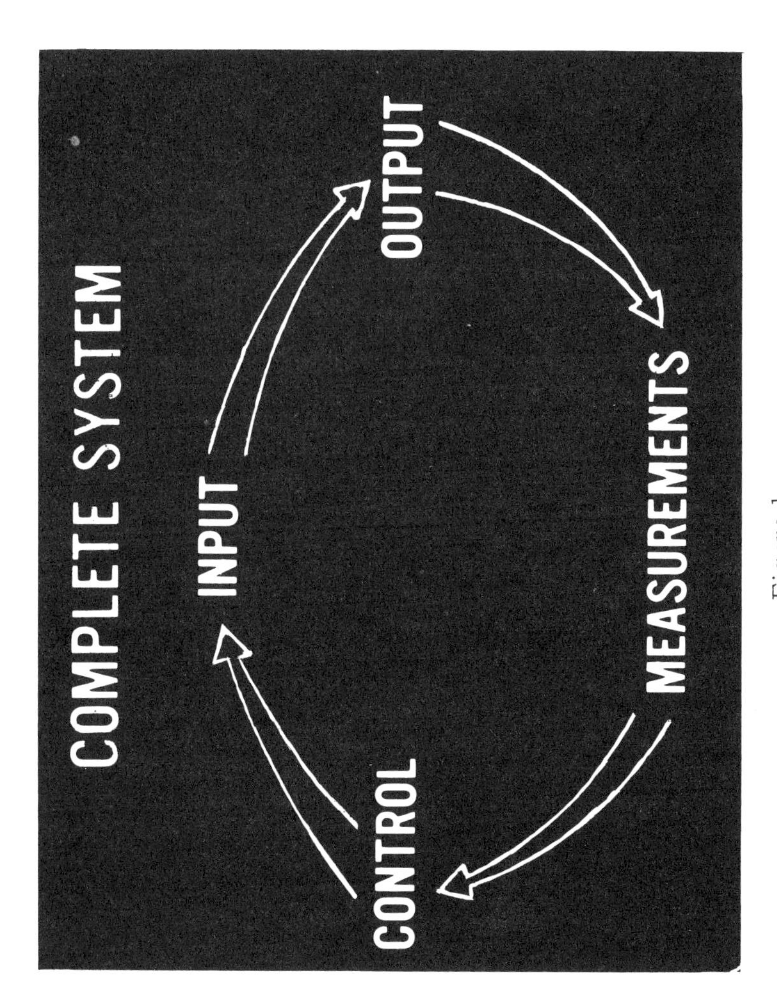

Figure 1
Process Control System

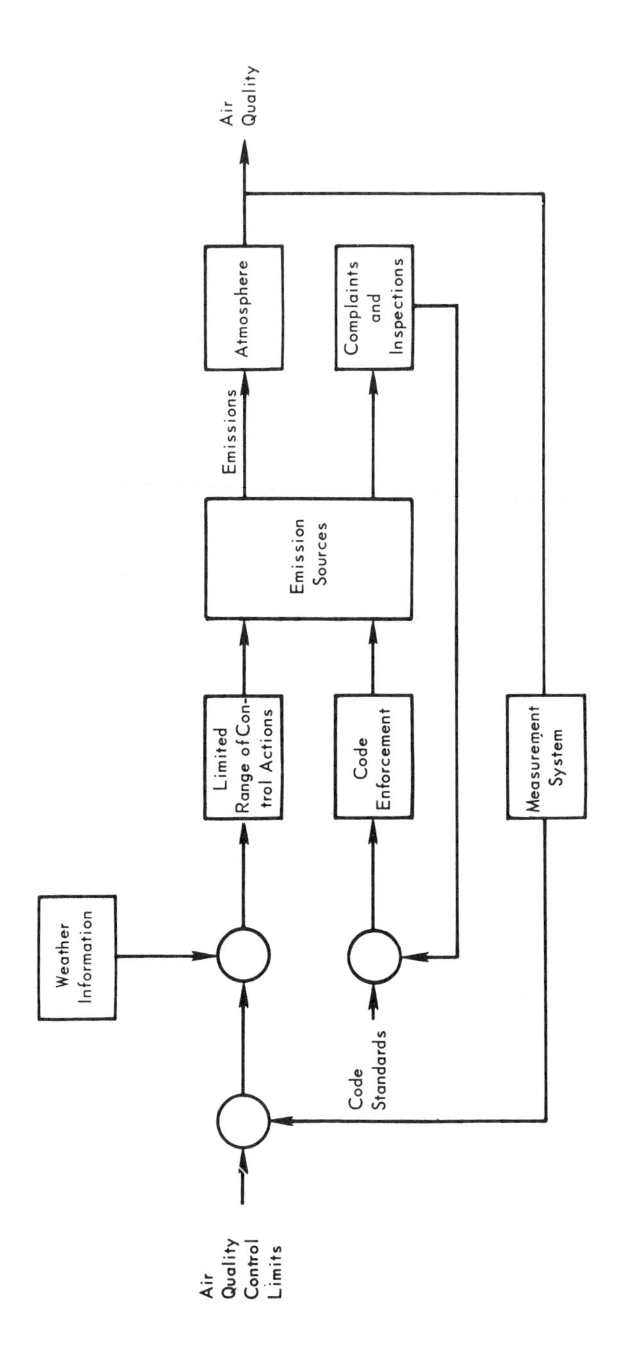

Figure 2
Traditional Urban System

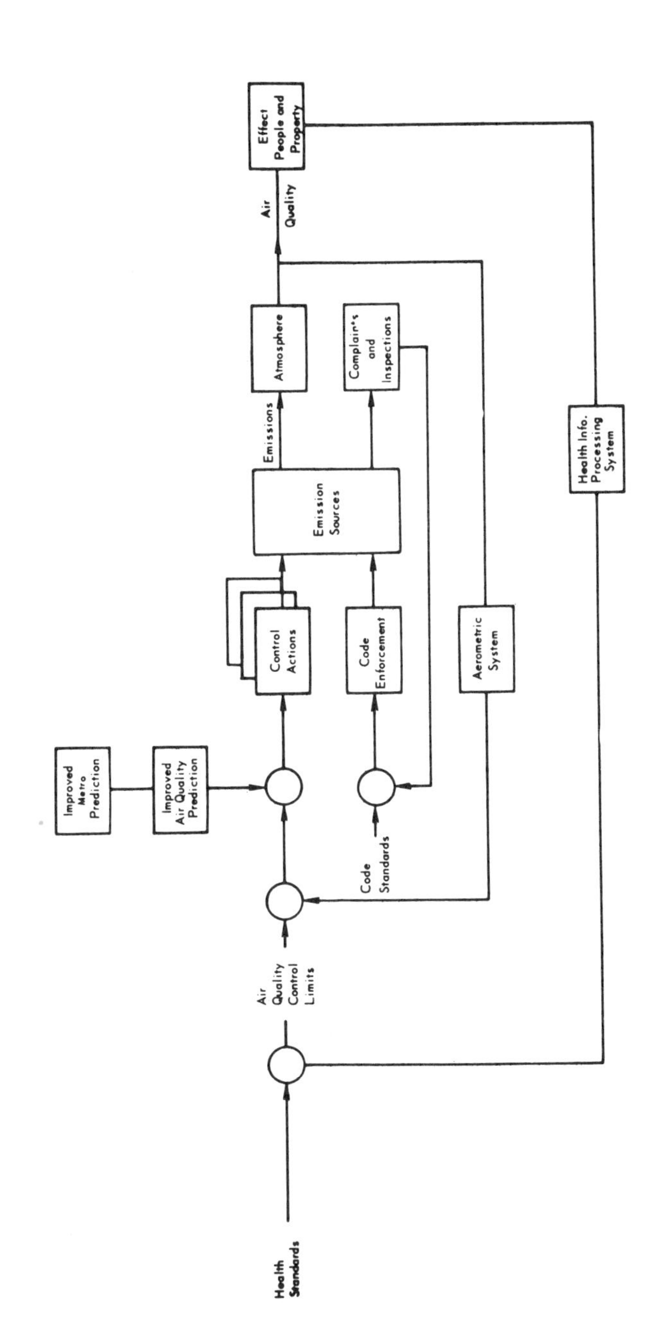

Figure 3
Fully Developed Urban System

people and their property provide the real significance of the measurement. This is particularly true of the effects on human health; it is the factor upon which the whole system is bottomed. Consequently, a health reporting sub-system was added so that the effects upon the population's health could be correlated with changes in the measured ambient air quality--and new health standards could be adopted to serve as the bases for limits on responding control actions directed at the emission sources.

The control of the process then takes the form of the following scenario: The pollution sources produce emissions, which when spread in the atmosphere, result in an air quality which is now monitored by the aerometric system. The control actions taken against the emitting sources are determined by the extent to which the air quality approaches the limits established by the code standards, now a function of the effects on the health and property of the community. The system can be taken a step farther. With improved weather forecasts, the effects of future weather conditions can be predicted, and control action taken in an anticipatory, rather than a corrective, mode. If the corrective action is timely, the air quality will never be allowed to deteriorate to a level that would endanger the community.

The components of the system have been rather generally described up to this point. Actually, they are numerous, difficult to identify and catalogue, and particularly difficult to organize. Figure 4 summarizes the controllable and non-controllable inputs identified in the IBM study for the state of Illinois some years ago. The various weather factors are grouped into general classifications of relevance: local, regional and synoptic. The pollutants are further identified and categorized in Figure 5. When such information relating to pollutants is received (generally in response to questionnaires) it must be organized into an emission inventory to provide the controllable source term in future control strategies and to identify sources against which abatement measures are to be enforced. Figure 6 shows such an arrangement. It should be noted that the pollution inventory is related to individual smokestacks or vents, and records concerned with the issuance of operating permits. Inspections and complaints relating to the stack, control equipment and process descriptions are included.

The output of the system, the pattern of emissions and the resulting air quality, can be detected and described in many ways. It can often be photographed or at times measured in terms of its effects such as the isopleths of pollution damage

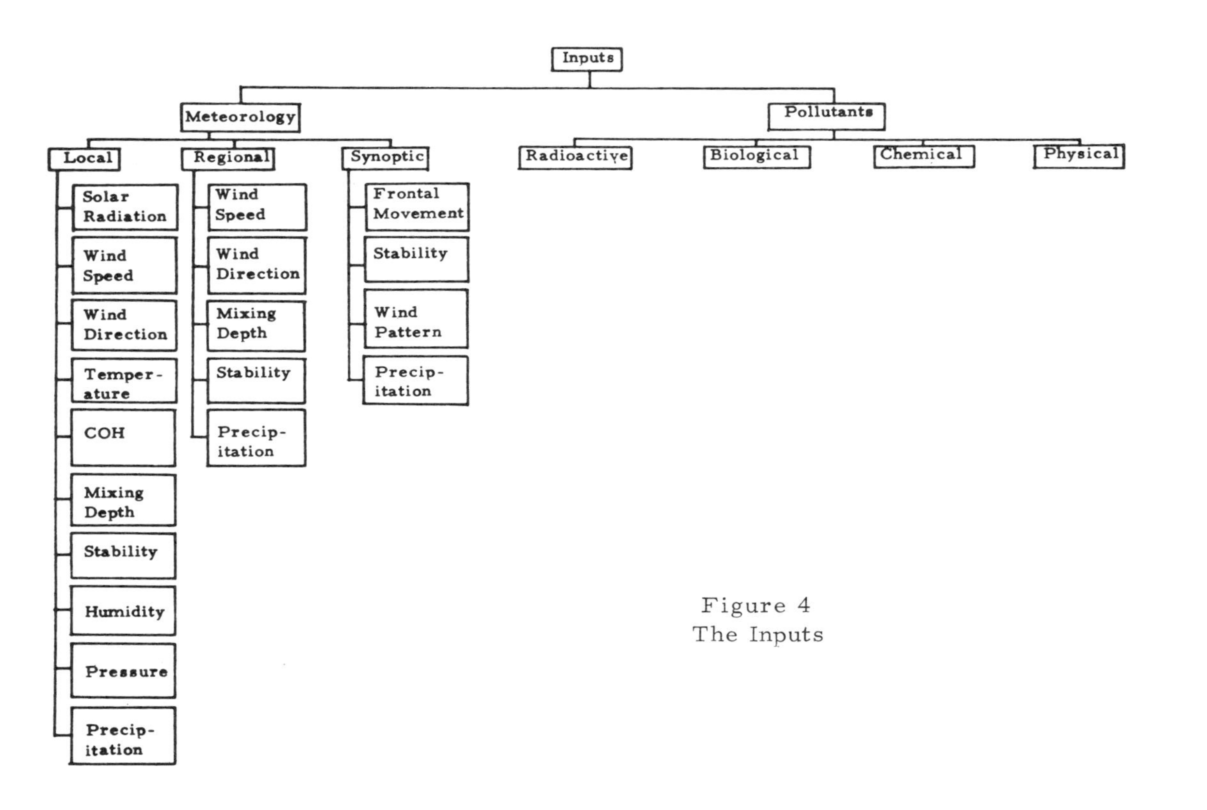

Figure 4
The Inputs

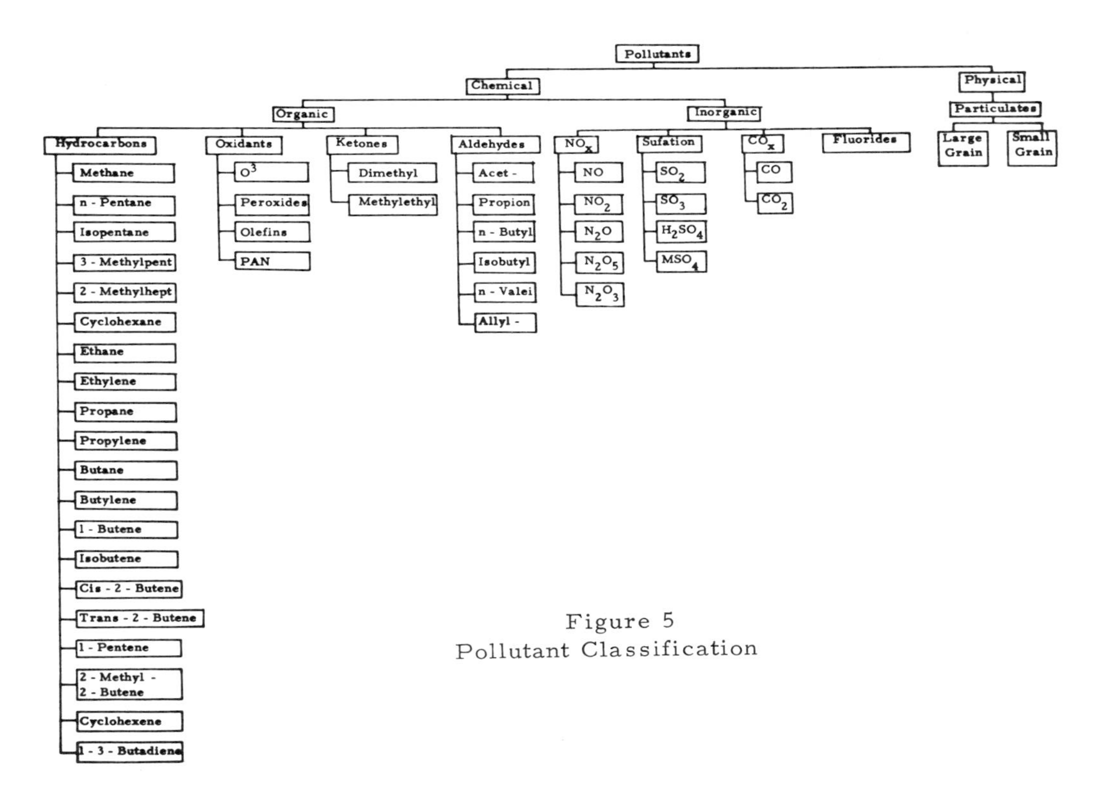

Figure 5
Pollutant Classification

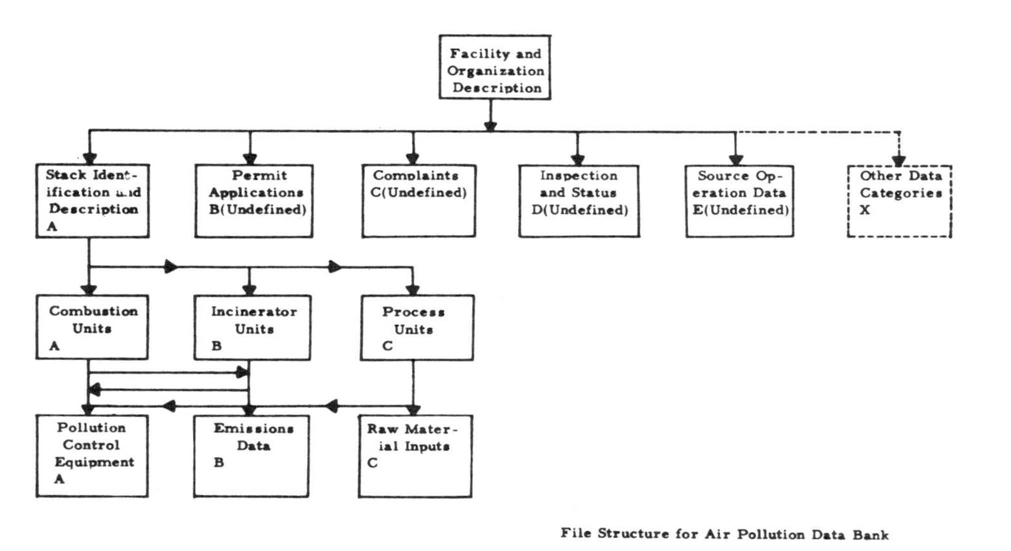

File Structure for Air Pollution Data Bank

Figure 6
Organization of the Emission Inventory

occurring in Amsterdam, depicted in Figure 7. More practically, the pollution pattern is measured by a system resembling that pictured in Figure 8. In this drawing a series of sensors remotely linked to a computing room by a data acquisition system is pictured. Based on the computations and predictions made there, certain decisions in combatting the rise in the pollution level would be made in the air pollution control center.

After the data is assembled and organized, correlations can be attempted. One of the most useful correlations would be the establishment of the relationship of various weather patterns with the meteorological stability associated with the build-up of pollution levels. For example, Figures 9 and 10 illustrate typical conditions and results of two controlling weather patterns. The effect of the controlling low with its concomitant instability, strong winds and general clearing is contrasted with the stability (associated with the anti-cyclonic high) that results in a general build-up of pollution in the northeastern corridor. The relationship between the pattern of pollution travel and buildup and the controlling weather is frequently described mathematically by meteorological diffusion models, based on general diffusion equations such as Sutton, Turner, et al, whereby expected levels of pollution from specified sources under various weather conditions can be determined. Modern mathematical models can be constructed for point and area sources. These models, with increasing precision, describe the pollution travel based on the thermodynamics properties of the emitting gas, the hydrodynamic aspects of its expansion, and utilizing the classical continuity questions, the overall pattern of expansion downwind.

After successfully constructing a data base describing the ingredients of the urban process system, effecting the necessary correlations, and finally constructing mathematical models of these interrelated functions, it is possible to turn to the problem of control of the process. As was earlier stated, control action must be brought against the only controllable pollution sources in the system, and it can take many forms. It is the objective of a systematic approach to the control problem to consider alternative control strategies, their effectiveness, ease of application and cost, before they are adopted or in the event of a multi-strategy approach, the order of their implementation. The overall approach to the control of the process is summarized in Figure 11. The evolution of the management approach takes the form developed in this paper. The

PATTERNS OF POLLUTANT DAMAGE TO VEGETATION NEAR AMSTERDAM

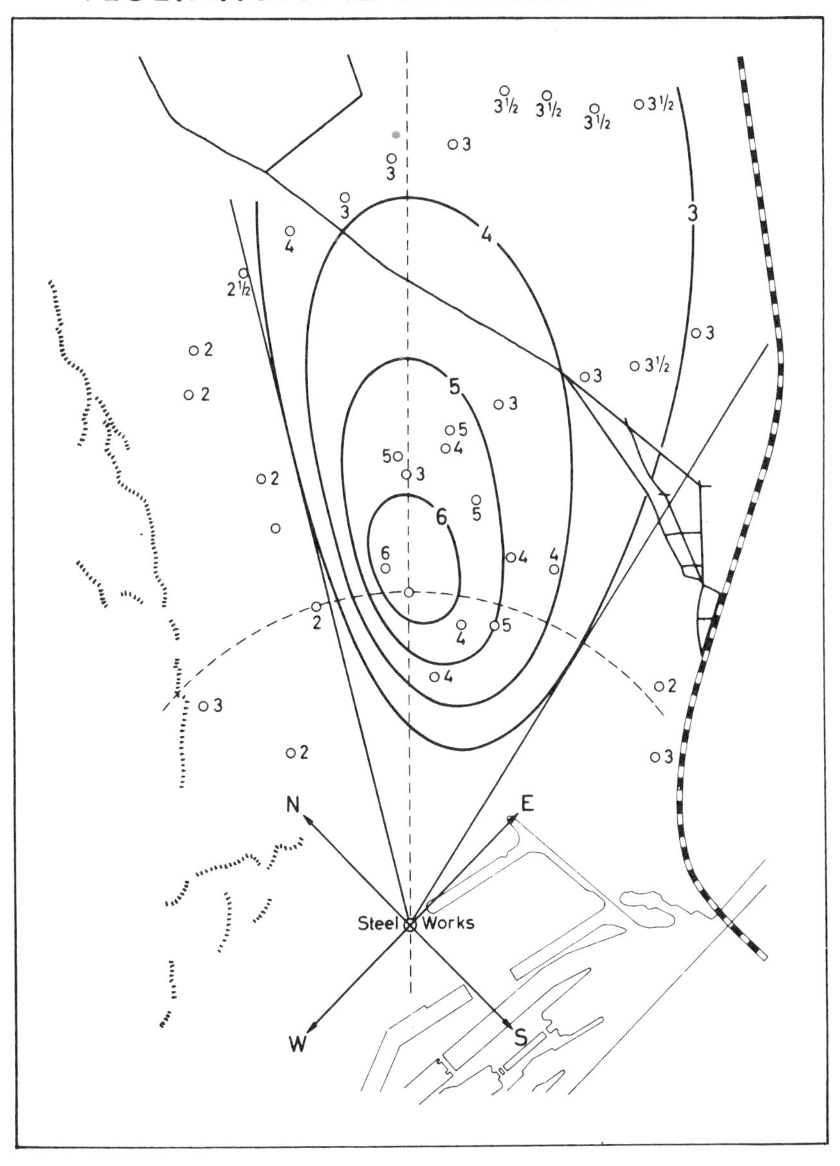

Figure 7

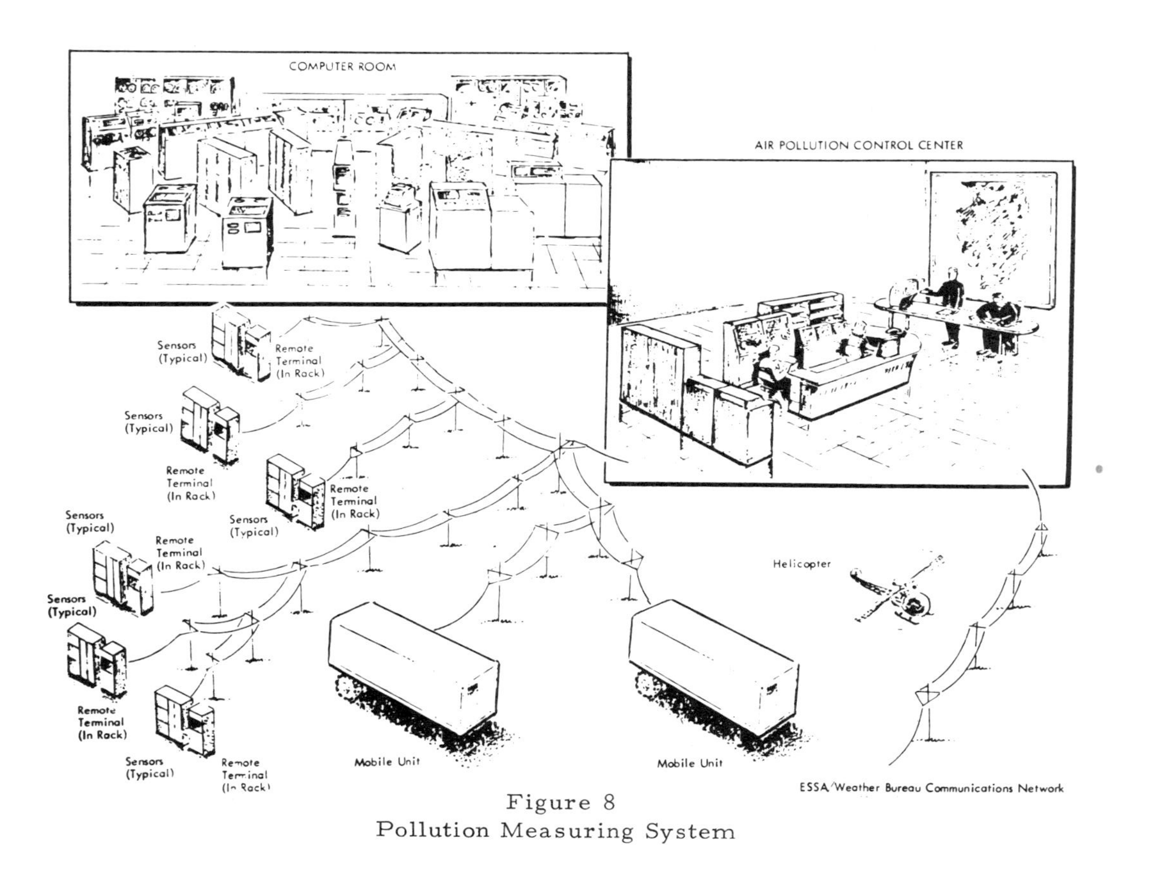

Figure 8
Pollution Measuring System

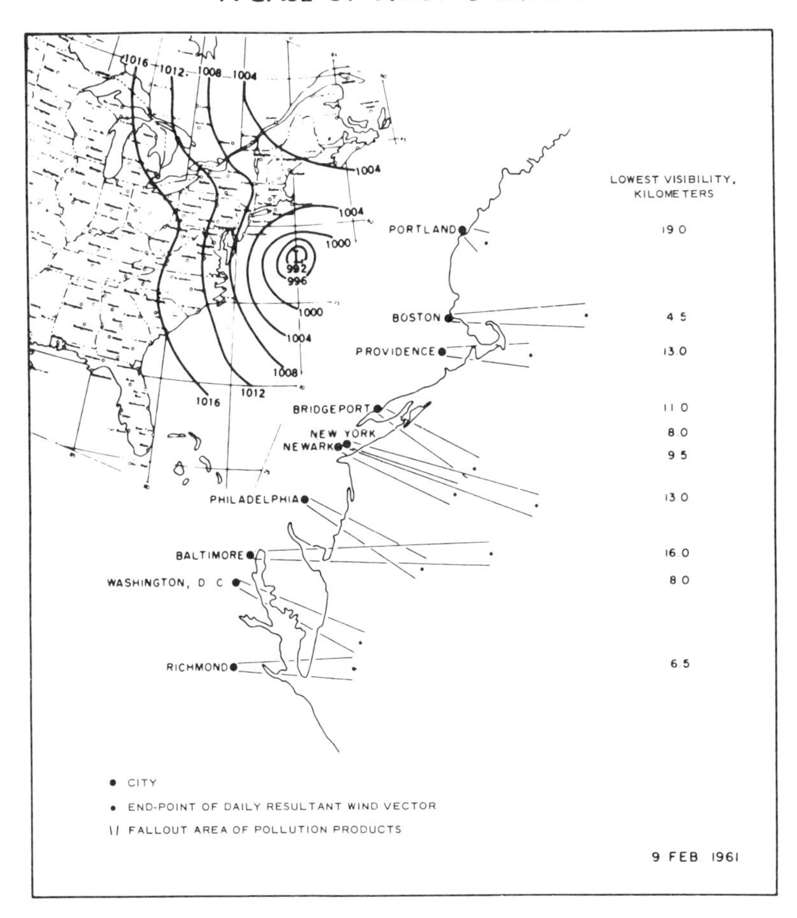

Figure 9
A Case of Strong Winds

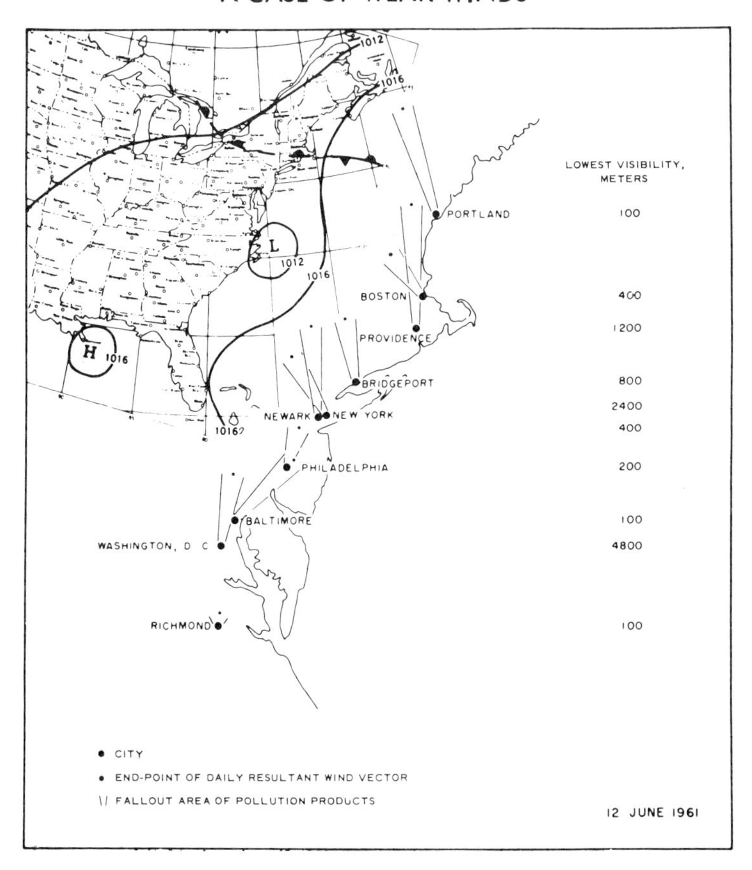

Figure 10
A Case of Weak Winds

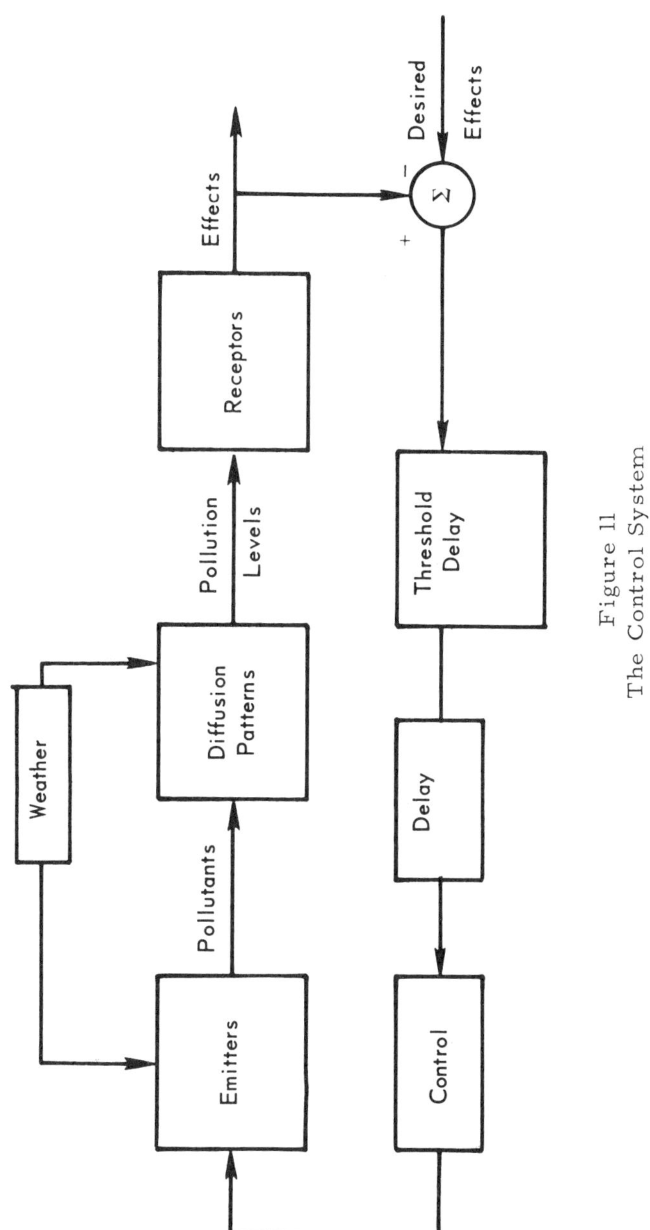

Figure 11
The Control System

elements of the system are identified, their dynamics defined, requirements generated, and decisions made and finally implemented to meet those requirements.

To implement an effective program, it is necessary to evaluate, on the basis of cost and effectiveness, the various control options available. To make such an evaluation, the components of the process system must be further defined. If the components are arranged as illustrated in Figure 11, it is seen that this system arrangement generally corresponds to that of the previously described "urban system." Two differences may be noted. The first is the role of the weather affecting both the amount of emission (since weather changes affect power and heating demand) and the diffusion pattern of the pollution. In this scheme feedback control is contemplated, based on the desired effects of the environmental quality. Correction (control) is made in accordance with the limits on effects, principally on the health of the community.

If, instead of using the emission inventory as the input and aggregate effects as the output, the time dependence of these components is introduced, the system evolves into one similar to that depicted in Figure 12. In this approach the emissions are reported on an hourly basis so that daytime vs. nighttime, weekday vs. weekend as well as seasonal variations can be noted, and as a result, an emission inventory model is developed. At the same time, a description of the meteorological variation with time (meteorology model) is developed, resulting in a time dependent diffusion description or diffusion model. A further description of the pollution absorption by the receptors as a function of time (the receptor model) provides the basis for an effects model. The control options are then costed and, based on the effects model, provide the basis for a cost/effects evaluation. From such a system, the following outputs can be identified: environmental quality as a function of time, effects as a function of time, and costs of various control options. With such time-dependent models available, it is now possible to return to Figure 12 and input the time relationships of environmental quality, effects and cost, and simulate each in the basic system to evaluate its cost/benefit. This extension to the system is depicted in Figure 13. Here the outer loop is unchanged from Figure 11. But the inner loop, composed of the time-dependent component models, provides the opportunity to determine from simulation the cost effectiveness of each control option before actually being inserted into the control situation. From such simulation the complete spectrum of control options can be simu-

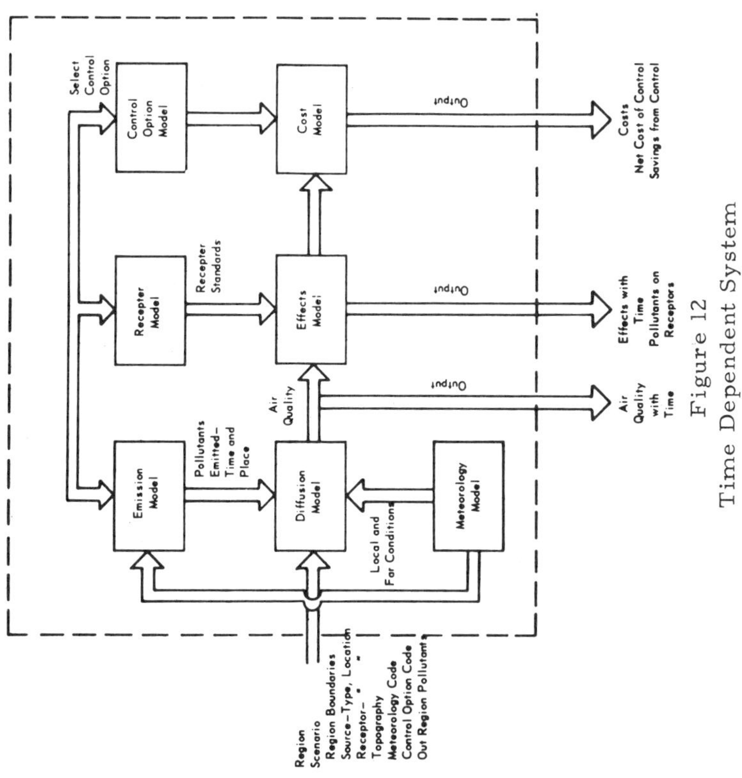

Figure 12
Time Dependent System

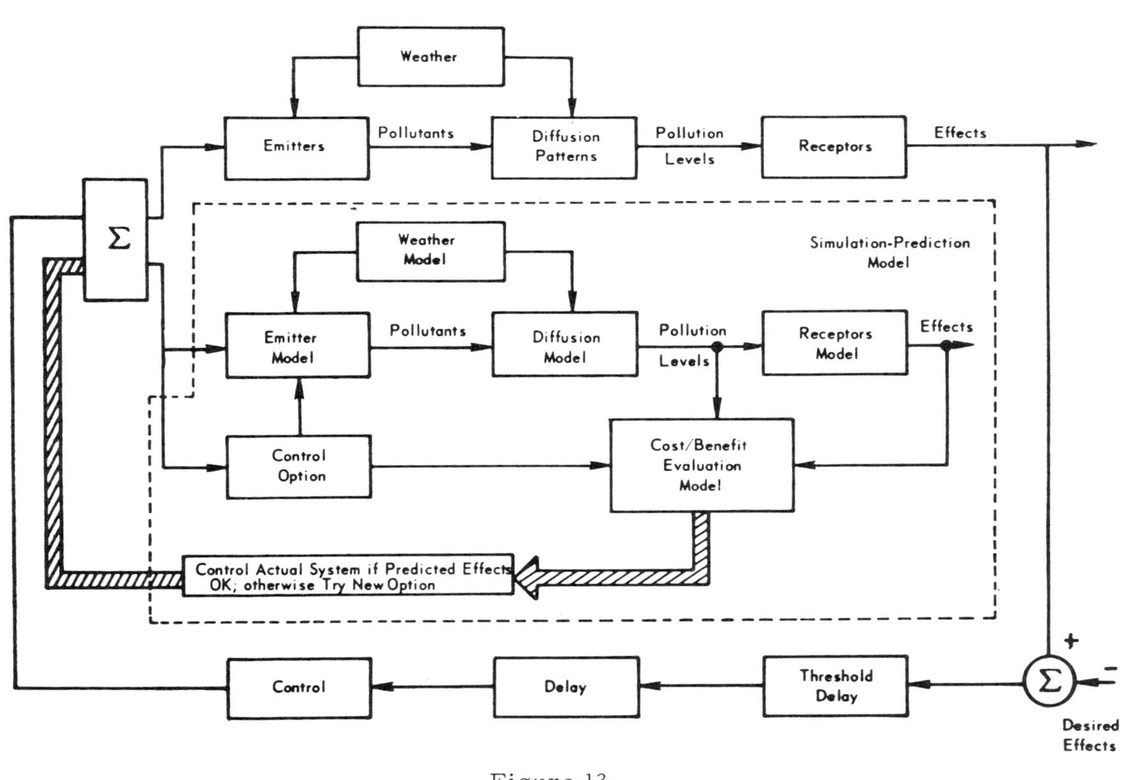

Figure 13
Feedback Control System

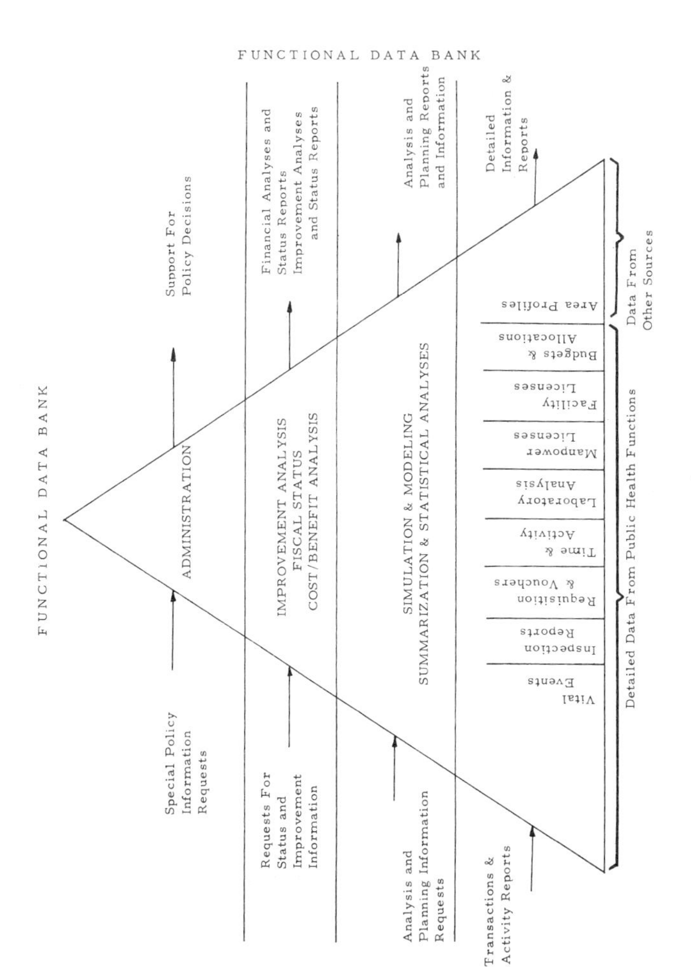
FUNCTIONAL DATA BANK
FUNCTIONAL DATA BANK
Special Policy Information Requests
Support For Policy Decisions
ADMINISTRATION
Requests For Status and Improvement Information
Financial Analyses and Status Reports Improvement Analyses and Status Reports
IMPROVEMENT ANALYSIS
FISCAL STATUS
COST/BENEFIT ANALYSIS
Analysis and Planning Information Requests
Analysis and Planning Reports and Information
SIMULATION & MODELING
SUMMARIZATION & STATISTICAL ANALYSES
Transactions & Activity Reports
Detailed Information & Reports
Vital Events
Inspection Reports
Requisition & Vouchers
Time & Activity
Laboratory Analysis
Manpower Licenses
Facility Licenses
Budgets & Allocations
Area Profiles
Detailed Data From Public Health Functions
Data From Other Sources

Figure 14

lated, evaluated and prioritized for use in actual pollution episodes.

To follow an approach such as is described in this paper, requires the capability for handling, reducing and organizing tremendous amounts of data in compiling an adequate data base. These data provide the basis for the systematic approach previously described. As illustrated in Figure 14, by simulation and modeling analysis of data an evaluation of the effectiveness of various control options can be made. Cost/benefit and fiscal analysis can then be undertaken to eventually lead to the final step in this systematic approach -- the decision of the control program to be adopted. Thus, as earlier suggested, the two most significant characteristics of modern technology, its ability to manage data and approach problems systematically, may be applied to a problem generally ascribed to this technology -- to aid rather than threaten man's survival in the modern technological world.

Chapter 20

POLUTION AND ECOLOGY

Y. Y. Haimes

Destruction of the ecology by pollution of the environment is undoubtedly one of the major common problems the world is faced with today. The assumption that our atmosphere, lands, and oceans have an infinite pollutant assimulative capacity has obviously proven to be wrong. President Nixon in his State of the Union Message last year said, "The great question of the seventies is, shall we surrender to our surroundings, or shall we make our peace with nature and begin to make preparations for the damage we have done to our air, our land and our water?" The answer to this challenge is inherent in the nature of man: never surrender; for with hard work a solution can always be found.

An old axiom says that half the way to solving a problem is by defining and understanding it. The systems approach follows this pattern by what is known as systems modeling and identification. Before we detail the systems approach to the solutions of the environmental problem, let us first define pollution and briefly survey its dimensions.

The author is Associate Professor, Division of Systems Engineering, Case Western Reserve University, Cleveland, Ohio.

First, what is pollution? Dr. Werner Stumm of the Harvard Division of Engineering and Applied Science in his talk to the American Chemical Society in Houston, 1969, gave a cybernetic view of natural systems controlling themselves with feedback mechanisms. Pollution is defined as an uncontrolled change in such a system. "An unpolluted ecological system is not necessarily an unchanging one; it is simply a system in which the cycling is controlled, not running out of control, in which energy flows are decreasing, not increasing; in which systems are developing increasingly complex checks and balances; in which diversification is increasing; in which instability is giving way to stability."[1]

These undesirable changes in the physical chemical or biological characteristics of the environment are generally classified under air pollution, water pollution, solid waste, noise and other. Air pollutants are solids dispersed in the atmosphere as smoke or other particulate matter, and gaseous and liquid materials that are air transported. Water pollutants are solids and gaseous and liquid materials suspended or dissolved in water [Ref. 14]. Solid waste is the presently unwanted residues of used natural or man-made resources, and of human activity, which are handled or managed in the solid state [Ref. 10]. What are the dimensions of these pollutants?

One major air pollutant is SO_2. "More than 68 billion pounds of SO_2 will pour into the U.S. atmosphere this year, 75% of it resulting from combustion of sulfur bearing fossil fuels [Ref. 9]. The National Air Pollution Control Administration gave the following statistics on the effect of SO_2 air pollution on the ecology: "....Adverse health effects were noted when 24-hr. average levels of sulfur dioxide exceeded 300 mg/m^3 (0.11 ppm) for three to four days....also when the annual mean level of sulfur dioxide exceeded 115 mg/m^3 (0.04 ppm)... Visibility reduction to about five miles was observed at 285 mg/m^3 (0.01 ppm), and adverse effects on materials at an annual mean of 345 mg/m^3 (0.12 ppm), also on vegetation at an annual mean of 85 mg/m^3 (0.03 ppm)". [Ref. 9]

In the water pollution scene, the need for more water is increasing as its quality is steadily degenerating. "Conservationists estimate that by 1980 daily consumption of water by U.S. industry, agriculture and municipalities will exceed our fresh

---

1 Werner Stumm, Harvard Division of Engineering & Applied Science, Harvard University, Cambridge, Massachusetts.

water resources of 600 billion gal/day." [Ref. 12] But our lakes are experiencing an accelerated eutrophication, i.e. an accelerated aging because of excessive growth of algae and other aquatic plants. Algae not only cause an aesthetic nuisance, such as serious taste and odor problems, it also adds organic matter that may deplete the water's oxygen content below the level required for fish and other aquatic life. [Ref. 12] Oil is also polluting our water; 1.5 million tons of oil are introduced into the oceans each year by ships and offshore drilling accidents. Almost 4.5 million tons of oil may be introduced into waterways and eventually the ocean as a result of emission and wasteful practices on land [Ref. 13].

Solid waste poses a third major threat to our environment. Today in the urban U.S. the annual accumulation of solid waste amounts to more than one ton per person, or over 200 million tons/year. To collect and dispose of these wastes we pay an excess of $4.5 billion, or an average figure of some $22/ton [Ref. 1]. We can expect that the quantity of solid waste generated per capita will increase in the future. The present $4.5 billion annual direct national cost for solid waste collection and disposal "does not include such important costs as the internal costs to industries and agriculture for solid waste management, householder and institutional costs for storage and handling of refuse, losses in property value due to inadequacies in collection and disposal of solid wastes, the value of potentially reusable fractions, or individual medical or loss-of-health costs from the various forms of pollution and inadequate control of vectors (flies, rats, mosquitos, etc.).... The 200 million tons per year of solid waste material represent a national resource and will in time be a major one. Return of fractions of solid waste to economical reuse must in the long run become common practice and must be a national objective." [Ref. 10]

The above examples indicate a clear trend in the ecological crisis we are faced with today. It is essential to identify the sources and origins of the crisis in order to reduce their effects and ultimately to eliminate them.

In his search for the "best" performance, man has often failed to consider and recognize the elements which are not directly related to his problem, and which are not only affected by, but affect his solution in the long run. This tradition has been followed for many reasons. Among them: lack of knowledge about the system under consideration and its environment; the magnitude, complexity and intractibility of the system as a whole; the unwillingness to pay the immediate price versus the longterm one; etc.

One astounding recent example of this limited outlook is the construction of the Aswan Dam in Egypt [Ref. 5, 2]. This multipurpose large scale project, which is aimed at generating hydroelectric power to promote industry, a better standard of living, and to provide water for the development of arid lands for agriculture, has failed to consider the environment. Some of the by-products of the Aswan Dam are the destruction of almost all the sardine fish in the eastern Mediterranean (estimated 97% reduction of the sardine catch), a greatly increased rate of schistosomiasis (a chronic debilitating disease dependent on snails) among Egyptians [Ref. 5, 2] stimulating the growth of aquatic weeds that greatly increase the rate of evaporation from the lake behind the dam, the generation of a demand for inorganic fertilizers followed by a salination in the irrigated region below the dam, etc. Obviously, a methodological approach, which considers the outcomes of the policies determined by the decision maker, is needed. The system approach is an attempt to study, analyze and evaluate all of the components which are related to, coupled with and otherwise affect the system solution.

The procedures that the systems analyst follows in his analysis include a definition of the objective, a systematic process of data gathering, an assumption of hypothesis, introduction of utility and value functions, an evaluation of the physical, technical, economical, political and other constraints and restrictions, a comparison of the effectiveness of the various alternative policies over the planning period, a selection of the best and most effective policy, and a forecasting of the future policies beyond the planning horizon. The systems analyst does all this in a scientific and coordinated way. His mathematical tools, techniques, and strategies are based on and arise from various disciplines. A broad and wide interdisciplinary knowledge is necessary because the best effective policy, often called the "optimal policy" can be reached only when all the factors affecting the system are considered. One of the major aids to the Systems Analyst is the mathematical model that he constructs. (A mathematical model is a set of equations that describes and represents the real system. This set of equations uncovers the various aspects of the problem, identifies the functional relationships among all components and elements of the system, establishes measures of effectiveness and constraints and thus indicates what data should be collected to deal with the problem quantitatively.)

Most mathematical models which attempt to represent urban, industrial, environmental or other engineering processes suffer from a lack of essential information regarding the process under consideration. Consequently the mathematical models developed are far from representing the real engineering process. This makes the analysis and "optimal decision" derived from the mathematical model not necessarily related to the real engineering processes it attempts to simulate and represent. Usually an improvement of the mathematical model takes place when additional information about the engineering process, such as observations on the process outputs, becomes available. This improvement and adaptation of the mathematical model is done via the identification of the system model parameters. Thus system identification plays an important role in the development of meaningful mathematical models which closely represent the real system. It is important to remember however, that the above observations on the system outputs generally become available only when the system has been operating on real time, under assumed optimal operational policy. This optimal operational policy should be determined in conjunction with the above system modeling and identification procedure.

The application of the above systems approach to environmental problem and pollution control generally lends itself to large scale complex and intractable mathematical models. Many ecological and biological phenomena are unexplained and covered with uncertainties and lack of knowledge. Consider for example the problem of pollution abatement and control of a water resource system. The physical, chemical, biological, ecological and other natural functional relationships should be taken into account. In addition the environmental, economical, social political and other non-commensurate aspects should also be considered. Such a large scale system model is obviously complex and of high dimensionality. In particular a conventional system analysis of the coupling and the interaction among the various elements of such a limnological system model is almost not feasible and may be impossible. It is evident that a modeling and optimization approach which leads to the desired reduction in dimensionality, is needed to successfully tackle parts of the complex system without losing track of the whole; and most important, an approach which leads to an overall optimal solution to the whole system. Decomposition and multilevel approach [Ref. 8] seems to be a promising answer to these requirements.

The concept of multilevel approach is based on the decomposition of large scale and complex systems, and then the reconstruction of the system into independent subsystems. This decentralized approach, by utilizing the concepts of strata, layers and echelons, enables the system analyst to analyze and comprehend the behavior of small parts of the system, called subsystems, in a lower level and to transmit the information obtained to fewer subsystems of a higher level. Whenever more decentralization is needed, the system is decomposed. This decomposition is accomplished by introducing new variables, which will be called pseudo variables, into the system. Then each subsystem is separately and independently optimized. This is called a first level solution. The subsystems are joined together by coupling variables which are manipulated by second level controllers in order to arrive at the optimal solution of the whole system. This coordination is called the second level solution. One way to achieve subsystem independence is by relaxing one or more of the necessary conditions for optimality and then satisfying this condition with the second level controller.

20-1 WATER RESOURCE POLLUTION MODELING

The modeling of water resources systems, where the quality of water is the theme, results essentially in models of high dimensions. This is due to the fact that "water quality" can not be expressed by a one dimensional state variable such as dissolved oxygen; moreover, there can be no unique representation of "water quality" since the water characteristics and properties are the result of a combination of factors, suspended material, organic material, temperature, biological oxygen demand, chlorides, phosphates, nitrates, algae, etc. Nevertheless, some measures of water characteristics have been established over the years. The more characteristics one considers, the more accurate a description of the water quality he gets and consequently, the more complex his system model becomes.

In the modeling and control of the pollution of water resources system (WRS), the various pollution inputs have to be considered. Each pollution input, municipal or industrial, has its own "quality" or characteristics and accordingly contributes to the quality of the total water resource system. It is the consideration of all these inputs to the WRS, where each input consists of a vector of pollution characteristics, that makes the modeling and control difficult and often intractable.

Haimes [Ref. 4] considered a general mathematical model of a water resource system. The cases where the pollution effluent is discharged directly into the river, into the lake, or into a bypass pipe leading to an advanced wastewater treatment (AWT) plant, each constituting a special case of that general model. The Water Resource System under consideration is decomposed into N subsystems. The pollution effluent input vector to each subsystem includes the water quantity and different water characteristics such as BOD, DO, pH, conductivity, temperature, algae, phosphates, nitrates, etc. Treatment cost functions and quality transition functions as well as system model constraints are introduced, where all functions can be nonlinear. It is assumed that a central authority for the WRS exists with AWT facility, and that each subsystem (polluter) may develop its own AWT facility should its development be economically justified by the optimization procedure. A system Lagrangian is formed to incorporate the system constraints and coupling. The Lagrangian is decomposed into N independent subsystems, and a two level optimization methodology is introduced. Each subsystem is independently and separately minimized at the first level assuming known Lagrange multipliers. At the second level, the total Lagrangian is maximized with respect to the Lagrange multipliers using optimal values for effluent inputs from all subsystems obtained from the first level. The Lagrange multipliers are prices imposed by the central authority for the pollution caused by the subsystems. These are prices for water improvements for use of the water and for wastewater discharges (or treatment). Each subsystem in turn is able to determine its own policy as to whether to pay the specified price by the central authority for causing the degradation of the quality of the WRS or to invest its money in developing a local AWT plant which in turn will improve the quality of its pollution effluent. An optimal policy for each subsystem can be reached while the total cost of improving the quality of the WRS by the central authority as a whole is minimized. Such an approach is both economically and politically desirable and feasible.

Kneese [Ref. 6, 7], is a major advocate of water quality control and management by imposition of a system of effluent charges. The imposition of a charge tends to encourage plants to make changes that in many cases reduce not only the volume of effluents, but the water intake. The advantages of such a system of effluent charges were noted in the Regional Water Quality Act of 1970 sponsored by Senator Proxmire [Ref. 10]:

1. It will enable us to make rapid strides in a relatively short time toward significantly improving the quality of our nation's waterways. The economic incentives not to pollute would be very strong. There would be almost an immediate impact due

to the natural desire of polluters to reduce their charges as soon as possible.

2. It would assign responsibility for pollution control to those who are responsible for the pollution. In other words, the polluters would pay for the damages caused by pollution, not the public at large.

3. It would contribute to the ultimate solution of the pollution problem; that is, reduction of waste production. It will provide the strongest possible incentives for the maximum use of our productive resources, and thus the reduction of waste.

4. It would provide revenue which could be utilized for further pollution control efforts in other sectors of the economy, most notably in eliminating domestic wastes produced by the municipalities, and, of course, to reduce the enormous size of the Federal Budget.

5. It will provide strong economic incentives for the creation of regional water management associations. These associations will provide the necessary coordination to make the comprehensive attack of a regions water pollution problems possible. They will also enable us to take advantage of significant economies of scale which can be realized in treatment efforts. This will sharply reduce the costs associated with water quality improvement.

This system of effluent charges has been successfully implemented in Germany, where the Ruhr and Esmer Gennosenschaften have for almost a century administered a program of environmental controls [Ref. 3]. This program includes areawide regulation geared to natural conditions, autonomous financing derived from user and effluent charges. Several approaches to a basin system have been made in the United States, such as in Otsego, Michigan and Springfield, Montana; but these efforts have been of the nature of voluntary federations that include an administrative superstructure without enforcement powers.

In order for such a system of effluent charges to be successfully implemented in the U.S., a freedom of choice should be offered to the polluter as to whether to treat his effluent by his own wastewater treatment plant or to pay the effluent charges imposed by the regional authority. In addition, in a free economy, the regional authority need not have complete knowledge of the cost function for each individual polluter; this will insure unbiased and fair charges to all polluters.

The multilevel approach in the modeling and control of the pollution of water resources systems has many advantages over other approaches. The following are the major advantages [Ref. 4]:

1. Enables each polluter to find his own optimal policy regarding the quality level of his pollution effluent, and provides a minimum cost of operation to the polluter and to the central authority.

2. The model can handle a vector of water quality characteristics, e.g., BOD, DO, nitrates, phosphates, pH, conductivity, temperature, algae, etc. This provides a better representation for water quality standards and control.

3. None of the system model functions need be linear. Thus a more flexible model representing the water quality characteristics may be developed and utilized.

4. Since at the first level each subsystem is separately optimized, there is no computational limitation on the number of polluters that the mathematical model can handle. This may increase, however, the computational time and add a coordination load to the second level. Furthermore, the model is capable of augmenting more than one polluter in one subsystem whenever this is physically, economically, or politically desirable.

5. This methodology is applicable to the cases where the pollution effluent is discharged directly into the river, into the lake, or into a bypass pipe leading to an advanced waste water treatment plant.

REFERENCES

1. Bowerman, F.R., "Solid Waste Disposal," Chemical Engineering, pp. 147-151, April, 1970.
2. Ellsaesser, H.W., "Air Pollution: Our Ecological Alarm and Blessing in Disguise," EOS, Transactions American Geophysical Union, Vol. 52, No. 3, pp. 92-100, 1971.
3. Federal Water Pollution Control Administration, U.S. Dept. of the Interior, (1970), The Economics of Clean Water.
4. Haimes, Y.Y., "Modeling and Control of the Pollution of Water Resources Systems via Multilevel Approach," Water Resources Bulletin, Vol. 7, No. 1, pp. 104-112, 1971.
5. Hardin, G., To Trouble A Star: The Cost of Intervention in Nature, Bulletin of The Atomic Science, Vol. 26, pp. 17-20, 1970.
6. Kneese, A.V., R.V. Ayers and R.C. D'Arge, (1970), Economics and the Environment, A Material Balance Approach, The Johns Hopkins Press, Baltimore, Maryland.

7. Kneese, A.V., and B.T. Bower, (1968), Managing Water Quality: Economics, Technology, Institutions. The Johns Hopkins Press, Baltimore, Maryland.
8. Mesarovic, M.D., D. Macko, and Y. Takahara, (1970), Theory of Hierarchical, Multilevel Systems, Academic Press, New York, N.Y.
9. National Air Pollution Control Administration, "Air Quality Criteria for Sulfur Oxides," Publication No. 50.
10. "Policies for Solid Waste Management," National Academy of Engineering, National Academy of Sciences, U.S. Department of Health, Education, and Welfare, Bureau of Solid Waste Management, 1970.
11. Proxmire, Introduction of the Regional Water Quality Act of 1970, (1969), Congressional Record-Senate, Proceedings and Debates of the 91st Congress, First Session, Vol. 115, No. 195, Washington, November 25, 1969.
12. Smith, W.W. and D. DiGregorio, "Advance Wastewater Treatment, An Overall Survey," Chemical Engineering, pp. 71-74, April 27, 1970.
13. The Plain Dealer Newspaper, Sunday August 2, 1970, article covering the preliminary report on the study of critical environmental problems sponsored by M. I. T.
14. Waste Management Control, Report to the Federal Council for Science and Technology by the Committee on Pollution, National Academy of Sciences, NRC, 1966, Publication 1400, Washington, D.C.

INDEX